Long-Term Potentiation

Volume 2

Long-Term Potentiation

Volume 2

edited by

Michel Baudry and Joel L. Davis

A Bradford Book
The MIT Press
Cambridge, Massachusetts
London, England

This book was set in Palatino by Asco Trade Typesetting Ltd., Hong Kong and was printed and bound in the United States of America.

Library of Congress Cataloging-in-Publication Data
(Revised for vol. 2)

Long-term potentiation.
"A Bradford book."
Vol. 1 based on the conference held in 1990.
Includes bibliographical references and indexes.
1. Memory—Congresses. 2. Neuroplasticity—Congresses.
I. Baudry, M. II. Davis, Joel L., 1942–
[DNLM: 1. Memory—physiology—congresses. 2. Neural Transmission—physiology—congresses. 3. Neurophysiology —congresses. WL102 L849 1990]
QP406.L66 1991 153.1'2 91-11092
ISBN 0-262-02330-X (v. 1)
ISBN 0-262-02370-9 (v. 2)

Contents

Preface

Most of the theories and models of learning and memory assume that learning is the result of activity-dependent modifications of synaptic efficacy along the pathways activated by various experiences. Ever since before the neuron doctrine was elaborated, the idea was put forward that these modifications take place at the synapses, the sites of neuron-to-neuron communication. According to these ideas, memories are represented in the brain as a collection of synaptic modifications waiting to be reactivated by the proper stimuli, be they internal or external. While theorists and network modelers have been satisfied with these general concepts, experimentalists have been searching for the nature and mechanisms underlying such synaptic modifications ever since the visualization of synapses in the brain. A gigantic step was made at the end of the 1960s and early 1970s by the discovery of the phenomenon of long-term potentiation (LTP), a long-lasting form of activity-dependent increase in excitatory postsynaptic potentials. Initially found in hippocampus, a brain structure known to be critically involved in memory formation from studies in amnestic patients and animals with hippocampal lesions, its existence in other brain structures also involved in memory formation is now well documented. The potential role of LTP as a memory storage device has since received considerable experimental and theoretical support but, as is the case for phenomena in search of a function, has generated endless debates and controversies and continues to fuel many intense discussions at various neuroscientific meetings. More recently, and almost surreptitiously, the phenomenon of long-term depression (LTD) has been discovered in various brain structures and, as will hopefully become apparent in this book, has slowly gained some official status. Initially observed in the cerebellum and discussed by both experimentalists and theorists as a possible mechanism for motor learning, it has now clearly been observed in hippocampus and neocortex and, because it appears to be the mirror image of LTP, is frequently associated with the latter in scientific debates.

The present volume reflects the current status of the studies directed at understanding the mechanisms underlying LTP and LTD and the roles that these mechanisms play in learning and memory. It will present the contributions of leaders in the field who summarize their most recent research, hopefully providing enough information for the reader to generate a critical conclusion. It also updates and complements a volume we edited two years ago and which presented what we perceived then to be critical issues. Several new themes have been added, including the possible role of nitric oxide and a more elaborate section on LTD. A remarkable characteristic about LTP is that it drives research in a multidisciplinary fashion (e.g., physiology, anatomy, biochemistry, behavior, genetics, computer sciences). Perhaps this should not be considered so unusual in a complex phenomenon attempting to model an even more complex one such as learning and memory. A good example is the question of the role of the gas nitric oxide (NO) in LTP. The NO studies demonstrate once again that evolution conserves and preserves chemical systems with good signaling properties—using these chemical signals in a variety of structures and functions—and also that neuroscientists are anxious to take on fashionable themes. In a rapidly evolving field it is not always easy to determine which themes will survive and which will provide the most fruitful source of ideas for the future. Hopefully, our selection is broad enough to satisfy a wide range of readers who have been interested in the field for a long time or for neophytes who want an introduction to it.

This volume was also the result of a meeting and of the debates and discussions it triggered in the fall of 1992. We would like to thank the local organizers, Drs. Serge Laroche and Yves Frégnac who obtained the ideal setting for this meeting on the grounds of the castle of the Commissariat de l'Energie Atomique at Gif-sur-Yvette. We would like to thank the authors of the chapters who engaged in spirited, courteous, and lively debates before, during, and after the presentations. Finally, we would like to thank the public and private organizations that provided the financial support for the meeting: the Office of Naval Research, the program Cogniscience of the CNRS, Rhone-Poulenc-Rohrer, and Wyeth Pharmaceuticals. Furthermore, we are indebted to Mlles Doyère and Rédini de Négro who made sure that the meeting ran smoothly and to the chairpersons and session rapporteurs who both stimulated and firmly guided our discussions.

M. B.
J. L. D.

Introduction

Long-term potentiation and long-term depression are now the leading candidates for cellular mechanisms of information storage in mammalian CNS. It is now widely believed that these phenomena are due to changes localized to synapses, and as long as the details of the mechanisms regulating synaptic efficacy are not fully understood, a number of fundamental questions will need to be answered: (1) Is the locus of expression of the changes in synaptic efficacy presynaptic or postsynaptic and, if it is presynaptic while the triggering mechanisms are postsynaptic, are there retrograde signals that could produce long-lasting changes in the characteristics of transmitter release? (2) If the locus of expression is postsynaptic, what are the mechanisms responsible for producing long-lasting changes in responsiveness to the neurotransmitter? (3) Are there different mechanisms of synaptic plasticity in different brain structures, and what are the relationships between mechanisms producing increase or decrease in synaptic efficacy? (4) What are the relationships between artificially induced modifications of synaptic efficacy and the phenomenon (or phenomena) of learning and memory, and what can we learn by implementing rules of synaptic plasticity in artificial networks that either do or do not mimic biological networks? This book is not intended to provide definitive answers to these questions but certainly represents the most current views and hypotheses developed by scientists deeply involved in finding these answers. This book also is an extension of a previously edited volume (*Long-term Potentiation: A Debate of Current Issues,* Baudry and Davis (eds.), MIT Press, 1991) and revisits some of its themes, as well as introducing new themes that have emerged during the past two years. Our choice of topics was guided by a desire to ensure a continuity in the presentation as well as the investigation and discussion of contemporary themes. Thus, those chapters that are written by the same authors will permit an estimation of the progress accomplished on some questions during the past two years, whereas new themes are reviewed by newcomers actively engaged in the topics we selected. As in the previous volume, the book is divided into five sections related to the general issues mentioned above, with several sections accompanied with a brief discussion analyzing and summarizing the main points of agreements/controversies.

I. Role of Nitric Oxide in LTP An important development during recent years has been the notion that there has to exist retrograde messengers which can transfer information from the postsynaptic side to the presynaptic side. The existence of retrograde messengers has received a lot of press coverage, and it is symptomatic that NO was selected Molecule of the Year for 1992 by *Science* magazine. A number of groups have suggested not only that NO is a retrograde messenger at a number of CNS synapses but also that NO plays a critical role in both LTP and LTD induction. Additional messengers have been proposed, such as arachidonic acid by Bliss and colleagues and more recently carbon monoxide by Kandel and Stevens. Because of space limitation, we limited the presentation to the discussion of the role of NO in LTP. Two opposite points of view are represented in this section, with two groups presenting convincing experimental evidence for and one group with similarly convincing evidence against. One of us (M. B.) attempted to provide a critical evaluation of the available evidence and to underline major problems raised by the hypothesis in favor of a critical role for NO in LTP induction. The commentary also proposes a series of criteria which should be met by putative retrograde messengers.

II. Expression and Maintenance Mechanisms Despite continuous progress in our understanding of the mechanisms involved in triggering LTP and possibly LTD, the mechanisms responsible for expressing and maintaining the increased synaptic responsiveness observed following the induction of LTP remain to be clearly identified. Our choice of topics again reflects the difficulties that scientists are facing in making progress on this question, and the various chapters summarize different experimental approaches and ideas. Although it seems well admitted that an increase in intracellular calcium is the trigger to activate the cascade of events leading to LTP as reviewed by Manabe et al., the duration of the calcium signal has been a matter of debate. A number of experiments indicate that the calcium signal is probably very brief, at least as far as free calcium is concerned. This does not necessarily imply that LTP takes place immediately as shown by the experiments of D. Muller, which clearly confirm the existence of a relatively long period of time (20–30 min) during which certain manipulations can disrupt previously established LTP. This agrees well with the notion that there exists a stabilization phase for LTP. R. Malinow continues his attempts at using quantal analysis techniques to resolve the question of the possible locus (or loci) of the modifications underlying LTP, and his new results indicate that both presynaptic and postsynaptic modifications are involved in LTP. This compromise would satisfy opposing factions in the LTP field and therefore would easily gain widespread approval in the neuroscience community. Finally, R. Malenka presents the results of a series of experiments which indicate that, depending on the conditions, activation of NMDA receptors can lead to either LTP or LTD, and thus provides a natural transition for the following section.

III. Relationships between LTP and LTD Increasing evidence indicates that long-lasting decrease in synaptic efficacy, generally defined as long-term depression (LTD), is observed in a number of pathways as a result of certain frequency of presynaptic activity. This form of synaptic plasticity has a relatively long history in the cerebellum, and F. Crepel presents a brief summary of the current views of the cellular mechanisms which are likely to underly it in this structure. Henri Korn and his collaborators then describe their elegant experiments using the Mauthner cell and stress a point which is often ignored, that potentiation or depression need not be restricted to excitatory synapses. In particular, they show that the synaptic efficacy of inhibitory synapses as well as those of excitatory synapses can be increased as a result of activity. Whether potentiation of an inhibitory synapse is functionally equivalent to depression of an excitatory synapse remains to be elucidated. The story in other brain structures is not as clear, and several chapters review experimental data that, while producing a convincing case for the existence of the phenomenon, leave open the issues of the mechanisms. In particular, Stanton et al. and Bindman et al. present their data concerning LTD in hippocampus and discuss possible cellular mechanisms involved in the triggering of this phenomenon. A. Artola develops the hypothesis for the existence of two thresholds for LTD and LTP based on changes in intracellular calcium concentrations. This idea is also present and discussed in the review by Frégnac and his collaborators.

IV. Relationships between LTP and Learning and Memory This issue was also debated in the previous volume but clearly remains a theme for endless discussions. Recent data provide quite convincing arguments that LTP is indeed a memory storage mechanism, and we thought it was important to continue to update the readers on these new developments. As reflected in the title of the review by Eichenbaum and Otto, "the hippocampus, LTP, and memory: Enhancing the connection," the use of olfactory learning has proven to be extremely fruitful to understand the links between LTP and learning. Two other reviews of the properties and functions of LTP in olfactory cortex by Haberly et al. and Roman et al. strengthen this idea and make the point that the relatively simple anatomy of the piriform cortex, its connections with the hippocampus, and its intimate links with olfaction provide unique opportunities to study the function of LTP in olfactory learning. In particular, the chapter by Roman et al. indicates that phenomena which look like LTP and LTD occur in piriform cortex during natural behavior and accompany learning processes.

V. Synaptic Plasticity and Computational Neurobiology The last section of the book deals with the roles of changes in synaptic efficacy in the functioning of different neuronal networks. As our understanding of the functions of most networks is still primitive, these questions have recently been extensively approached through the use of computer simulations of realistic biological neuronal networks. This theme was also present in the previous volume but, as it represents a rapidly evolving field which is of interest not only for neuroscientists but also for computer scientists and applied mathematicians, we considered it worthwhile to present two different views and approaches of groups attempting to provide a global understanding of hippocampal function. Gilbert Chauvet and Ted Berger review their approach, which combines a mathematical model derived from physics theories with experimental data generated from studies of hippocampal function. The rationale for such an approach is that the use of an n-level field theory allows the integration of data from the molecular up to the system level and that the dynamics of the system can be studied once the underlying structures and the rules of plasticity of the system are elucidated. On the other hand, Richard Granger, Gary Lynch, and their collaborators provide an update of their original analysis of the computational operations performed by the piriform cortex/hippocampal circuitries and propose some new hypotheses concerning cortical encoding of information.

Again, it is clear that this volume will not provide the ultimate answers to several of the current issues in the field of LTP/LTD. What it does provide, however, is a clear picture of the progress made during the past two years. If we have raised more questions than answers, we believe this to be the best evidence why this dynamic area of the biological sciences continues to fascinate scientists interested in issues of learning and memory.

I The Role of Nitric Oxide in Long-Term Potentiation

1 Nitric Oxide as a Physiological Messenger in Long-Term Potentiation and Memory Formation: Electrophysiological and Behavioral Evidence

Christelle Bon, Martine Lemaire, Odile Piot, Michel Reibaud, Jean-Marie Stutzmann, Adam Doble, and Georg Andreas Böhme

Nitric oxide (NO) is one of the most surprising messenger molecules produced by mammalian cells. This small, diffusible radical was first shown to be the endothelial-derived relaxing factor (EDRF) responsible for the relaxation of vascular smooth muscles in response to bradykinin (Ignarro et al., 1987; Palmer et al., 1987). Further work identified L-arginine as the source of NO (Palmer et al., 1988) and extended our knowledge of the physiological function of NO to the inhibition of platelet aggregation and adhesion (Moncada, 1992). Studies in immunology showed that NO is also produced from L-arginine by activated macrophages (Marletta et al., 1988) and is involved in their tumoricidal (Stuehr and Nathan, 1989) and antimicrobial activities (Nathan and Hibbs, 1991).

The ascent of NO to the brain followed the finding by John Garthwaite's laboratory that the activation of *N*-methyl-D-aspartate (NMDA) receptors results in the production of EDRF/NO in the cerebellum (Garthwaite et al., 1988). NO is produced from the oxidation of one of the guanidino nitrogen atoms of L-arginine and acts on guanylate-cyclase to increase cyclic GMP levels in target cells (Garthwaite, 1991). This oxidation is catalysed by an NO synthase (NOS) that forms L-citrulline as a coproduct, a feature used to monitor NOS activity in brain preparations (Bredt and Snyder, 1989). Both cGMP and L-citrulline formation are inhibited in an L-arginine–sensitive manner by L-arginine analogs that are substituted on the guanidino moiety (Garthwaite et al., 1989; Bredt and Snyder, 1989).

NOS was purified from rat and human cerebella as a soluble flavoprotein with a monomer molecular weight of 160 kDa. This enzyme contains FAD and FMN and its activity is dependent on NADPH. In both species, NOS enzymatic activity appears to be both calcium/calmodulin- and tetrahydrobiopterin-dependent (Bredt and Snyder, 1990; Schmidt and Murad, 1991; Schmidt et al., 1992). Molecular cloning of brain NOS revealed sequence homology with cytochrome P-450 reductase (Bredt et al., 1991), and recent evidence also indicates that brain NOS contains a P-450 type intramolecular heme moiety (Klatt et al., 1992; McMillan et al., 1992; Stuehr and Ikeda-Saito, 1992).

Long-term potentiation (LTP) in the hippocampus is one of the most thoroughly studied mechanisms of synaptic plasticity, a process by which the brain is believed to acquire, store, or retrieve information (for extensive reviews, see Landfield and Deadwyler, 1988; Baudry and Davis, 1991; Bliss and Collingridge, 1993). At the synapses between Schaffer collaterals and CA1 pyramidal cells, LTP is observed after brief tetanic stimulation of the afferent pathway and involves activation of NMDA receptor channels leading to postsynaptic calcium influx (Collingridge et al., 1983; Lynch et al., 1983; Asher and Nowak, 1988, Malenka et al., 1988). The maintenance of an enhanced synaptic transmission in LTP is thought to be related, at least in part, to presynaptic events (Bliss et al., 1986; Bekkers and Stevens, 1990; Malinow and Tsien, 1990). This suggests the existence of retrograde messengers mediating post-to-presynaptic signaling during LTP. Arachidonic acid was the first of such putative agents identified as being able to increase synaptic strength in the hippocampus (Williams et al., 1989). However, its effect appears too slow to account for the rapid installation of LTP. The hypothesis emerged that the small, hydrophobic radical NO that is produced in response to NMDA-receptor activation might be involved in early post-tetanic phases. A theoretical basis for this hypothesis was provided by computer simulation of brain functioning showing that a short-lived and highly diffusible molecule, such as NO, could influence the strength of neighboring synapses (Gally et al., 1990). In the work summarized herein using pharmacological tools that inhibit or mimic local NO formation, we provide experimental evidence that endogenous NO formation is indeed functionally involved in LTP. We also report that blocking NO production in the intact animal impairs learning in hippocampal-dependent learning tasks, a result supporting the existence of a functional link between LTP and some forms of memory.

ELECTROPHYSIOLOGICAL STUDIES

LTP was studied in vitro, in rat hippocampal slices, at the level of the synapse between Schaffer-collateral/commissural fibers and CA1 pyramidal cells. Synaptic efficacy was recorded extracellularly by monitoring every 5 sec the slope of the field excitatory postsynaptic potentials (EPSPs) in the stratum radiatum. All drugs were applied through the superfusion bath. Tetanic stimulation consisted of two bursts (1 sec duration) of high-frequency (100 Hz) stimuli delivered at twice the baseline voltage. The hippocampal slices were maintained at 32°C during the experiments. Other details are as described elsewhere (Bon et al., 1992).

NOS Inhibitors Block LTP Induction

The strong tetanic stimulation consistently elicited LTP in control slices, but was unable to enhance the EPSP slopes in slices tetanized in the presence of a low concentration of L-Ng-nitro-arginine (L-NA), a potent NOS inhibitor (East and Garthwaite, 1990; Moore et al., 1990). In contrast, slices treated with D-NA, the enantiomer of L-NA, consistently exhibited LTP, showing that the blockade was stereoselective. The effect of L-NA could be prevented by pre-incubating the slices with an excess of L-arginine, but not D-arginine, suggesting that the inhibitory effect of L-NA was specific to NOS. L-arginine alone had no obvious effect on the magnitude of LTP, although it does enhance NMDA-induced increases in cGMP levels in the cerebellum (Garthwaite et al., 1989). Figure 1.1 summarizes these findings.

To confirm and extend these observations, we compared the effects of various concentrations of L-NA, its methyl ester derivative L-NAME, and L-Ng-monomethyl-arginine (L-NMMA), an early identified guanidino-substituted arginine analog able to block NO-mediated responses (Moncada, 1992). L-NA and L-NAME blocked LTP in a concentration-dependent manner, with half-maximal inhibitory concentrations in the nanomolar range. L-NMMA, however, appeared to be much weaker than the nitro-substituted derivatives in its ability to prevent LTP (figure 1.2). This difference is somewhat surprising since L-NA and L-NMMA exhibit similar potencies

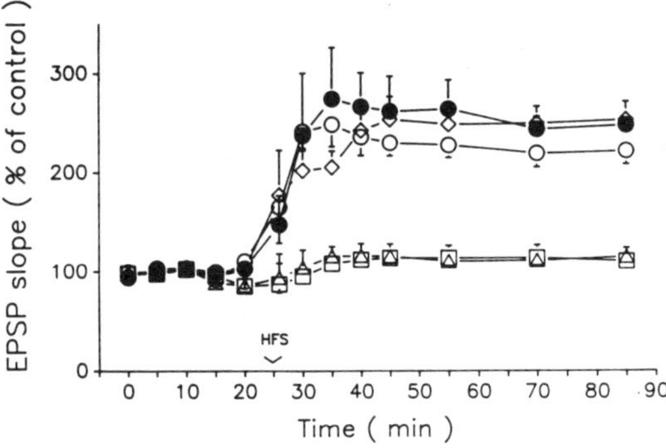

Figure 1.1 Blockade of LTP by the NOS inhibitor L-Ng-nitro-arginine. Changes in synaptic strength for untreated slices (close circles, n = 7) and slices exposed for 20 min (starting from time 10 min) to 0.1 μM L-NA (squares, n = 6) or 0.1 μM D-NA (open circles, n = 7). Also shown are the effects of 0.1 μM L-NA on slices superfused for 2 hr with medium containing 100 μM L-arginine (diamonds, n = 6) or 100 μM D-arginine (triangles, n = 4). Each point represents the mean ± SEM field EPSP slope plotted against time in minutes. Values were normalized relative to the mean of three measurements taken 5 min apart at the beginning of the recording. High-frequency stimulation (HFS) was delivered at the time marked by the arrow. (Data from Böhme et al., 1991)

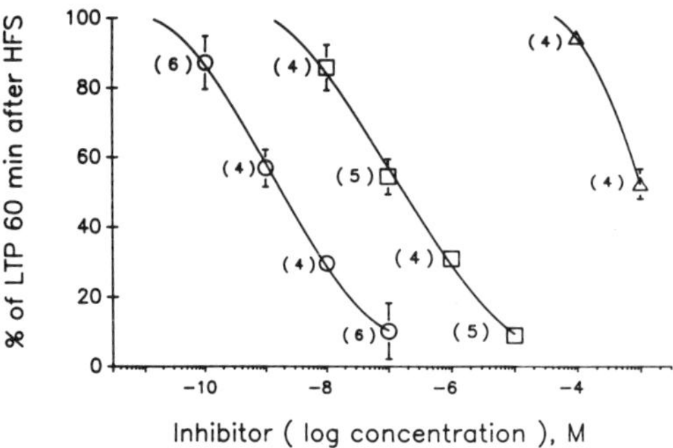

Figure 1.2 Concentration-response curves for inhibitory effects of various NOS inhibitors. Percentage of LTP in treated slices was calculated from the normalized field EPSP slope values (1 hr after tetanic stimulation) with respect to untreated control slices from the same brains as follows: $\%\text{LTP} = 100 - [(\text{EPSP}_{\text{control}} - \text{EPSP}_{\text{treated}})/(\text{EPSP}_{\text{control}} - 100) \times 100$. Numbers in brackets indicate the number of treated slices for each concentration of the inhibitor. Circles, L-NA; squares, L-NAME; triangles, L-NMMA. (Reprinted from Bon et al., 1992 with permission of Oxford University Press. © 1992 European Neuroscience Association)

on the purified cerebellar enzyme (Schmidt et al., 1991). There is, however, evidence that L-NA inhibits components of NMDA-induced cGMP formation in cerebellar and hippocampal slices (East and Garthwaite, 1990, 1991) and in striatal neurons in culture (Marin et al., 1992) several orders of magnitude more potently than the monomethylated derivative. Whether this is due to the existence of isoforms of NOS particularly sensitive to the nitro-derivative remains to be established.

L-NA also affected LTP in hippocampal slices prepared from rats previously treated intraperitoneally with the NOS inhibitor. To ensure a maximal blockade of the enzyme, the inhibitor was given twice daily for 4 days before preparing the slices (Dwyer et al., 1991). L-NA (25–100 mg/kg i.p.) blocked LTP dose-dependently; doses of 100 mg/kg were necessary for complete blockade of LTP (Böhme et al., 1993). These findings suggest that the NOS inhibitor L-NA given systemically is able to penetrate the brain to block hippocampal LTP.

Several other laboratories working independently have also observed the inhibitory effect of guanidino-substituted arginine analogs on LTP. Bath applications of micromolar concentrations of L-NA or L-NMMA have been found to prevent LTP induced by tetanic stimulation or by pairing depolarization with presynaptic stimulation during intracellular recordings with NOS inhibitor–containing electrodes (O'Dell et al., 1991; Schuman and Madison, 1991). The involvement of guanylate cyclase as a molecular target for NO in LTP is suggested by the finding that dibutyryl-cGMP partially reverses the blockade of LTP by L-NAME (Haley et al., 1992).

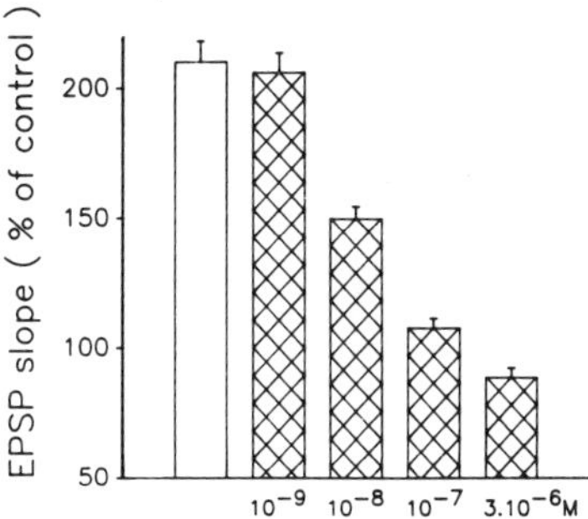

Figure 1.3 Blockade of LTP by the NO-scavenger hemoglobin. Mean ± SEM field EPSP slope (1 hr after tetanic stimulation) for pooled control slices (n = 14, open bar) and slices tetanized 15 min after bath application of reduced bovine hemoglobin (hatched bars: n = 4–6 slices per concentration). (Data from Bon et al., 1992)

There is also evidence that an ADP-ribosyltransferase may fulfill this role (Schuman et al., 1992; Duman et al., 1993).

Further arguments in favor of a role for endogenous NO production in LTP came from studies with hemoglobin. This voluminous protein potently binds NO and suppresses NO-mediated responses in vascular (Martin et al., 1985) and brain tissue (Garthwaite et al., 1988). As also reported by others, we observed that bath applications of oxyhemoglobin blocked LTP. As with the NOS inhibitors, the effect of hemoglobin appeared to be concentration-dependent (figure 1.3). Since hemoglobin does not penetrate neurons, the scavenging of NO happens in the extracellular space. This observation alone, however, does not demonstrate that the effect of NO is presynaptic. The high affinity of NO for hemoglobin could possibly create an extracellular sink, preventing NO from reaching a critical intracellular level. Recent evidence indicates that the potentiating effect of NO in hippocampal slices occurs only upon simultaneous afferent activity, a finding consistent with the idea that NO acts presynaptically (Zhuo et al., 1993).

NO Donors Mimic LTP

To test whether exogenous NO is able to influence synaptic transmission, we used various chemicals that conveniently release this compound in brain slices (Southam and Garthwaite, 1991). Bath application of millimolar concentrations of sodium nitroprusside (SNP), a substance that spontaneously releases NO, or hydroxylamine, which releases NO once it is metabolized, potentiates synaptic transmission. These drugs increased the EPSP slopes to 150–200% of baseline values, a level where they remained for at least 1 hr (Böhme et al., 1991; Bon et al., 1992). The similarity of this

NO-induced potentiation with electrically induced LTP was tested with occlusion experiments. When tetanic stimulation was delivered to slices potentiated by NO donors, only a transient potentiation was observed, indicating that LTP was not additive with the NO donor–induced potentiation. Conversely, when slices already potentiated by tetanus application were exposed to SNP or hydroxylamine, no further potentiation was seen (Böhme et al., 1991; Bon et al., 1992). A similar mutual occlusion had previously been observed for potentiation induced by arachidonic acid in vivo (Williams et al., 1989).

These results are consistent with the idea that NO is involved in the installation of LTP. If NO plays the role of a retrograde messenger, the potentiation brought about by the NO donors should be independent of NMDA-receptor activation. To test this hypothesis, we examined the effects of NO donors in the presence of NMDA-receptor antagonists. As shown in figure 1.4, hydroxylamine applied in the presence of a concentration of D,L-aminophosphonovaleric acid (APV) that fully blocked LTP retained its ability to enhance synaptic transmission in a nonadditive manner with tetanus-induced LTP.

BEHAVIORAL LEARNING TESTS

The role of NO in learning tasks requiring spatial memory formation was assessed by training rats to explore an eight-arm radial maze (Olton and Samuelson, 1976; Olton, 1987). Food-deprived rats were trained every day

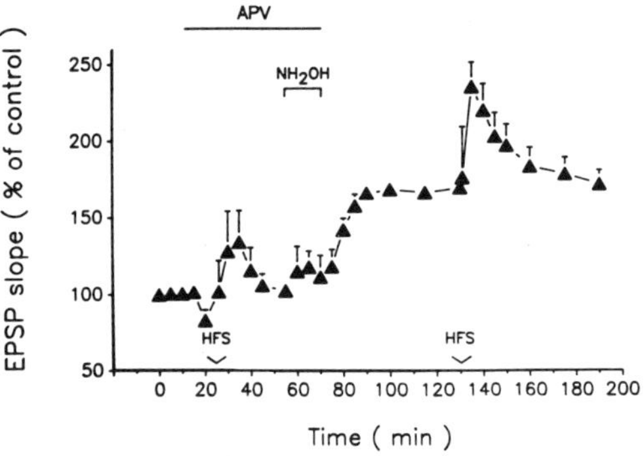

Figure 1.4 Potentiation of synaptic transmission by hydroxylamine. Mean ± SEM field EPSP slope values for n = 6 slices. APV (10 μM) was applied to the slices during the time indicated by the uppermost bar above the graph. High-frequency stimulation (HFS) delivered at time 25 min increased synaptic strength only transiently. The NO donor hydroxylamine (1 mM) was subsequently applied for 15 min in the continuing presence of APV. Note the stable enhancement of synaptic strength 20 min after hydroxylamine washout. A second HFS applied to these potentiated slices at time 130 min did not result in further potentiation.

for 12 days to collect bait placed at the arm extremities. The capacity of the animals to learn about the spatial environment of the maze was measured as the progressive decline in the number of reentries in already visited arms. Rats were pretreated intraperitoneally with a suspension of 100 mg/kg L-NA twice daily for 4 days (as necessary to fully block LTP) or 100 mg/kg D-NA, and then once daily for the remainder of the experiment 90 min before each training session. The mean number of reentries diminished within each group over the 2-week training period, showing that the animals had learned the task. However, rats treated with 100 mg/kg L-NA learned significantly more slowly than did the controls. In contrast, D-NA did not affect radial-maze performances, and total exploration time was not significantly affected by L-NA or D-NA (figure 1.5). All the arms were visited and all the bait was consumed during each session, indicating that treatment with the inhibitor did not affect motivation. L-NA also impaired radial maze learning when water was used as the positive reinforcing stimulus (data not shown).

The effect of the NOS inhibitor on olfactory memory formation was assessed by using a test of social recognition based on the fact that rats explore familiar juvenile conspecifics for a shorter time than unfamiliar

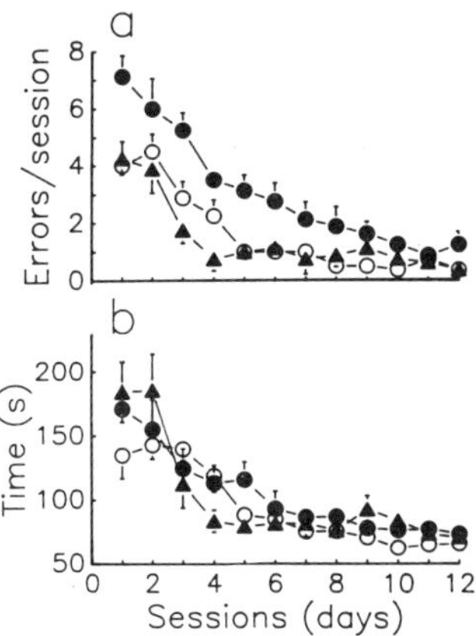

Figure 1.5 Impairment of spatial learning by the NOS inhibitor L-NA. Performances of food-deprived rats in an eight-arm radial maze are represented as (a) the mean number of reentries in already visited arms (= errors) per session and (b) time (seconds) spent visiting the maze on each day of learning. Vertical bars indicate SEM, if greater than symbol size. Open circles, controls (n = 8); closed circles, L-NA 100 mg/kg (n = 8); triangles, D-NA 100 mg/kg (n = 8). Two-way analysis of variance revealed that the overall number of reentries decreased during the training [$F(11, 154) = 22.1$ and 19.3 for L-NA and D-NA, respectively, $P < .0001$] and that L-NA, but not D-NA, treated animals made significantly more errors than the controls [$F(1, 154) = 47.7$, $P < .0001$ and 0.48, $P > .4$ for L-NA and D-NA, respectively]. The same analysis of the time-to-visit data revealed no significant effect of either treatment [$F(1, 154) = 1.5$ and 0.9 for L-NA and D-NA, respectively, $P > .2$]. (Data from Böhme et al., 1993)

Table 1.1 Impairment of rat olfactory memory by repeated pretreatment with L-NA[a]

Treatment	Investigation duration (sec)	
	1st contact (T_0)	2nd contact (T_{30})
Vehicle	77.4 ± 7.4	39.6 ± 7.8[b]
L-NA	82.2 ± 9.8	78.7 ± 13

[a] 100 mg/kg i.p. Duration of social investigation for two consecutive 5 min exposures 30 min apart. Data are means ± SEM (n = 10 couples).
[b] Statistically different from first contact values, $P < .001$, paired t-test.

ones. The recognition derives from olfactory cues and can be measured as the decrease in the length of time a male adult rat spends investigating a male juvenile during the second of two brief encounters separated by 30 min. This parameter is known to be sensitive to drugs influencing cognitive processes (Dantzer et al., 1987; Perio et al., 1989; Lemaire et al., 1992). We exposed adult rats, that had been repeatedly pretreated with 100 mg/kg i.p. L-NA as above, twice to the same juveniles at 30-min intervals to test whether they could recognize their odors. The duration of investigation at the first contact was not affected by the NOS inhibitor, suggesting that the animals' sense of smell was not altered. However, the L-NA–treated adults investigated the juveniles during the second contact for as long as during the first contact, while the vehicle-treated controls showed a decrease by about one half of investigation duration (table 1.1). This suggests that L-NA impaired the olfactory recognition of the juveniles.

These experiments suggest that NO is involved in spatial and olfactory learning. Other investigators found that inhibition of NO synthesis impairs conditioned eyeblinking in rabbits (Chapman et al., 1992) and taste avoidance learning in chicks (Hölscher and Rose, 1992). NO appears therefore to be involved in the formation of some kinds of memory. NO also plays a major role in the control of peripheral processes, such as blood pressure regulation, and systemic administration of L-NA induces marked hypertension (Moncada, 1992). The absence of motor deficit in the radial maze and the absence of changes in the amount of social interaction at the first contact after chronic treatment with L-NA argue against an interference of these peripheral effects with our interpretation of the learning tests. Moreover, certain other forms of memory such as shock-avoidance learning in rats seems not to be affected by L-NA at doses up to 300 mg/kg (Böhme et al., 1993).

That identical doses of L-NA disrupt learning and alter synaptic plasticity in the hippocampus provides support for the existence of a link between LTP and memory formation. Spatial and olfactory learning are known to be sensitive to hippocampal lesions (Morris et al., 1982; Staubli et al., 1984; Eichenbaum et al., 1989). Both forms of memory can be pharmacologically impaired by infusing NMDA antagonists into the brain (Staubli et al., 1989; Davis et al., 1992), and there is a correlation between doses of NMDA antagonists impairing spatial learning and those blocking LTP (Davis et al., 1992). There is also a linear relationship between LTP and learning of a classical tone/shock conditioning paradigm (Laroche et al., 1989; Doyère and Laroche, 1992). Two recent studies, in which gene knockout techniques were used to alter the expression of key enzymes in the LTP process, further support the hypothesis. Mutant mice lacking either the alpha isoform of calcium-calmodulin-dependent kinase II (Silva et al., 1992a,b) or the *fyn* tyrosine kinase (Grant et al., 1992) simultaneously lost their capacity both to express normal LTP and to learn spatial tasks.

CONCLUSIONS

Seven different laboratories have now provided evidence in favor of the involvement of NO in hippocampal LTP (O'Dell et al., 1991; Schuman and Madison, 1991; Gribkoff and Lum-Ragan, 1992; Haley et al., 1992; Izumi et al., 1992; Li et al., 1992; Musleh et al., 1993). Other observations suggest that NO is involved in long-term depression in the cerebellum (Crepel and Jaillard, 1990; Ito and Karachot, 1990; Shibuki and Okada, 1991), although evidence obtained with brain slice preparations has not been reproduced with Purkinje cell cultures (Linden and Connor, 1992). NO donors have also been shown to influence synaptic efficacy in the rat superior cervical ganglion (Briggs, 1992). Therefore, NO is probably a common mediator for synaptic processes supporting plasticity in different regions of the central nervous system. The role of NO may be rather complex, since experimental conditions can largely influence the effects observed. Some investigators report that the effect of NOS inhibitors in the hippocampus is temperature-dependent, since L-NA blocked LTP in both the CA1 area and dentate gyrus at room temperature (24°C) but failed to do so in the CA1 area at 29°C (Errington et al., 1991; Li et al., 1992). Others suggest the existence of NOS inhibitor–sensitive and –insensitive forms of LTP as revealed by varying the strength of tetanic stimulation (Gribkoff and Lum-Ragan, 1992). NOS inhibitors can also paradoxically restore LTP in the CA1 area under conditions in which endogenous NO is released tonically through prior activation of NMDA receptors (Izumi et al., 1992), an effect that reflects the negative feedback inhibition NO exerts on NMDA receptor function (Lei et al., 1992; Manzoni et al., 1992).

The cellular origin of NO during LTP has been questioned. NADPH-diaphorase staining, a histochemical reaction thought to trace NOS, shows this enzyme to be absent from the CA1 pyramidal cell bodies (Vincent and Hope, 1992). However, detailed immunohistochemical and in situ hybridization studies indicate the presence of NOS in a population of non-pyramidal interneurons that extend their processes widely throughout the CA1 pyramidal cell body area (Dawson et al., 1992; Schmidt et al., 1992a; Valtschanoff et al., 1992; Vincent and Hope, 1992). These local cells may be the actual source of NO during LTP. Alternatively, other enzymes may also participate in the generation of NO, so that its production would not necessarily be restricted to NOS-positive cells. Consistent with this idea, it has recently been shown that L-hydroxy-arginine, a stable intermediate in the oxidation of L-arginine by NOS (Stuehr et al., 1991), is converted to NO and L-citrulline by liver microsome P-450s and that this conversion is sensitive to neither L-NA nor L-NMMA (Boucher et al., 1992).

In conclusion, the endogenous production of NO appears to be essential to the LTP process in the CA1 area of the hippocampus. The highly original properties of NO allow it to mediate tridimensional signaling in a sphere of brain tissue defined by its high diffusibility, but short half-life (Edelman and Gally, 1992). This makes NO particularly well suited to play the role of a retrograde messenger, the existence of which is suspected in the presynaptic hypothesis of LTP.

REFERENCES

Ascher, P., and Nowak, L. (1988) Electrophysiological studies of NMDA receptors. *Trends Neurosci.* 10:284–288.

Baudry, M., and Davis, J. (eds) (1991) *Long-term Potentiation: A Debate of Current Issues.* Cambridge, Mass.: MIT Press.

Bekkers, J., and Stevens, C. F. (1990) Presynaptic mechanism for long-term potentiation in the hippocampus. *Nature* 346:724–729.

Bliss, T. V. P., and Collingridge, G. L. (1993) A synaptic model of memory: long-term potentiation in the hippocampus. *Nature* 361:31–39.

Bliss, T. V. P., Douglas, R. M., Errington, M. L., and Lynch, M. A. (1986) Correlation between long-term potentiation and release of endogenous amino acids from dentate gyrus of anaesthetized rats. *J. Physiol.* 377:391–408.

Böhme, G. A., Bon, C., Stutzmann, J. M., Doble, A., and Blanchard, J. C. (1991) Possible involvement of nitric oxide in long-term potentiation. *Eur. J. Pharmacol.* 199:379–381.

Böhme, G. A., Bon, C., Lemaire, M., Reibaud, M., Piot, O., Stutzmann, J. M., Doble, A., and Blanchard, J. C. (1993) Altered synaptic plasticity and memory formation in nitric oxide synthase inhibitor treated rats. *Proc. Natl. Acad. Sci. USA* 90:9191–9194.

Bon, C., Böhme, G. A., Doble, A., Stutzmann, J. M., and Blanchard, J. C. (1992) A role for nitric oxide in long-term potentiation. *Eur. J. Neurosci.* 4:420–424.

Boucher, J.-L., Genet, A., Vadon, S., Delaforge, M., Henry, Y., and Mansuy, D. (1992) Cytochrome P450 catalyses the oxidation of Nω-hydroxy-L-arginine by NADPH and O_2 to nitric oxide and citrulline. *Biochem. Biophys. Res. Commun.* 187:880–886.

Bredt, D. S., and Snyder, S. H. (1989) Nitric oxide mediates glutamate-linked enhancement of cGMP levels in the cerebellum. *Proc. Natl. Acad. Sci. USA* 86:9030–9033.

Bredt, D. S., and Snyder, S. H. (1990) Isolation of nitric oxide synthetase, a calmodulin-requiring enzyme. *Proc. Natl. Acad. Sci. USA* 87:682–685.

Bredt, D. S., Hwang, P. M., Glatt, C. E., Lowenstein, C., Reed, R. R., and Snyder, S. H. (1991) Cloned and expressed nitric oxide synthase structurally resembles cytochrome P-450 reductase. *Nature* 351:714–718.

Briggs, C. A. (1992) Potentiation of nicotinic transmission in rat superior cervical sympathetic ganglion: Effect of cyclic GMP and nitric oxide generators. *Brain Res.* 573:139–146.

Chapman, P. F., Atkins, C. M., Allen, M. T., Haley, J. E., and Steinmetz, J. E. (1992) Inhibition of nitric oxide synthesis impairs two different forms of learning. *NeuroReport* 3:567–570.

Collingridge, G. L., Kehl, S. L., and McLennan, H. (1983) Excitatory amino acids in synaptic transmission in the Schaffer collateral-commissural pathway of the rat hippocampus. *J. Physiol.* (*Lond.*) 334:33–46.

Crepel, F., and Jaillard, D. (1990) Protein kinases, nitric oxide and long-term depression of synapses in the cerebellum. *NeuroReport* 1:133–136.

Dantzer, R., Bluthe, R. M., Koob, G. F., and Le Moal, M. (1987) Modulation of social memory in male rats by neurohypophyseal peptides. *Psychopharmacology* 91:363–368.

Davis, S., Butcher, S. P., and Morris, R. G. M. (1992) The NMDA receptor antagonist D-2-amino-5-phosphonopentanoate (D-AP5) impairs spatial learning and LTP in vivo at intracerebral concentrations comparable to those that block LTP in vitro. *J. Neurosci.* 12:21–34.

Dawson, T. M., Dawson, V. L., and Snyder, S. H. (1992) A novel neuronal messager molecule in brain: The free radical, nitric oxide. *Ann. Neurol.* 32:29–311.

Doyère, V., and Laroche, S. (1992) Linear relationship between the maintenance of hippocampal long-term potentiation and retention of an associative memory. *Hippocampus* 2:39–48.

Duman, R. S., Terwilliger, R. Z. and Nestler, E. J. (1993) Alteration in nitric-oxide stimulated endogenous ADP-ribosylation associated with long-term potentiation in rat hippocampus. *J. Neurochem.* 61:1542–1545.

Dwyer, M. A., Bredt, D. S., and Synyder, S. H. (1991) Nitric oxide synthase: Irreversible inhibition by L-Ng-nitroarginine in brain in vitro and in vivo. *Biochem. Biophys. Res. Commun.* 176:1136–1141.

East, S. J., and Garthwaite, J. (1990) Nanomolar Ng-nitroarginine inhibits NMDA-induced cyclic GMP formation in rat cerebellum. *Eur. J. Pharmacol.* 184:311–313.

East, S. J., and Garthwaite, J. (1991) NMDA receptor activation in rat hippocampus induces cyclic GMP formation through the L-arginine-nitric oxide pathway. *Neurosci. Lett.* 123:17–19.

Edelman, G. M., and Gally, J. A. (1992) Nitric oxide: Linking space and time in the brain. *Proc. Natl. Acad. Sci. USA* 89:11651–11652.

Eichenbaum, H., Mathews, P., and Cohen, N. J. (1989) Further studies of hippocampal representation during odor discrimination learning. *Behav. Neurosci.* 103:1207–1216.

Errington, M. L., Li, Y.-G., Matthies, H., Williams, J. H., and Bliss, T. V. P. (1991) The nitric oxide synthase inhibitor Nω-nitro-L-arginine reduces the magnitude of long-term potentiation in the dentate gyrus but not in area CA1 of the hippocampus. *Soc. Neurosci. Abstr.* 17:951.

Gally, J. A., Read Montague, P., Reeke G. N., Jr., and Edelman, G. M. (1990) The NO hypothesis: Possible effects of a short-lived, rapidly diffusible signal in the development and function of the nervous system. *Proc. Natl. Acad. Sci. USA* 87:3547–3551.

Garthwaite, J. (1991) Glutamate, nitric oxide and cell-cell signalling in the nervous system. *Trends Neurosci.* 14:60–67.

Garthwaite, J., Charles, S. L., and Chess-Williams, R. (1988) Endothelium-derived relaxing factor release on activation of NMDA receptors suggests role as intercellular messenger in the brain. *Nature* 336:385–388.

Garthwaite, J., Garthwaite, G., Palmer, R. M. J., and Moncada, S. (1989) NMDA receptor activation induces nitric oxide synthesis from arginine in rat brain slices. *Eur. J. Pharmacol.* (*Mol. Pharmacol.*) 172:413–416.

Grant, S. G. N., O'Dell, T. J., Karl, K. A., Stein, P. L., Sorieno, P., and Kandel, E. R. (1992) Impaired long-term potentiation, spatial learning and hippocampal development in *fyn* mutant mice. *Science* 258:1903–1910.

Gribkoff, V. K., and Lum-Ragan, J. T. (1992) Evidence for nitric oxide synthase inhibitor-sensitive and insensitive hippocampal synaptic potentiation. *J. Neurophysiol.* 68:639–642.

Haley, J. E., Wilcox, G. L., and Chapman, P. F. (1992) The role of nitric oxide in hippocampal long-term potentiation. *Neuron* 8:211–216.

Hölscher, C., and Rose, S. P. R. (1992) An inhibitor of nitric oxide synthesis prevents memory formation in the chick. *Neurosci.* 145:165–167.

Ignarro, L. J., Byrns, R. E., Buga, G. M., and Wood, K. S. (1987) Endothelium-derived relaxing factor from pulmonary artery and vein possesses pharmacologic and chemical properties identical to those of nitric oxide radical. *Circ. Res.* 61:866–879.

Ito, M., and Karachot, L. (1990) Messengers mediating long-term desensitization in cerebellar Purkinje cells. *NeuroReport* 1:129–132.

Izumi, Y., Clifford, D. B., and Zorumski, C. F. (1992) Inhibition of long-term potentiation by NMDA-mediated nitric oxide release. *Science* 257:1273–1276.

Klatt, P., Schmidt, K., and Mayer, B. (1992) Brain nitric oxide synthase is a haemoprotein. *Biochem. J.* 288:15–17.

Landfield, P. W., and Deadwyler, S. A. (1988). *Long-term Potentiation: From Biophysics to Behavior.* New York: Alan R. Liss.

Laroche, S., Doyère, V., and Bloch, V. (1989). Linear relation between the magnitude of long-term potentiation in the dentate gyrus and associative learning in the rat: A demonstration using commissural inhibition and local infusion of an *N*-methyl-D-aspartate receptor antagonist. *Neuroscience* 28:375–386.

Lei, S. Z., Pan, Z. H., Aggarwal, S. K., and Chen, H. S. V. (1992) Effect of nitric oxide production on the redox modulatory site of NMDA receptor-channel complex. *Neuron* 8:1087–1099.

Lemaire, M., Piot, O., Roques, B. P., Böhme, G. A., and Blanchard, J.-C. (1992) Evidence for an endogenous cholecystokininergic balance in social memory. *NeuroReport* 3:929–932.

Li, Y.-G., Errington, M. L., Williams, J. H., and Bliss, T. V. P. (1992) Temperature-dependent block of LTP by NO synthase inhibitor L-NARG. *Soc. Neurosci. Abstr.* 18:343.

Linden, D., and Connor, J. A. (1992) Long-term depression of glutamate currents in cultured cerebellar Purkinje neurons does not require nitric oxide signalling. *Eur. J. Neurosci.* 4:10–15.

Lynch, G., Larson, J., Kelso, S., Barrionuevo, G., and Schottler, F. (1983) Intracellular injections of EGTA block induction of hippocampal long-term potentiation. *Nature* 305:719–721.

Malenka, R. C., Kauer, J. A., Zucker, R. S., and Nicoll, R. A. (1988) Postsynaptic calcium is sufficient for potentiation of hippocampal synaptic transmission. *Science* 242:81–84.

Malinow, R., and Tsien, R. W. (1990) Presynaptic enhancement shown by whole-cell recordings of long-term potentiation in hippocampal slices. *Nature* 346:177–180.

Manzoni, O., Prezeau, L., Marin, P., Deshager, S., Bockaert, J., and Fagni, L. (1992) Nitric oxide-induced blockade of NMDA receptors. *Neuron* 8:653–662.

Marin, P., Lafon-Cazal, M., and Bockaert, J. (1992) A nitric oxide synthase activity selectively stimulated by NMDA receptors depends on protein kinase C activation in mouse striatal neurons. *Eur. J. Neurosci.* 4:425–432.

Marletta, M. A., Yoon, P. S., Iyengar, R., Leaf, C. D., and Wishnok, J. S. (1988) Macrophage oxidation of L-arginine to nitrite and nitrate: Nitric oxide is an intermediate. *Biochemistry* 27:8706–8711.

Martin, W., Villani, G. M., Jothianandan, D., and Furchgott, R. F. (1985) Selective blockade of endothelium-dependent and glyceryl trinitrate-induced relaxation by hemoglobin and by methylene blue in the rabbit aorta. *J. Pharmacol. Exp. Ther.* 232:708–716.

McMillan, K., Bredt, D. S., Hirsch, D. J., Snyder, S. H., and Clark, J. E. (1992) Cloned, expressed rat cerebellar nitric oxide synthase contains stoichiometric amounts of heme, which binds carbon monoxide. *Proc. Natl. Acad. Sci. USA* 89:11141–11145.

Moncada, S. (1992) The L-arginine: Nitric oxide pathway. *Acta Physiol. Scand.* 145:201–227.

Moncada, S., Higgs, E. A., Hodson, H. F., Knowles, R. G., Lopez-Jaramillo, P., McCall, T., Palmer, R. M. J., and Radomski, M. W. (1991) The L-arginine: Nitric oxide pathway. *J. Cardiovasc. Pharmacol.* 17:S1–S9.

Moore, P. K., Al-Swayeh, O. A., Chong, N. W. S., Evans, R. A., and Gibson, A. (1990) L-Ng-nitro arginine (L-NOARG), a novel, L-arginine-reversible inhibitor of endothelium-dependent vasodilatation in vitro. *Br. J. Pharmacol.* 99:408–412.

Morris, R. G. M., Garrud, P., Rawlins, J. N. P., and O'Keefe, J. (1982) Place navigation impaired in rats with hippocampal lesions. *Nature* 297:681–683.

Musleh, W. Y., Shahi, K., and Baudry, M. (1993) Further studies concerning the role of nitric oxide in LTP induction and maintenance. *Synapse* 13:370–375.

Nathan, C. F., and Hibbs, J. B., Jr. (1991) Role of nitric oxide synthesis in macrophage antimicrobial activity. *Curr. Opin. Immunol.* 3:65–70.

O'Dell, T. J., Kandel, E. R., and Grant, S. G. N. (1991) Long-term potentiation in the hippocampus is blocked by tyrosine kinase inhibitors. *Nature* 353:558–560.

Olton, D. (1987) The radial arm maze as a tool in behavioral pharmacology. *Physiol. Behav.* 40:793–797.

Olton, D. S., and Samuelson, R. J. (1976) Remembrance of places passed: Spatial memory in rats. *J. Exp. Psychol.* 2:97–115.

Palmer, R. M. J., Ferrige, A. G., and Moncada, S. (1987) Nitric oxide release accounts for the biological activity of endothelium-derived relaxing factor. *Nature* 327:524–526.

Palmer, R. M. J., Ashton, D. S., and Moncada, S. (1988) Vascular endothelial cells synthesize nitric oxide from L-arginine. *Nature* 333:664–666.

Perio, A., Terranova, J. P., Worms, P., Bluthe, R. M., Dantzer, R., and Biziere, K. (1989) Specific modulation of social memory in rats by cholinomimetic and nootropic drugs, by benzodiazepine inverse agonists, but not by psychostimulants. *Psychopharmacology* 97:262–268.

Schmidt, H. H. H. W., and Murad, F. (1991) Purification and characterization of human NO synthase. *Biochem. Biophys. Res. Commun.* 181:1372–1377.

Schmidt, H. H. H. W., Pollock, J. S., Nakane, M., Gorsky, L. D., Förstermann, U., and Murad, F. (1991) Purification of a soluble isoform of guanylyl cyclase-activating-factor synthase. *Proc. Natl. Acad. Sci. USA* 88:365–369.

Schmidt, H. H. H. W., Smith, R. M., Nakane, M., and Murad, F. (1992a) Ca^{2+}/calmodulin-dependent NO synthase type I: A biopteroflavoprotein with Ca^{2+}/calmodulin-independent diaphorase and reductase activities. *Biochemistry* 31:3243–3249.

Schmidt, H. H. H. W., Gagne, G. D., Nakane, M., Pollock, J. S., Miller, M. F., and Murad, F. (1992b) Mapping of neural nitric oxide synthase in rat suggests frequent colocalization with NADPH diaphorase but not with soluble guanylyl cyclase, and novel paraneural functions for nitrinergic signal transduction. *J. Histochem. Cytochem.* 40:1439–1456.

Schuman, E. M., and Madison, D. V. (1991) A requirement for the intercellular messenger nitric oxide in long-term-potentiation. *Science* 254:1503–1506.

Schuman, E. M., Meffert, M. K., Schulman, H., and Madison, D. V. (1992) A potential role for an ADP-ribosyltransferase (ADPRT) in hippocampal long-term potentiation (LTP). *Soc. Neurosci. Abstr.* 18:761.

Shibuki, K., and Okada, D. (1991) Endogenous nitric oxide release required for long-term synaptic depression in the cerebellum. *Nature* 349:326–328.

Silva, A. J., Stevens, C. F., Tonegawa, S., and Wang, Y. (1992a) Deficient long-term potentiation in alpha-calcium-calmodulin kinase II mutant mice. *Science* 257:201–206.

Silva, A. J., Paylor, R., Wehner, J. M., and Tonegawa, S. (1992b) Impaired spatial learning in alpha-calcium-calmodulin kinase II mutant mice. *Science* 257:206–211.

Southam, E., and Garthwaite, J. (1991) Comparative effects of some nitric oxide donors on cyclic GMP levels in rat cerebellar slices. *Neurosci. Lett.* 130:107–111.

Staubli, U., Ivy, G., and Lynch, G. (1984) Hippocampal denervation causes rapid forgetting of olfactory information in rats. *Proc. Natl. Acad. Sci. USA* 81:5885–5887.

Staubli, U., Thibault, O., DiLorenzo, M., and Lynch, G. (1989) Antagonism of NMDA receptors impairs acquisition but not retention of olfactory memory. *Behav. Neurosci.* 103:54–60.

Stuehr, D., and Ikeda-Saito, M. (1992) Spectral characterization of brain and macrophage nitric oxide synthases cytochrome P-450 like hemeproteins that contain a flavin semiquinone radical. *J. Biol. Chem.* 267:20547–20550.

known that glutamate efflux from hippocampus increases after the induction of LTP (e.g., Dolphin et al., 1982; Bliss et al., 1986). This increase in efflux is prevented by the manipulations that prevent LTP. Since glutamate is the transmitter at the potentiated synapses, it is reasonable to suppose that the terminals may be the source of the glutamate efflux, and that the increase in this efflux represents enhanced transmitter release during LTP. However, it cannot be ruled out that this glutamate arises from another depolarization-released store.

The second line of evidence has used the techniques associated with quantal analysis, analysis of variance, or failure analysis (cf. Bekkers and Stevens, 1990; Malinow and Tsien, 1990; Malinow, 1991; Larkman et al., 1992; Malgaroli and Tsien, 1992; but see Foster and McNaughton, 1991; Manabe et al., 1992). This evidence is controversial because there is frank disagreement about the results and interpretations in the field. The differences in results and interpretations may also result because of a lack of crucial knowledge about the functioning of this particular synapse, because some of the underlying assumptions to these analyses cannot be tested, and because the characteristics of the synapses in the hippocampus are not ideal for performing this analysis. Nonetheless, some of these data have provided evidence that suggests a presynaptic component to LTP, although as noted, there is some disagreement. It may be that mechanisms on both sides of the synapse come into play (Malinow, 1991; Kullmann and Nicoll, 1992). However, if any part of the LTP expression mechanisms contain a presynaptic component, then there must be a way for that component to know that the postsynaptic induction mechanisms have been induced. It is in this case that a retrograde messenger becomes necessary.

PROPERTIES OF A RETROGRADE MESSENGER

There are several general classes of mechanism that could subserve the functions of a retrograde signal during LTP induction. The nature of this messenger could be chemical, ionic, electrical, or even mechanical. A mechanical messenger might take the form of a large extracellular matrix assembly that is able to move in response to postsynaptic LTP induction. In the crudest sense, this could be called the pull-chain hypothesis, conjuring up an image of the postsynaptic cell pulling the chain of the presynaptic terminal to turn on LTP expression. An electrical messenger might take the form of a field that is produced by the postsynaptic cell and which transduces an effect there. An ionic signal could take the form of calcium being actively pumped from the postsynaptic cell. The resulting rise in extracellular calcium could lead to an increase in transmitter release. Finally, a chemical messenger would take the form of a substance that

Stuehr, D., and Nathan, C. (1989) A macrophage product responsible for cytostasis and respiratory inhibition in tumor target cells. *J. Exp. Med.* 169:1543–1555.

Stuehr, D. J., Soo Kwon, N., Nathan, C. F., Griffith, O. W., Feldman, P. L., and Wiseman, J. (1991) Nω-hydroxy-L-arginine is an intermediate in the biosynthesis of nitric oxide from L-arginine. *J. Biol. Chem.* 266:6259–6263.

Valtschanoff, J. G., Kharazia, V. N., and Weinberg, R. J. (1992) Neurons containing nitric oxide synthase in rat hippocampus. *Soc. Neurosci. Abstr.* 18:639.

Vincent, S. R., and Hope, B. T. (1992) Neurons that say NO. *Trends Neurosci.* 46:755–784.

Williams, J. H., Errington, M. L., Lynch, M. A., and Bliss, T. V. P. (1989) Arachidonic acid induces a long-term activity-dependent enhancement of synaptic transmission in the hippocampus. *Nature* 341:739–742.

Zhuo, M., Small, S. A., Kandel, E. R., and Hawkins, R. D. (1993) Nitric oxide and carbon monoxide produce activity-dependent long-term synaptic enhancement in hippocampus. *Science* 260:1946–1950.

2 Involvement of Nitric Oxi Potentiation

Daniel V. Madison and Erin M. Sc

The idea that a retrograde messenger resides in the ble for inducing or maintaining long-term potentia has been widely discussed for several years. The hy retrograde messenger, a signal that can communica ward across a synapse, has arisen because the induc mechanisms of LTP may be found on opposite sides of the most widely accepted and least controversial as at least a major part of the induction mechanisms resides cell. One of the more controversial hypotheses is that expression mechanism for LTP may reside in the pre manifest as an increase in per-action potential transmitt found to be true that a part of LTP expression lies in the then it becomes necessary to propose that the inductio postsynaptic cell are communicated to the presynaptic would travel opposite, or retrograde, to the normal in across the synapse.

Several lines of evidence have served to establish that cri mechanisms are located in the postsynaptic cell. In general, that have demonstrated this have taken the form of manip affect only the postsynaptic cell and block the induction of LT of such experiments are injection of EGTA (Lynch et al., 198 calcium chelators (e.g., Malenka et al., 1988) into postsynapti injection of protein kinase inhibitors into postsynaptic cells (M al., 1989; Malinow et al., 1989). These manipulations prevent being induced, but generally do not reverse already induced LTP showing that they are interrupting only the induction processes.

Evidence for a presynaptic locus of the expression processes ing LTP, those processes that actually make the synapse transm strongly, is far less certain. Two lines of evidence have been used to in favor of presynaptic expression of LTP. In the first, it has long diffuses from the postsynaptic cell during LTP induction (Gally et al., 1990).

A retrograde messenger operating during the formation or expression of LTP must have the following properties: (1) It must be necessary, but not necessarily sufficient, for the production of LTP; (2) it must be produced by the postsynaptic cell during LTP induction or maintenance; (3) the signal should be transmitted from the postsynaptic to the presynaptic cell; and (4) it must act on the presynaptic cell to transduce an increase in per-action potential transmitter release. There may be other subrequirements as well, such as a need for cofactors such as concurrent presynaptic activity, but the most basic requirements remain these four. There is no reason, a priori, for choosing one type of general mechanism over another, but attention has focused on a chemical messenger as the most likely and most easily testable possibility. The two candidates that have drawn the most interest are arachidonic acid (Williams et al., 1989; O'Dell et al., 1991) and nitric oxide (NO; see Moncada et al., 1989; Gally et al., 1990) Work in our laboratory and others has provided evidence that NO may serve as a retrograde messenger during the formation of LTP (e.g., Böhme et al., 1991, O'Dell et al., 1991, Schuman and Madison, 1991; Haley et al., 1992).

NO AS A POTENTIAL RETROGRADE MESSENGER IN LTP

We have performed studies to test whether NO can meet any of the criteria for a retrograde messenger. To do so, we have taken advantage of pharmacological probes (Bredt and Snyder, 1989; Shibuki and Okada, 1991) that are known to prevent NO production by its synthesizing enzyme, nitric oxide synthase (NOS). Experiments were performed in in vitro hippocampal slices from rats. Field excitatory postsynaptic potentials (EPSPs) were evoked in the stratum radiatum of hippocampal slices by test stimuli delivered to the Schaffer collaterals. To determine if NO was necessary for the production of LTP, we bathed slices in artificial cerebrospinal fluid (ACSF) containing the NOS inhibitor L-methyl arginine (L-Me-Arg). Experiments in this series were always done using two stimulating electrodes positioned to give two independent Schaffer collateral pathways onto the same population of postsynaptic CA1 cells. In this way, we could use each slice for its own internal control. One of the pathways, selected at random (by coin toss), was tetanized in the absence of inhibitor, and the resulting LTP was recorded for at least 1 hr. Slices failing to potentiate in this control condition were discarded. In slices showing LTP, inhibitor was then applied, followed by tetanic stimulation to the other path. Following this second tetanic stimulation, post-tetanic potentiation appeared as normal, and an early, slowly decaying potentiation was recorded. However,

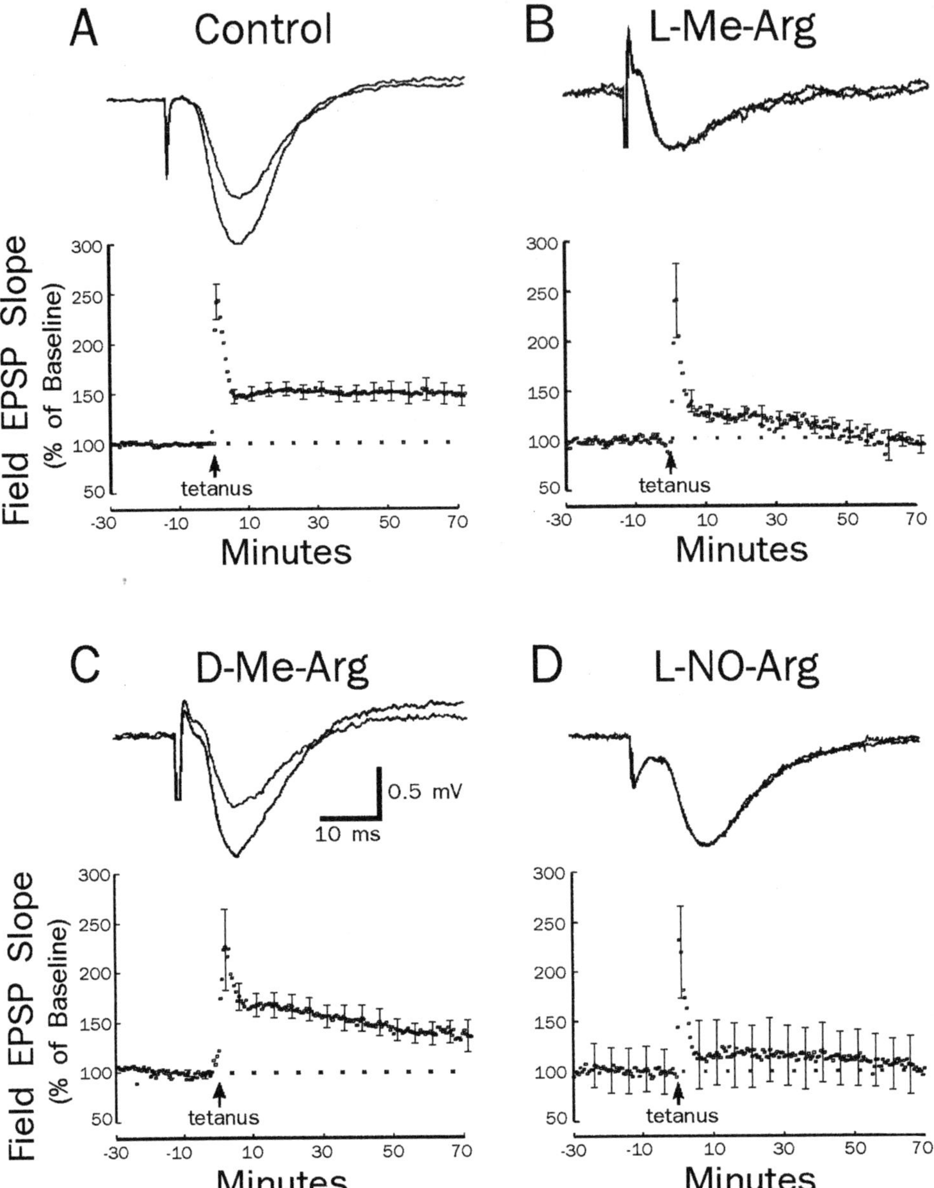

Figure 2.1 Extracellular application of the NOS inhibitor L-Me-Arg prevents tetanus-induced LTP of the field EPSP. (*A*) Graph showing ensemble averages for all control experiments. Tetanic stimulation resulted in a significant enhancement of the field EPSP slope in control conditions. *Inset*: Two representative EPSPs from a slice bathed in control ACSF, recorded 10 min before and 60 min after tetanic stimulation. (*B*) Graph showing ensemble average for all L-Me-Arg experiments. No significant LTP occurred in slices bathed in L-Me-Arg. *Inset*: Two representative EPSPs from a slice bathed in ACSF containing L-Me-Arg (100 μM), recorded 10 min before and 60 min after tetanic stimulation. (*C*) Graph showing ensemble average for all D-Me-Arg experiments. Significant LTP occurred in D-Me-Arg-treated slices. *Inset*: Two representative EPSPs from a slice bathed in ACSF containing D-Me-Arg (100 μM), recorded 10 min before and 60 min after tetanic stimula-

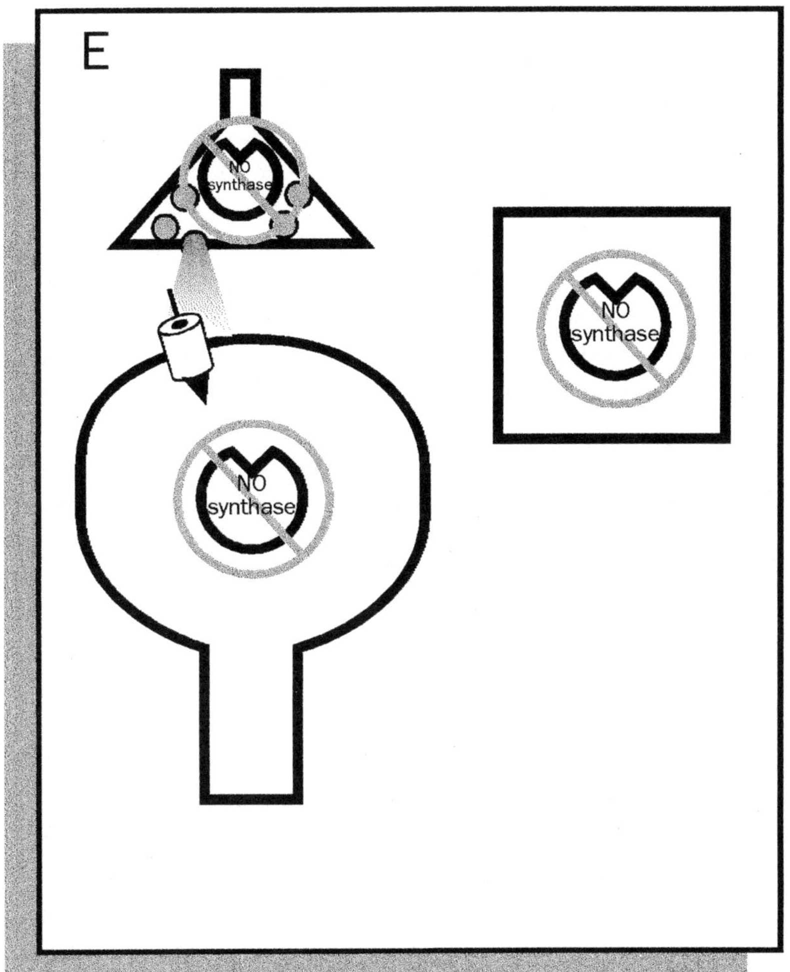

tion. Note: the amount of potentiation produced in L-Me-Arg-experiments was significantly less than that occurring in control ($P < .001$) and D-Me-Arg experiments ($P < .005$), which did not differ significantly from one another. (*D*) Graph showing ensemble average for all NO-Arg experiments. No significant LTP occurred in slices bathed in NO-Arg (100 μM), recorded 10 min before and 60 min after tetanic stimulation. No significant LTP occurred in the presence of NO-Arg. (*E*) Diagram illustrating our interpretation of this experiment. Since NOS inhibitors were bath applied, NOS was potentially inhibited both in presynaptic and postsynaptic cells, as well as other compartments in the slice such as glia. (Adapted from Schuman and Madison, 1991) by an hour after the tetanic stimulation, the size of the EPSP in the test pathway had decayed to baseline (figure 2.1; Schuman and Madison, 1991). As a group, control paths showed persistent potentiation of the EPSP (148.2 $\pm$ 8.9% at 1 hr post tetanus; mean % of baseline, $\pm$SEM, n = 15; figure 2.1) lasting at least 1 hr. In contrast, the paths tested in the presence of L-Me-Arg failed to produce a persistent enhancement of the EPSP (103.2 $\pm$ 6.5%; n = 16; figure 2.1). Application of the inactive isomer of methyl arginine, N^G-methyl-D-arginine (D-Me-Arg; 100 μM), did not prevent tetanus-induced LTP (158.6 $\pm$ 19.7%; n = 6; figure 2.1). These results fulfill the first requirement for a retrograde messenger by suggesting that NOS activity and NO are necessary for the production of LTP.

L-Me-Arg is a competitive inhibitor of NOS, competing with endogenous substrate, L-arginine (Palmer et al., 1988; Sakuma et al., 1988; Knowles et al., 1989) for the substrate site on the enzyme (Bredt and Snyder, 1989). We found that we were able to reverse the L-Me-Arg block of LTP by the addition of excess L-arginine to the ACSF. Slices initially incubated in ACSF containing the inhibitor, L-Me-Arg, failed to display LTP (103.0 $\pm$ 1.5%; n = 4, figure 2.2). When L-arginine was then added to the bath in the continued presence of L-Me-Arg, subsequent tetanic stimulation of the same presynaptic axons resulted in clear potentiation of the EPSP (138.5 $\pm$ 13.2%; figure 2.2). Thus, the inhibition of LTP by L-Me-Arg can be reversed by excess arginine. These data support the conclusion that the site of action of L-Me-Arg is the NOS.

The second requirement, if NO is acting as a retrograde messenger, is that it be produced by the postsynaptic pyramidal cell. In situ hybridization (Bredt et al., 1991) and immunohistochemical studies (Bredt et al., 1990) indicate that neuronal NOS is expressed in many brain areas. However, these studies, (Bredt et al., 1991), have failed to identify CA1 pyramidal neurons, which stain positive for NOS in the CA1 region of the hippocampus. So far, only one brain NOS has been identified, although a distinct inducible NOS is expressed in macrophages (Xie et al., 1992). The apparent lack of NOS in CA1 pyramids may be explained either as the actual absence of NOS in these cells, or because there may be a different isoform of the NOS in CA1 pyramidal neurons. Alternatively, NOS may be expressed at levels too low to be detected in the above assays. Interestingly, a new immunohistochemical study now provides evidence that NOS is indeed present in CA1 pyramidal cells, albeit at lower levels than in some other neurons (Wendland et al., 1993).

We performed experiments to test physiologically for the presence of a necessary NOS activity in CA1 pyramidal cells. This was done by inhibiting NOS activity exclusively in a postsynaptic cell while attempting

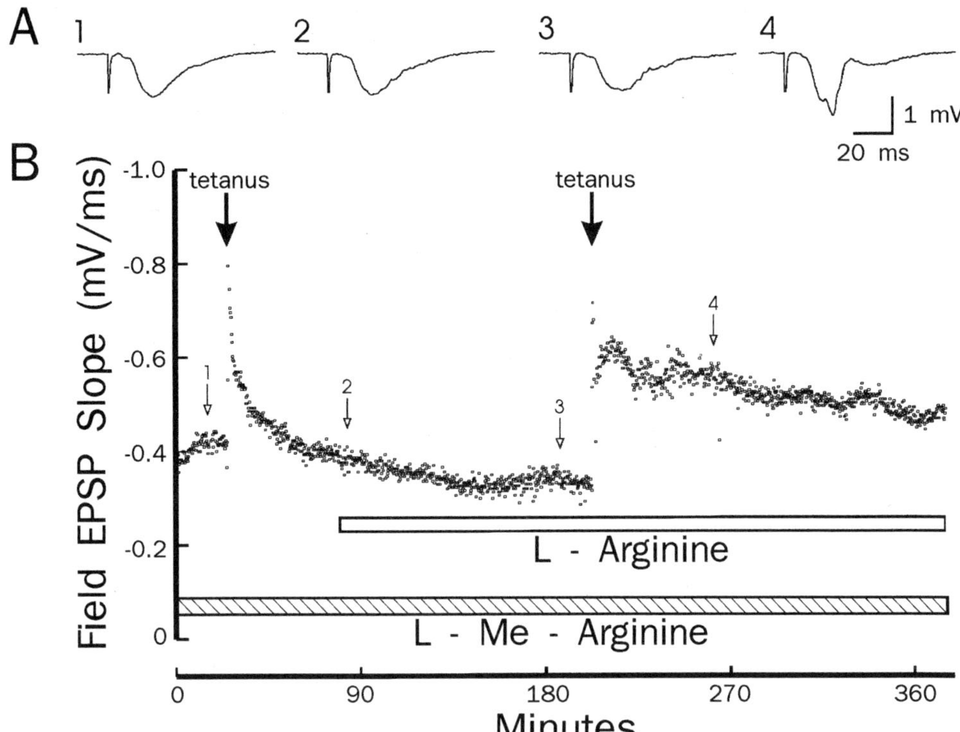

Figure 2.2 L-Arginine reverses the L-Me-Arg block of LTP. Because the conversion of L-arginine to NO is inhibited competitively by L-Me-Arg, production of NO should be restored by addition of an excess of L-arginine. Tetanic stimulation was delivered twice in this experiment, once in the presence of L-Me-Arg (100 μM), and again after the addition of L-arginine (1 mM). (*A*) Field EPSPs recorded in stratum radiatum. EPSPs illustrated were collected 15 min before (1), and 60 min after (2) the first tetanic stimulation (in L-Me-Arg); 15 min before (3), and 60 min after (4) the second tetanic stimulation (after addition of L-arginine). (*B*) Graph showing the slope of recorded field EPSPs over the entire time course of the experiment. Each point represents the rising slope of a single, nonaveraged EPSP. EPSPs were evoked at a rate of 2/min; data points for all EPSPs collected during the experiment are shown. Bars below the data indicate the period of application of L-Me-Arg (hatched bar) and L-arginine (open bar). Numerals over open arrows indicate the time at which the records in *A* were collected. Closed arrows indicate the time of delivery of tetanic stimulations. In all experiments, long-term potentiation, measured 60 min after tetanic stimulation, was blocked by L-Me-Arg (pre: $.43 \pm .07$ mV/msec, mean $\pm$SEM; post: $.45 \pm .08$, ns) but addition of L-arginine restored the potency of tetanic stimulation to induce LTP (pre: $.35 \pm .06$; post: $.48 \pm .07$, $P < .05$). Note: the amount of potentiation observed in the presence of L-Me-Arg + L-arginine was significantly greater than that observed in the presence of L-Me-Arg alone ($P < .05$). (From Schuman and Madison, 1991)

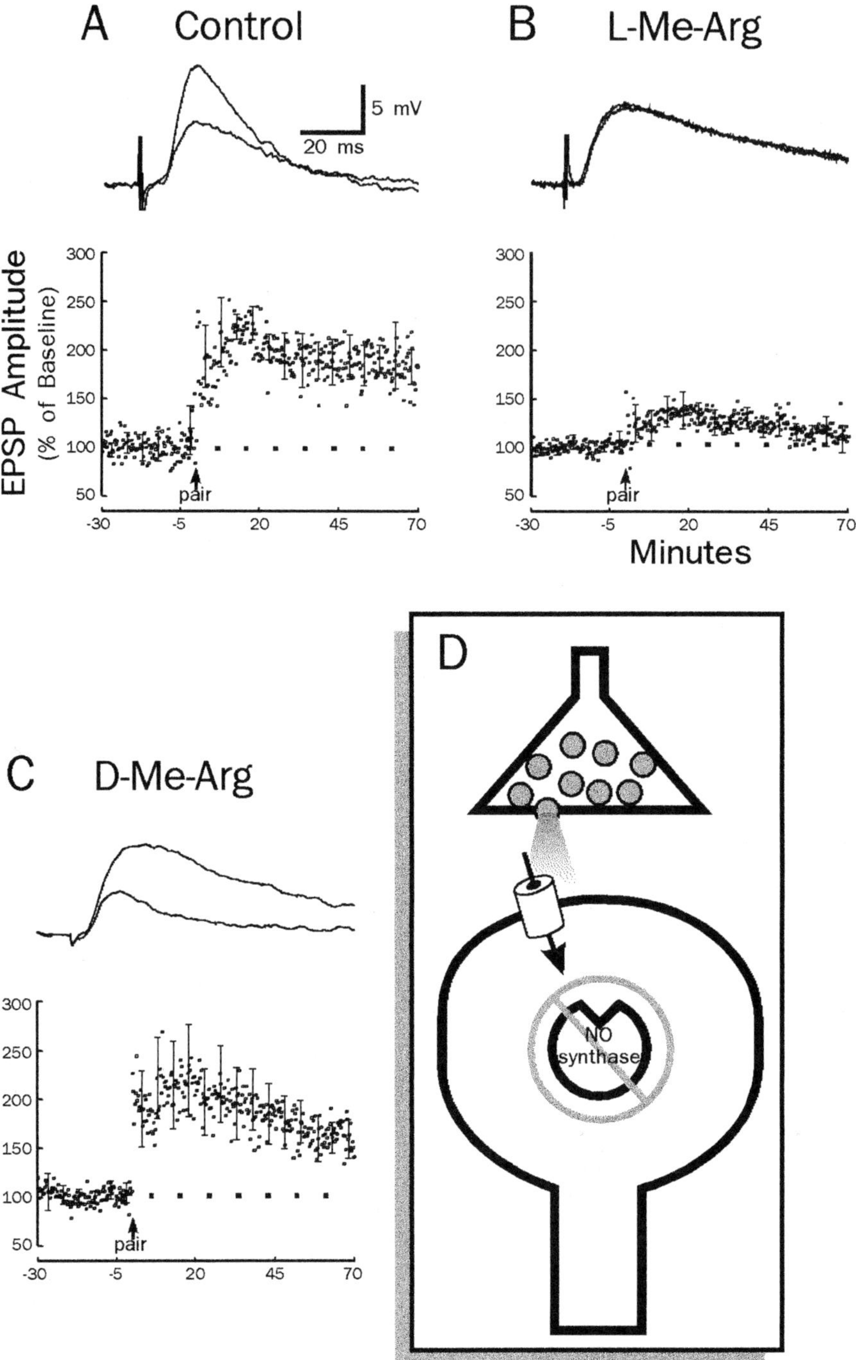

Figure 2.3 Injection of L-Me-Arg into postsynaptic cells blocks long-term potentiation. This result shows that the NO production necessary for inducing LTP occurs in the postsynaptic cell. LTP was induced in these experiments by pairing steady postsynaptic depolarization with 1 Hz stimulation of Schaffer collaterals. (*A, top*) Representative intracellular EPSPs recorded from the same control cell 15 min before and 60 min after pairing.

to induce LTP. L-Me-Arg was injected into single CA1 pyramidal cells by placing the inhibitor into the filling solution of an intracellular recording electrode. After impalement of the cell, the recording was maintained for at least an hour to allow for the diffusion of the inhibitor into the cell. After this period, we attempted to induce LTP by pairing constant postsynaptic depolarization with low-frequency stimulation of presynaptic axons. In control (no inhibitor in electrode), this resulted in a robust and long-lasting potentiation of the intracellular EPSP (potentiation: 190.3 ± 24.3% of baseline, n = 7; figure 2.3). However, when postsynaptic neurons were filled with L-Me-Arg, the pairing protocol elicited a similar decaying potentiation as seen with tetanus and bath-applied inhibitor, but did not elicit potentiation lasting for a hour (109.5 ± 4.5% potentiation n = 15; figure 2.3, ns). Cells filled with the inactive isomer, D-Me-Arg, exhibited LTP similar to control (172.2 ± 16.7%, n = 6; figure 2.3).

As a further test for the postsynaptic location of NOS, we performed an additional experiment. Hippocampal slices were bathed in 100 μM of L-Me-Arg at a concentration that blocks tetanus-induced LTP. A single CA1 pyramidal cell was then impaled with an electrode containing 10 mM L-arginine. EPSPs resulting from Schaffer collateral stimulation were monitored both with the intracellular electrode, and also with an extracellular field electrode placed in the same vicinity. Upon tetanic stimulation, no significant LTP was recorded in the field potentials, reflecting the effectiveness of L-Me-Arg in blocking NOS in the bulk of the tissue. However, LTP was induced in the single arginine-filled cell (figure 2.4). This result implies that while NOS was inhibited in the entire slice, NOS function was restored by competition of the injected excess substrate with the inhibitor.

(*Bottom*) Graph showing ensemble average for all control experiments. In control conditions, pairing resulted in a significant enhancement of the intracellular EPSP (pre: 3.76 ± .42 mV [10 min before pairing]; post: 6.72 ± .49 mV [60 min after pairing]; $P < .001$; mean ± SEM) (*B, top*) Representative intracellular EPSPs recorded from a cell injected with L-Me-Arg. Shown are two superimposed EPSPs taken 15 min before and 60 min after pairing. (*Bottom*) Graph showing ensemble average for all L-Me-Arg-injected neurons. In contrast to control experiments, no significant LTP occurred in these cells (pre: 3.38 ± .25 mV; post: 3.77 ± .31 mV; ns). (*C, top*) Representative intracellular EPSPs recorded from a D-Me-Arg–injected cell. Shown are two superimposed EPSPs taken 15 min before and 60 min after pairing. (*Bottom*) Graph showing ensemble average for all D-Me-Arg-injected neurons. D-Me-Arg–injected cells exhibited significant pairing-induced LTP (pre: 3.29 ± .42 mV; post: 5.61 ± .73 mV, $P < .01$). Note: the amount of potentiation produced in L-Me-Arg-injected cells was significantly less than that observed in controls ($P < .001$) or D-Me-Arg–injected cells ($P < .001$), which did not differ significantly from one another. (*D*) Diagram illustrating our interpretation of this experiment. Since NOS inhibitors were applied to the intracellular space of a single postsynaptic cell, this is evidence that the NOS critical for LTP is found in the postsynaptic cell. (Adapted from Schuman and Madison, 1991).

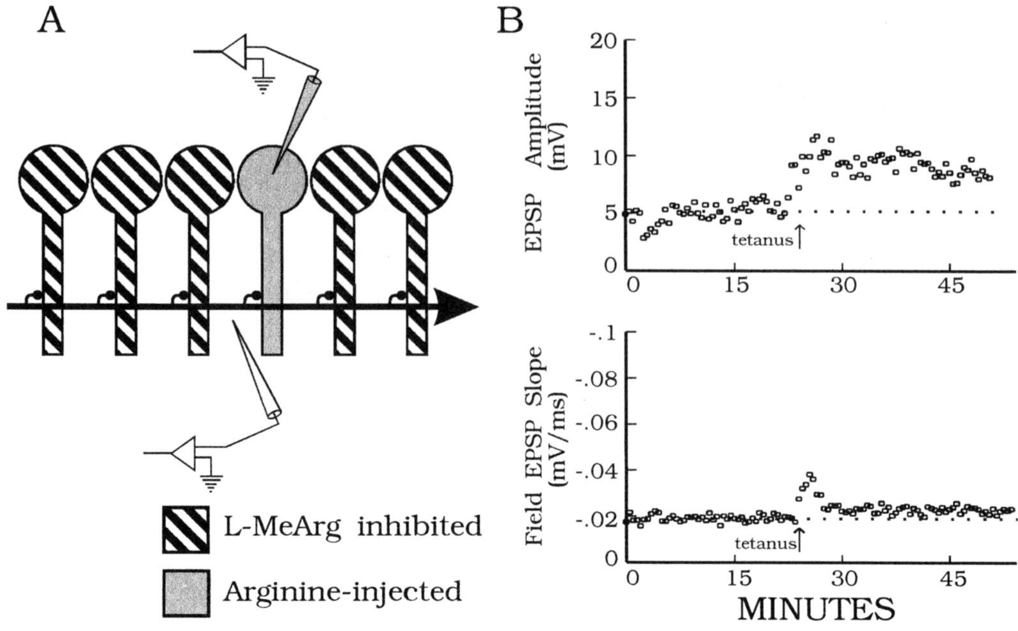

Figure 2.4 L-Arginine injected into a postsynaptic cell can reconstitute that cell's ability to support LTP in the presence of bath-applied NOS inhibitor. (*A*) Diagram of the experimental arrangement. A hippocampal slice was bathed in 100 μM L-Me-Arg, while a single CA1 pyramidal cell was injected with 10 mM L-arginine. EPSPs were recorded via the intracellular electrode. Recordings of field EPSPs were made simultaneously with an extracellular electrode placed in stratum radiatum adjacent to the intracellular electrode. (*B, top*) Recordings of EPSPs made with the intracellular microelectrode containing 2 M K-methylsulfate and 10 mM L-arginine. Note that LTP occurred in response to tetanic stimulation (100 Hz/1sec, 1 train at test intensity). (*Bottom*) Simultaneous recording of field EPSPs made with the extracellular recording electrode (3 M NaCl). The entire experiment was conducted in bath-applied L-Me-Arg (100 μM). Note that no significant LTP occurred in the extracellular recording reflecting the efficacy of the NOS inhibitor in preventing LTP in the slice, while LTP did occur in the single arginine-injected cell. This is further evidence that the NOS that is critical for LTP is found in postsynaptic CA1 pyramidal cells.

This is further evidence that the NOS that is critical for producing LTP is found in the postsynaptic cell. These data suggest that NOS activity in the postsynaptic cell is critical for the production of LTP.

The third criterion specifies that a retrograde messenger of the chemical variety would likely need to pass outside the postsynaptic cell en route to its presynaptic target. To address the possibility for NO, we tested if extracellular application of hemoglobin attenuates LTP. Hemoglobin binds and inactivates NO but itself does not readily cross cell membranes (Traylor and Sharma, 1992). Slices incubated in solution containing hemoglobin failed to exhibit significant LTP (115.8 ± 6.2%, n = 10; figure 2.5) compared with matched controls (159 ± 10.7%, n = 10; figure 2.5). The simplest interpretation of this finding is that hemoglobin, present outside

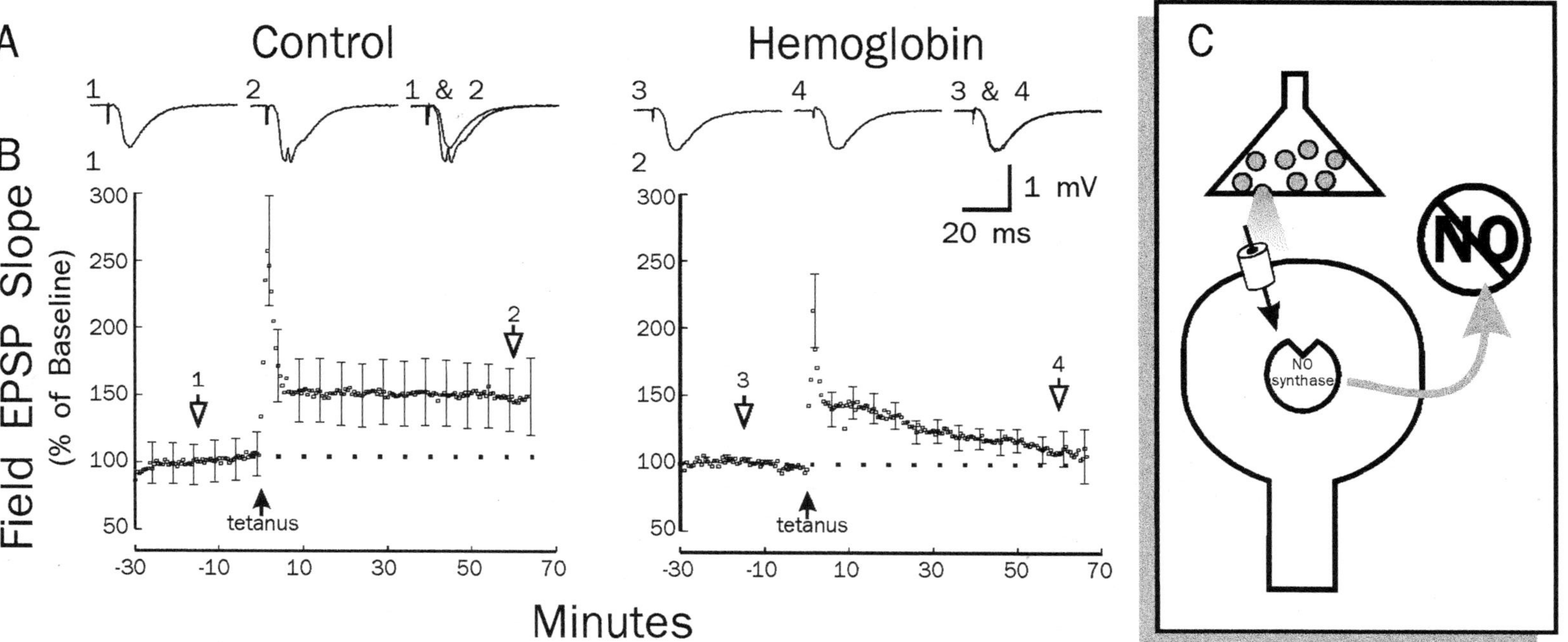

Figure 2.5 Extracellular hemoglobin attenuates long-term potentiation. Since hemoglobin binds NO and cannot readily cross cell membranes, this result suggests that NO must pass outside the cell en route to its target. Control and hemoglobin records in this figure were obtained in the same slices. EPSPs were evoked by stimulation of two independent groups of Schaffer collateral fibers, one tetanized before, and the other after addition of hemoglobin. (*A, top*) Field EPSPs from a slice bathed in control ACSF collected 15 min before (1), and 60 min after (2) tetanic stimulation. (*Bottom*) Graph showing ensemble averages for all control experiments. Tetanic stimulation resulted in a significant enhancement of the field EPSP in control conditions (pre: $.63 \pm .05$ mV/msec; post: $1.04 \pm .14$ mV; $P < .005$; mean $\pm$ SEM). (*B, top*) Field EPSPs from a slice bathed in ACSF containing hemoglobin (100 μM) collected 15 min before (3) and 60 min after (4) tetanic stimulation. (*Bottom*) Graph showing ensemble averages for all hemoglobin experiments. No significant LTP occurred in slices after bathing in hemoglobin (pre: $.58 \pm .08$ mV/msec; post: $.68 \pm .12$ mV/msec; ns). Note: the amount of LTP observed in control pathways was significantly greater than that observed in hemoglobin-treated pathways ($P < .01$). (*C*) Diagram illustrating our interpretation of this experiment. Attenuation of LTP by extracellular hemoglobin is consistent with the need for NO to travel out of the postsynaptic cell en route to its target. (Adapted from Schuman and Madison, 1991)

the cell, scavenges extracellular NO and prevents it from reaching its ultimate site of action, perhaps in the presynaptic terminal.

DOES NO HAVE A ROLE IN INDUCTION OR MAINTENANCE OF LTP?

As a retrograde messenger NO could, in theory, act either in the induction mechanisms of LTP or in the mechanisms that maintain potentiation. If NO participates solely in the induction of LTP, then it should be produced for only a brief period after tetanic stimulation. Conversely, if LTP is maintained by the continuous presence of NO, then its production must persist, since NO itself has such a short half-life. The most straightforward way to resolve this issue would be to use NOS inhibitors or NO chelators that could reach their site of action in hippocampal slices without delay. The agents in present use take, we believe, on the order of an hour to reach effective concentration in the slice, and thus give too poor a time resolution to address this question adequately. An inhibitor capable of penetrating the tissue quickly could be applied at varying times before or after LTP induction to define the time window during which NO is active in producing LTP.

One result, illustrated in figure 2.2, does, however, suggest that NO is important for LTP induction rather than maintenance. Since NO has a very short life, a continuous presence of NO, as would be required for LTP maintenance, would require the continual activity of NOS. The addition of excess arginine clearly can compete inhibitor off the enzyme and restore the function of the NOS. If NOS were continuously active following a tetanic stimulation, then addition of excess arginine after tetanic stimulation (in the presence of L-Me-Arg) would result in the production of NO and the expression of LTP. Clearly, it takes a second tetanic stimulation in the presence of the arginine to reactivate the NOS and to produce LTP. This suggests to us that the enzyme is not active for long periods after tetanic stimulation, but rather that NO is produced for only a brief period of time. Such a short burst of NO would be more consistent with its having a role in the short-lived induction mechanisms of LTP rather than the persistent maintenance mechanisms.

Another aspect of our data that may have bearing on a role for NO in induction or maintenance of LTP is that some potentiation does occur immediately after tetanic stimulation or pairing in the presence of L-Me-Arg or hemoglobin. The presence of this initial, nonpersistent potentiation seems more consistent with NO having a role in the maintenance of LTP, rather than the induction. This potentiation decays to baseline within an hour of induction, unlike LTP, and is reminiscent of the slowly decaying potentiation that is observed in several conditions, including the presence of protein kinase inhibitors (Malinow et al., 1988). The appearance of this decaying potentiation does not, however, necessarily give any useful information about which phase of LTP requires NO. This decaying potentiation may represent a distinct process that is induced in parallel with LTP, but is not otherwise related. Alternatively, decaying potentiation may represent nonconsolidated LTP that was not made persistent because of a missing NO-related component in the induction steps. A hypothetical example of the latter could be proposed under a scheme where LTP is expressed as a result of the activity of a protein kinase. In this hypothetical scheme, this kinase would be activated by tetanic stimulation, action would be slowly reversed by enzyme deactivation and phosphatase activity, unless the kinase were made constitutively active (e.g., by proteolytic cleavage by a brief NO-related event). Under this idea, persistent potentiation (LTP) and decaying potentiation would have the same expression mechanisms, but the degree of persistence would be regulated by the brief presence of NO. The important point here is that while the presence or absence of NO would influence how long LTP was maintained, NO's role itself would be confined to the induction stages of LTP.

TARGETS OF NO

The ultimate site of NO action during LTP is not known. A target of NO in many tissues is a soluble guanylyl cyclase (Murad et al., 1978; Garthwaite et al., 1988; Bredt and Snyder, 1989). NO binds to a heme moiety in guanylyl cyclase, resulting in activation of the cyclase and production of cyclic GMP (Murad et al., 1978; see also Mulsch et al., 1988). Activation of NMDA receptors, either by exogenous agonist (East and Garthwaite, 1991) or by high-frequency synaptic stimulation (Chetkovich and Sweatt, 1992) in hippocampal slices results in elevations in cGMP levels, which are sensitive to NOS inhibitors. However, extracellular application of membrane-permeant analogs of cGMP do not appear to enhance synaptic transmission when applied alone (Haley et al., 1992) or together with presynaptic activity (Schuman et al., 1992). Inhibitors of cGMP-dependent protein kinase applied before high-frequency stimulation do not block LTP (Schuman et al., 1992). Thus, while LTP induction procedures clearly promote NO-sensitive rises in cGMP, the actual contribution of cGMP to LTP production remains to be determined.

NO has also been suggested to activate an cytosolic ADP-ribosyltransferase (Brune and Lapetina, 1989, 1990). ADP-ribosyltransferases covalently modify their substrate proteins by attaching ADP-ribose moieties from NAD^+ or $NADP^+$ (Ueda and Hayaishi, 1985; Jacobson et al., 1990;

see also Moss et al., 1980, 1986; Oka et al., 1984; Smith et al., 1985). NO has been reported to stimulate the ADP-ribosylation of several proteins including glyceraldehyde 3′ phosphate dehydrogenase (Dimmeler et al., 1992; Kots et al., 1992; Zhang and Snyder, 1992), actin (Matsuyama and Tsuyama, 1991), and neuronal GTP-binding proteins (Matsuoka et al. 1989; Duman et al., 1991; see also Tanuma et al., 1988). Preliminary results suggest that ADP-ribosyltransferase activity is required for LTP and that NO can stimulate the ADP-ribosylation of hippocampal proteins in vitro (Schuman et al., 1992). However, it remains to be shown that NO-induced increases in synaptic strength are mediated by ADP-ribosyltransferase activity. Also to be determined is whether tetanic stimulation results in the ADP-ribosylation of specific proteins. Thus, the contribution of NO-stimulated ADP-ribosylation to LTP is presently unknown, but an LTP induction pathway including NO and ADP-ribosylation appears to be promising.

CONCLUSION

If any part of the expression of LTP occurs in the postsynaptic neuron, there must exist a postsynaptically generated retrograde signal that travels to the presynaptic neuron during LTP induction. Nitric oxide is a viable candidate for such a signal. Our own experimental evidence, and that of others, indicates that NO satisfies several of the criteria for a retrograde messenger in LTP induction. However, at this time, it is still not proven that NO serves this role. It still must be shown that NO has a target in the presynaptic terminal, and its interaction with this target leads to an increase in transmitter release following LTP induction. Emerging, but as of yet mostly unpublished data, indicate that the efficacy of NOS inhibitors may depend on experimental conditions such as temperature or other factors. Our view is that this information will support a complex, but as yet poorly understood, role for NO in LTP. We will address the role of experimental temperature on NOS inhibitor efficacy in detail in later publications. In our most recent work, we have replicated our earlier NOS inhibitor findings, with the experiments being performed blind. Our future experiments will concentrate in part on identifying the target receptor for NO in LTP, and in finding the location of this target. This information will assist in the determination of the actual nature of the role of NO in long-term potentiation.

REFERENCES

Bekkers, J. M., and Stevens, C. F. (1990) Presynaptic mechanism for long-term potentiation in the hippocampus. *Nature* 346:724–729.

Bliss, T. V. P., Douglas, R. M., Errington, M. L., and Lynch, M. A. (1986) Correlation between long-term potentiation and release of endogenous amino acids from dentate gyrus of anaesthetized rats. *J. Physiol.* 377:391–408.

Böhme, G. A., Bon, C., Stutzmann, J. M., Doble, A., and Blanchard, J. C. (1991) Possible involvement of nitric oxide in long-term potentiation. *Eur. J. Pharmacol.* 199:379–381 .

Bredt, D. S., and Snyder, S. H. (1989) Nitric oxide mediates glutamate-linked enhancement of cGMP levels in the cerebellum. *Proc. Natl. Acad. Sci. USA* 86:9030–9033.

Bredt, D. S., Hwang, P. H., and Snyder, S. H. (1990) Localization of nitric oxide synthase indicating a neural role for nitric oxide. *Nature* 347:768–770.

Bredt, D. S., Glatt, C., Hwang, P. M., Fotuhi, M., Dawson, T. M., and Snyder, S. H. (1991) Nitric oxide synthase protein and mRNA are discretely localized in neuronal populations of the mammalian CNS together with NADPH diaphorase. *Neuron* 7:615–624.

Bruene, B., and Lapetina, E. G. (1989) Activation of a cytosolic ADP-ribosyltransferase by nitric oxide generating agents. *J. Biol. Chem.* 264:8455–8458.

Bruene, B., and Lapetina, E. G. (1990) Properties of a novel nitric oxide-stimulated ADP-ribosyltransferase. *Arch. Biochem. Biophys.* 279:286–290.

Chetkovich, D. M., and Sweatt, J. D. (1992) LTP-inducing tetanic stimulation causes a nitric-oxide mediated increase in cGMP in hippocampal area CA1. *Soc. Neurosci. Abstr.* 18:761.

Dimmeler, S., Lottspeich, F., and Brune, B. (1992) Nitric oxide causes ADP-ribosylation and inhibition of glyceraldehyde-3-phosphate dehydrogenase. *J. Biol. Chem.* 267:16771–16774.

Dolphin, A. C., Errington, M. L., and Bliss, T. V. P. (1982) Long-term potentiation of the perforant path in vivo is associated with increased glutamate release. *Nature* 297:496–498.

Duman, R. S., Terwilliger, R. Z., and Nestler, E. J. (1991) Endogenous ADP-ribosylation in brain: Initial characterization of substrate proteins. *J. Neurochem.* 57:2124–2132.

East, S. J., and Garthwaite, J. (1991) NMDA receptor activation in rat hippocampus induces cyclic GMP formation through the L-arginine nitric oxide pathway. *Neurosci. Lett.* 123:17–19.

Foster, T. C., and McNaughton, G. L. (1991) Long-term enhancement of CA1 synaptic transmission is due to increased quantal size, not quantal content. *Hippocampus* 1:79–91.

Gally, J. A., Montague, P. R., Reeke, G. N., and Edelman, G. M. (1990) The NO hypothesis: Possible effects of a short-lived, rapidly diffusible signal in the development and function of the nervous system. *Proc Natl. Acad. Sci. USA* 87:3547–3551.

Garthwaite, J., Charles, S. L., and Chess-Williams, R. (1988) Endothelium-derived relaxing factor release on activation of NMDA receptors suggests a role as intracellular messenger in the brain. *Nature* 336:385–388.

Haley, J. E., Wilcox, G. L., and Chapman, P. F. (1992) The role of nitric oxide in long-term potentiation. *Neuron* 8:211–216.

Jacobson, M. K., Loflin, P. T., Aboul-Ela, N., Mingmuang, M., Moss, J., and Jobson, E. L. (1990) Modification of plasma membrane protein by ADP-ribose in vivo. *J. Biol. Chem.* 265:10825–10828.

Knowles, R. G., Palacios, M., Palmer, R. M. G., and Moncada, S. (1989) Formation of nitric oxide from L-arginine in the central nervous system: A transduction mechanism for stimulation of soluble guanylate cyclase. *Proc. Natl. Acad. Sci. USA* 86:5159–5162.

Kots, A. Y., Skurat, A. V., Sergienko, E. A., Bulgarina, T. V., and Severin, E. S. (1992) Nitroprusside stimulates the cysteine-specific mono(ADP-ribosylation) of glyceraldehyde-3-phosphate dehydrogenase from human erythrocytes. *FEBS Lett.* 300:9–12.

Kullmann, D. M., and Nicoll, R. A. (1992) Long-term potentiation is associated with increases in quantal content and quantal amplitude. *Nature* 357:240–244.

Larkman, A., Hannay, T., Stratford, L., and Jack, J. (1992) Presynaptic release probability influences the locus of long-term potentiation. *Nature* 360:70–73.

Lynch, G., Larson, J., Kelso, S., Barrionuevo, G., and Schottler, F. (1983) Intracellular injections of EGTA block induction of hippocampal long-term potentiation. *Nature* 305: 719–721.

Malenka, R. M., Kauer, J. A., Zuker, R. J., and Nicoll, R. A. (1988) Postsynaptic calcium is sufficient for potentiation of hippocampal synaptic transmission. Science 242:81–84.

Malenka, R. C., Kauer, J. A., Perkel, D. J., Mauk, M. D., Kelly P. T., Nicoll, R. A., and Waxham, M. N. (1989) An essential role for postsynaptic calmodulin and protein kinase activity in long-term potentiation. *Nature* 340:554–557.

Malgaroli, A., and Tsien, R. W. (1992) Glutamate-induced long-term potentiation of the frequency of miniature synaptic currents in cultured hippocampal neurones. *Nature* 357: 134–139.

Malinow, R. (1991) Transmission between pairs of hippocampal slice neurons: Quantal levels, oscillations and LTP. *Science* 252:722–724.

Malinow R., and Tsien, R. W. (1990) Presynaptic enhancement shown by whole-cell recordings of long-term potentiation in hippocampal slices. *Nature* 346:177–180.

Malinow, R., Madison, D. V., and Tsien, R. W. (1988) Persistent protein kinase activity underlies long-term potentiation. *Nature* 335:820–824.

Malinow, R., Schulman, H., and Tsien, R. W. (1989) Inhibition of postsynaptic PKC and CAMKII blocks induction but not expression of LTP. *Science* 245:862–866.

Manabe, T., Renner, P., and Nicoll, R. A. (1992) Postsynaptic contribution to long-term potentiation revealed by the analysis of miniature synaptic currents. *Nature* 355:50–55.

Matsuoka, I., Sakuma, H., Syuto, B., Moriishi, K., Kubo D., and Kurihara, K. (1989) ADP-ribosylation of 24-26 kDA GTP-binding proteins localized in neuronal and non-neuronal cells by botulinum neurotoxin. *J. Biol. Chem.* 264:706–712.

Matsuyama, S., and Tsuyama, S. (1991) Mono-ADP-ribosylation in brain: Purification and characterization of ADP-ribosyltransferases affecting actin from rat brain. *J. Neurochem.* 57:1380–1387.

Moncada, S., Palmer, E. M. J., and Higgs, E. A. (1989) The biological significance of nitric oxide formation from L-arginine. *Biochem. Pharmacol.* 38:1709.

Moss, J., Stanley, S. J., and Watkins, P. A. (1980) Isolation and properties of an NAD- and guanidine-dependent ADP-ribosyltransferase from turkey erythrocytes. *J. Biol. Chem.* 255: 5838–5840.

Moss, J., Oppenheimer, N. J., West, R. E., and Stanley, S. J. (1986) Amino acid specific ADP-ribosylation: substrate specificity of an ADP-ribosylarginine hydrolase from turkey erythrocytes. *Biochemistry* 25:5408–5414.

Mulsch, A., Busse, R., Liebau, S., and Forstermann, U. (1988) LY83583 interferes with the release of endothelium-derived relaxing factor and inhibits soluble guanylate cyclase. *J. Pharmacol. Exp. Ther.* 247:283–288.

Murad, F., Mittal, C. K., Arnold, W. P., Katsuki, S., and Kimura, H. (1978) Guanylate cyclase: Activation by azide, nitro compounds, nitric oxide, and hydroxyl radical and inhibition by hemoglobin and myoglobin. *Adv. Cyclic Nuc. Res.* 9:145–158.

O'Dell, T. J., Hawkins, R. D., Kandel, E. R., and Arancio, O. (1991) Tests on the roles of two diffusible substances in LTP: Evidence for nitric oxide as a possible early retrograde messenger. *Proc. Natl. Acad. Sci. USA* 88:11285–11289.

Oka, J., Ueda, L., Hayaishi, O., Komura, H., and Nakanishi, K. (1984) ADP-ribosyl lyase. Purification, properties, and identification of the product. *J. Biol. Chem.* 259:986–995.

Palmer, R. M. J., Rees, D. D., Ashton, D. S., and Moncada, S. (1988) L-Arginine is the physiological precursor for the formation of nitric oxide in endothelium-dependent relaxation. *Biochem. Biophys. Res. Commun.* 153:1251–1256.

Sakuma, I., Stuehr, D., Gross, S. S., Nathan, C., and Levi, R. (1988) Identification of arginine as the precursor of endothelium-derived relaxing factor. *Proc. Natl. Acad. Sci. USA* 85: 1011–1020.

Schuman, E. M., and Madison, D. V. (1991) A requirement for the intercellular messenger nitric oxide in long-term potentiation. *Science* 254:1503–1506.

Schuman, E. M., Meffert, M. K., Schulman, H., and Madison, D. V. (1992) A potential role for an ADP-ribosyltransferase in long-term potentiation. *Soc. Neurosci. Abstr.* 18:761.

Shibuki, K., and Okada, D. (1991) Endogenous nitric oxide release required for long-term synaptic depression in the cerebellum. *Nature* 349:326–328.

Smith, K. P., Benjamin, R. C., Moss, J., and Jacobson, M. K. (1985) Identification of enzymatic activities which process protein bound mono(ADP-ribose). *Biochem. Biophys. Res. Commun.* 126:136–142.

Tanuma, S., Kawashima, K., and Endo, H. (1988) Eukaryotic mono (ADP-ribosyl)transferase that ADP-ribosylates regulatory G_i protein. *J. Biol. Chem.* 263:5485–5489.

Traylor, T. G., and Sharma, V. S. (1992) Why NO? *Biochemistry* 31:2847–2850.

Ueda, K., and Hayaishi, O. (1985) ADP-ribosylation. *Annu. Rev. Biochem.* 54:73–100.

Wendland, B., Schweizer, F. E., Ryan, T. A., Nakane, M., Murad, F., Scheller, R. H., and Tsien, R. W. (1993) Existence of nitric oxide synthase in rat hippocampus. *Proc. Natl. Acad. Sci. (USA)*, in press.

Williams, J. H., Errington, M. L., and Bliss, T. V. P. (1989) Arachidonic acid induces a long-term activity-dependent enhancement of synaptic transmission in the hippocampus. *Nature* 341:739–742.

Xie, Q., Cho, H. J., Calaycay, J., Mumford, R. A., Swiderek, K. M., Lee, T. D., Ding, A., Troso, T., and Nathan, C. (1992) Cloning and characterization of inducible nitric oxide synthase from mouse macrophages. *Science* 256:225–228.

Zhang, J., and Snyder, S. H. (1992) Nitric oxide stimulates auto-ADP-ribosylation of glyceraldehyde-3-phosphate dehydrogenase. *Proc. Natl. Acad. Sci. USA* 89:9382–9385.

3 Nitric Oxide Synthase Inhibition In Vivo: Lack of Effect on Hippocampal Synaptic Enhancement or Spatial Memory

Carol A. Barnes, Bruce L. McNaughton, David S. Bredt, Christopher D. Ferris, and Solomon H. Snyder

Experiments performed on the in vitro hippocampal slice or cell culture preparations suggest that nitric oxide (NO) may translate induction events into expression of long-term enhancement (LTP/LTE) of synapses in hippocampal subfield CA1 (Böhme et al., 1991; O'Dell et al., 1991; Schuman and Madison, 1991; Haley et al., 1992). NO synthesis is presumably triggered postsynaptically through the activation of NMDA receptors, which leads to calcium influx, calmodulin binding, and nitric oxide synthase (NOS) activation (Garthwaite et al., 1988, 1989; Fujimori and Pan-Hou, 1991). The release of NO from postsynaptic neurons appears to increase transmitter release from terminals near the site of NO release (Schuman and Madison, 1991); however, such a presynaptic mechanism for the expression of LTE has been the subject of controversy (Bekkers and Stevens, 1990; Malinow and Tsien, 1990; Foster and McNaughton, 1991; Manabe et al., 1992). In the hippocampus, the highest levels of NOS are present in the fascia dentata (FD), although the CA1 pyramidal cells stain for the NADPH-diaphorase isoform, which has been found to have a neuronal distribution coincident with NOS (Bredt and Snyder, 1990; Bredt et al., 1991; Dawson et al., 1991).

Two questions were addressed in the experiments presented here: first, does in vivo treatment with L-N^G-nitroarginine (NO_2Arg), an irreversible inhibitor of NOS (Dwyer et al., 1991), have the same effect on LTE in awake freely behaving rats as has been observed in vitro? and second, is spatial learning ability disrupted in the same NOS-depleted rats from which electrophysiological measurements are also obtained? The specific hypothesis under investigation is that if NO is critical in the pathway that leads to stable LTE, and if an LTE-like process normally subserves information storage, then depletion of NOS should result in defective LTE and spatial behavior. This approach is analogous to experiments that have examined the effects of blockade of the NMDA receptor, which results both in blockade of LTE induction and disruption of spatial behavior (e.g., Morris et al., 1986; Morris, 1989).

Under surgical anesthesia, male F344 rats (9 months) received bilateral implants of electrodes for stimulation and recording of perforant path synaptic potentials in FD (see McNaughton et al., 1986 for details). Following a 1-week recovery period, a 14-day experimental protocol was initiated, during which evoked synaptic potentials (figure 3.1) were recorded daily from both hemispheres, using low-frequency (0.1 Hz) test stimuli (100–500 μA, 200 μsec diphasic pulses).

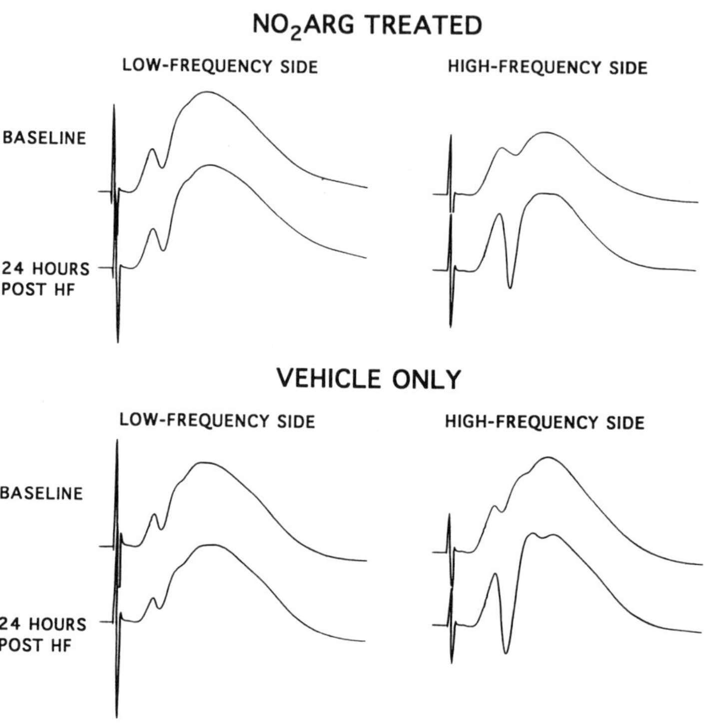

Figure 3.1 Representative electrically evoked synaptic potentials recorded bilaterally from the fascia dentata of conscious, unrestrained rats following single test stimuli delivered to the perforant path of the corresponding hemisphere. EPSP magnitude was computed from the amplitude of the synaptic wave at a fixed latency from its onset, prior to the appearance of the negative-going population spike (granule cell discharge). The population spike magnitude was taken as the area under a tangent line fitted between the onset and offset of the population spike. Following systemic administration of NO_2Arg, an irreversible inhibitor of NOS, high-frequency (HF) stimuli were delivered unilaterally to the perforant path. Vehicle-only animals were treated identically. For quantification of the induction of LTE by high-frequency stimulation, fractional changes of the EPSP and spike measures (compared to baseline) were computed as the corresponding ratio minus 1.0. There was no effect of NOS inhibition either on the baseline responses, or on the induction of long-term synaptic enhancement.

On day 6 of the procedure, daily i.p. injections of NO_2Arg (20 or 50 mg/kg in phosphate buffered saline) [n = 7 rats] or vehicle [n = 6 rats] were begun. On days 10 and 11, after enough time for the depletion of NOS (Dwyer et al., 1991), spatial learning was evaluated using the Morris water task (Morris, 1981). On both days rats were given 14 training trials (7 blocks of 2 trials), with an intertrial interval of 30 sec, and an interblock interval of about 15 min. Latency to find the submerged platform and path length were monitored with the aid of a video tracking system (HVS Systems). During training, the platform was maintained in a fixed location beneath the surface of the pool of water made opaque by powdered chalk, and could be located only by distal spatial cues. On the fifteenth (final) trial of each day, the platform was removed, and the proportion of time spent in the training quadrant was assessed during a 60-sec probe trial.

On days 12 and 13, unilateral high-frequency (25 msec trains of 400 Hz repeated 10 times at 0.1 Hz, repeated twice per day at 2-hr intervals) was administered to induce LTE in both control and NOS-depleted rats. No injections were given days 12–14. The contralateral hemisphere for each group served as an internal control, in which single test stimuli were delivered at low frequency, which produced no LTE. Twenty-four hours after the last high-frequency stimulation treatment, final magnitudes of the responses were recorded, and hippocampi were rapidly dissected from individual animals for enzymatic assay (Bredt and Snyder, 1990). The tissue samples were coded, and the code was broken only after the analyses were complete.

Fractional change (response magnitude after high-frequency [or control] stimulation, minus baseline responses before treatment, divided by baseline values) of the EPSP and population spike indicated that NO_2Arg treatment alone produced no effect on baseline synaptic transmission. The electrophysiological measurements (days 7–12) were always assessed approximately 22 hr after any injection, and thus we would not have detected immediate effects that may have occurred following injection on evoked responses parameters. The amplitude of the EPSP and population spike, measured immediately after the last high-frequency session and 24 hr later, were compared with the control hemispheres between groups using a two-factor (hemisphere by NO_2Arg treatment), repeated-measures analysis of variance (with $\alpha = 0.05$). Although the enzymatic analysis indicated at least 90% inhibition of NOS, there was no effect of NOS depletion on either control responses or the enhancement of the EPSP or population spike (figure 3.2A,B). There was also no effect on spatial learning (figure 3.2C).

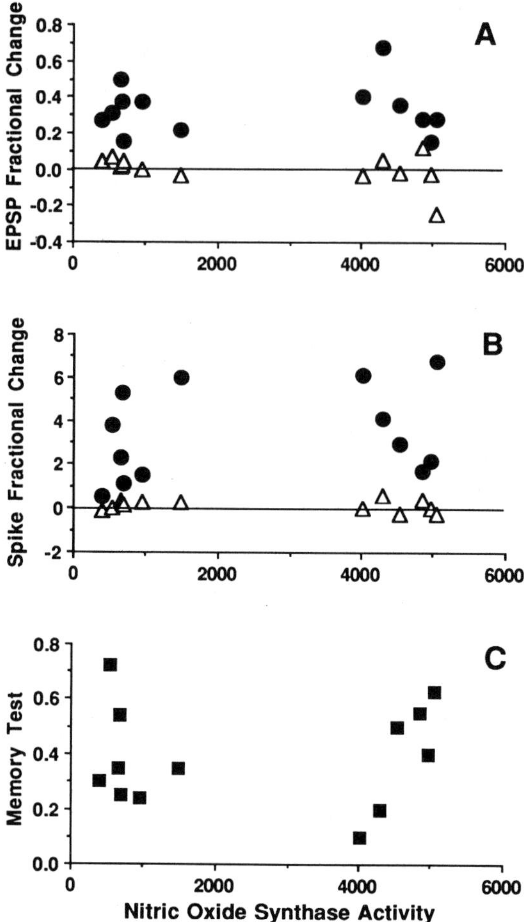

Figure 3.2 Relationship between measured levels of NOS activity and physiological and behavioral variables. (*A*) Fractional change of excitatory postsynaptic potential (EPSP) in high-frequency hemispheres (circles) and control hemispheres (triangles). (*B*) Corresponding measures for the population spike. (*C*) Spatial memory score during probe trials following training on the Morris water task, in which animals are required to escape to a submerged platform located in a large pool of water. Test score reflects the proportion of time the animals spent swimming in the pool quadrant where the platform had been located during training. Both groups performed significantly better than chance, but there was no effect of NOS inhibition.

The present results are not consistent with those obtained by Chapman et al. (1992), who found deficits in water task and conditioned eyeblink response learning following 75 mg/kg doses of NO_2Arg. One difference between the procedures utilized is that training occurred 1 hr after injections in the Chapman et al. (1992) experiment, whereas testing in the present experiment occurred 22 hr after injection. Another difference was that the water task training was given over 2 days in the present study and 7 days in the other experiment. Because enzymatic analysis was not performed in the Chapman et al. (1992) experiment, nor were electrophysiological recordings obtained, it is not possible to know whether NOS depletion was different than that obtained here, nor is it possible to know, even if they did obtain behavioral deficits, whether LTE would have also been disrupted.

A number of possible interpretations are left open by these results. First, the majority of NOS activity may be localized in the vasculature and sensitive to systemic NO_2Arg, whereas neuronal NOS may be immune. Second, LTE may be expressed through a fundamentally different mechanism in FD than in CA1, and may not be NOS-dependent. Third, it is quite possible that there are two NMDA receptor-dependent component processes to the expression of LTE, one which is NOS-dependent and involves increased transmitter release, and one which is NOS-independent and involves postsynaptic changes in transmitter sensitivity. The relative balance of these two components may vary with experimental conditions. Clearly, synaptogenesis in developing tissue, or possible sprouting reactions following injury, must involve some presynaptic component, whereas it is possible that the plasticity required for learning in the intact hippocampus is predominantly of the postsynaptic variety. The present results suggest that the NO_2Arg-sensitive component of LTE, which appears to predominate under some in vitro conditions, does not reflect the same process that predominates under in vivo conditions, and whose blockade is correlated with spatial learning impairment.

ACKNOWLEDGEMENTS

Supported by AGO3376, MH00897, and the Office of Naval Research. We thank T. W. Abel, L. T. Church, S. Davis, and D. L. Korol for assistance, and G. S. Wenk for helpful comments on the manuscript.

REFERENCES

Bekkers, J. M., and Stevens, C. F. (1990) Presynaptic mechanisms for long-term potentiation in the hippocampus. *Nature* 346:724–729.

Böhme, G. A., Bon, C., Stutzman, J.-M., Doble, A., and Blanchard, J.-C. (1991) Possible involvement of nitric oxide in long-term potentiation. *Eur. J. Pharmacol.* 199:379–381.

Bredt, D. S., and Snyder, S. H. (1990) Isolation of nitric oxide synthetase, a calmodulin-requiring enzyme. *Proc. Natl. Acad. Sci. USA* 87:682–685.

Bredt, D. S., Glatt, C. E., Hwang, P. M., Fotuhi, M., Dawson, T. M., and Snyder, S. H. (1991) Nitric oxide synthase protein and mRNA are discretely localized in neuronal populations of the mammalian CNS together with NADPH diaphorase. *Neuron* 7:615–624.

Chapman, P. F., Atkins, C. M., Allen, M. T., Haley, J. E., and Steinmetz, J. E. (1992) Inhibition of nitric oxide synthesis impairs two different forms of learning. *NeuroReport* 3:567–570.

Dawson, T. M., Bredt, D. S., Fotuhi, M., Hwang, P. M., and Snyder, S. H. (1991) Nitric oxide synthase and neuronal NADPH diaphorase are identical in brain and peripheral tissues. *Proc. Natl. Acad. Sci. USA* 88:7797–7801.

Dwyer, M. A., Bredt, D. S., and Snyder, S. H. (1991) Nitric oxide synthase: Irreversible inhibition by L-N^{G}-nitrioarginine in brain in vitro and in vivo. *Biochem. Biophys. Res. Commun.* 176:1136–1141.

Foster, T. C., and McNaughton, B. L. (1991) Long-term enhancement of CA1 synaptic transmission is due to increased quantal size, not quantal content. *Hippocampus* 1:79–91.

Fujimori, H., and Pan-Hou, H. (1991) Effect of nitric oxide on L-[^{3}H]glutamate binding to rat brain synaptic membranes. *Brain Res.* 554:355–357.

Garthwaite, J., Charles, S. L., and Chess-Williams, R. (1988) Endothelium-derived relaxing factor release on activation of NMDA receptors suggests role as intercellular messenger in the brain. *Nature* 336:385–388.

Garthwaite, J., Garthwaite, G., Palmer, R. M. J., and Moncada, S. (1989) NMDA receptor activation induces nitric oxide synthesis from arginine in rat brain slices. *Eur. J. Pharmacol.* 172:413–416.

Haley, J. E., Wilcox, G. L., and Chapman, P. F. (1992) The role of nitric oxide in hippocampal long-term potentiation. *Neuron* 8:211–216.

Malinow, R., and Tsien, R. W. (1990) Presynaptic changes revealed by whole-cell recordings of long-term potentiation in rat hippocampal slices. *Nature* 346:177–180.

Manabe, T., Renner, P., and Nicoll, R. A. (1992) Long-term potentiation revealed by the analysis of miniature synaptic currents. *Nature* 355:50–55.

McNaughton, B. L., Barnes, C. A., Rao, G., Baldwin, J., and Rasmussen, M. (1986) Long-term enhancement of hippocampal synaptic transmission and the acquisition of spatial information. *J. Neurosci.* 6:563–571.

Morris, R. G. M. (1981) Spatial localization does not require the presence of local cues. *Learn. Motiv.* 12:239–261.

Morris, R. G. M. (1989) Synaptic plasticity and learning: Selective impairment of learning in rats and blockade of long-term potentiation in vivo by the *N*-methyl-D-aspartate receptor antagonist AP5. *J. Neurosci.* 9:3040–3057.

Morris, R. G. M., Anderson, E., Lynch, G. S., and Baudry, M. (1986) Selective impairment of learning and blockade of long-term potentiation by an *N*-methyl-D-aspartate receptor antagonist, AP5. *Nature* 319:774–776.

O'Dell, T. J., Hawkins, R. D., Kandel, E. R., and Arancio, O. (1991) Tests on the roles of two diffusible substances in LTP: Evidence for nitric oxide as a possible early retrograde messenger. *Proc. Natl. Acad. Sci. USA* 88:11285–11289.

Schuman, E. M., and Madison, D. V. (1991) A requirement for the intercellular messenger nitric oxide in long-term potentiation. *Science* 254:1503–1506.

4 NO or NOT: Fact or Artifact?

Michel Baudry

Over the past 3 or 4 years, a number of experimental data have suggested that the diffusible gas nitric oxide (NO) plays the role of retrograde messenger at various synapses (Bredt and Snyder, 1992; Snyder, 1992). In particular, it has been argued that NO synthesis and release is a critical element of the mechanisms underlying LTP in hippocampus (Böhme et al., 1991; O'Dell et al., 1991; Schuman and Madison, 1991; Haley et al., 1992) and LTD in cerebellum (Shibuki and Okada, 1991; see Crepel et al., chapter 9). As discussed in the preceding chapters, this conclusion would appear to fit well with the requirement for a calcium influx in both LTP and LTD and with the calcium-calmodulin stimulation of nitric oxide synthase (NOS). In the case of LTP, it has been further hypothesized that NO is synthesized in and released from postsynaptic structures and triggers some biochemical cascade in the presynaptic terminal which results in an increased transmitter release. In the case of LTD, it remains unclear where NO is synthesized, as there is no clear evidence for the presence of NOS in Purkinje cells, the site of convergence of the two stimuli that trigger LTD (see Crepel et al., chapter 9). While this scheme for LTP has been widely publicized (in fact, NO was selected the Molecule of the Year for 1992 by *Science* magazine) and would appear to be well supported by published evidence, we would like to present a critical discussion of the experimental data and to raise a number of problems that are generally ignored or vaguely addressed by the proponents of the NO hypothesis.

ARE THE FACTS TOTALLY CONVINCING?

The NO hypothesis is based in part on the results of experiments showing that inhibitors of NO synthesis, such as L-*N*-methyl-arginine (LNMA) or L-nitro-arginine (L-NA), as well as NO scavengers, such as hemoglobin (Hb), block the induction of LTP in hippocampal slices and in intact animals. While several laboratories have reported positive results with these compounds (see Bon et al., chapter 1, and Madison and Schuman, chapter 2; Mizutani et al., 1993), other laboratories have had difficulties in replicating these findings, and Gribkoff and Lum-Ragan observed NOS inhibitor–sensitive and –insensitive forms of LTP (Gribkoff and Lum, 1992). In particular, Bliss has raised the issue that, although these drugs were able to block LTP induction in hippocampal slices maintained at 24°C, they were ineffective when the temperature of the experiment was 29°C. He has further tested the effects of these drugs in hippocampal slices prepared from animals of different ages and showed that in slices from adult rats, these drugs were not effective at either temperature. In intact animals, L-nitro-arginine (L-NA) had little effect on LTP in the dentate gyrus, although Hb (20 μM) did block LTP (Bliss, personal communication). Recently, the laboratory of McNaughton and Barnes, in collaboration with that of Snyder, observed that animals treated with L-NA and exhibiting a massive depletion in NOS capacity were still able to produce LTP in the dentate gyrus in response to high-frequency stimulation delivered to the perforant path (Barnes et al., chapter 3).

The other argument supporting the role of NO in LTP is provided by the fact that NO-generating molecules, such as sodium nitroprusside or hydroxylamine, produce an LTP-like phenomenon when perfused in hippocampal slices (Böhme et al., 1991; O'Dell et al., 1991; Zhuo et al., 1993). However, Bliss and colleagues have observed that release of NO from cage compounds had no effect on synaptic transmission. Similar negative results were obtained by the laboratory of John Garthwaite (Carol Boulton, personal communication).

As is generally the case in the LTP literature, it remains difficult to determine the origin of the discrepancy in the experimental data obtained in separate laboratories. Several factors are probably important to account for the outcome of LTP studies: the temperature at which the slices are maintained, which possibly relates to NO solubility, the age of the animals, which could affect drug disposition, and the experimental conditions (duration of incubation, rate of perfusion, etc.). Independent of these factors, an initial conclusion is that no consensus exists concerning the effects of NO-related drugs on LTP induction. Recent results again from McNaughton's laboratory might provide some clues concerning the origin of the discrepancies. He and his collaborators made the surprising observation that, in hippocampal slices, repetitive stimulation, even with trains at frequency as modest as 10 stimulations at 2 Hz, results in a long-lasting lowering of the threshold for antidromic activation of the Schaffer-commissural pathways when slices are maintained at room temperature. This effect was not observed when slices were maintained at 32°C. Furthermore, the modification of axonal excitability resulting from repetitive stimulation was totally abolished by NOS inhibitors (McNaughton et al., 1994). These results raise the intriguing possibility that several of the data reported in the literature concerning the role of NO in LTP might be due to (1) an alteration in axonal excitability induced by the manipulation used to produce LTP and (2) an inhibition of this effect by NOS inhibitors.

ADDITIONAL PROBLEMS OF THE NO HYPOTHESIS

At least three additional problems complicate the interpretation of experimental results and cast serious doubts as to the generality of the role of NO in LTP induction. They are related to the questions of the specificity of the drugs used to probe the involvement of NO, of the origin of NO and of the regulation of NMDA receptor via an oxydation-reduction reaction.

Specificity of the NOS Inhibitors or of NO Scavengers

A critical point when probing NO involvement in LTP induction is that the pharmacological agents commonly used to inhibit NOS or to scavenge NO after its release do not affect another critical step in LTP induction.

The initial steps in LTP induction are now well understood and involve the activation of non-NMDA receptors (or AMPA receptors) to produce fast EPSPs, the activation of $GABA_B$ receptors to produce a transient disinhibition (by reducing GABA release) and the temporal summation of depolarization to release NMDA receptor channels from the voltage-dependent Mg^{2+} blockade, and finally the influx of calcium in postsynaptic structures and the activation of a cascade of calcium-dependent processes (Baudry and Lynch, 1993). It has been clearly demonstrated that interfering with any of these steps is sufficient to prevent LTP induction. In a recent study, we showed that the drugs commonly used to probe the NO hypothesis produce a significant inhibition of the temporal summation of depolarization, which is generated by the trains of high-frequency stimulation used to induce LTP (Musleh et al., 1993). In fact, the extent of inhibition was similar to that produced by APV, an inhibitor of NMDA receptor, and the inhibition was not additive with that produced by APV. Although we did not try to determine the precise mechanism by which these drugs inhibit temporal summation, it is not likely that is it related to NO synthesis or release, steps that are supposed to take place after the influx of calcium through the NMDA receptor-channel.

Localization of NOS

Numerous reports have now been published regarding the distribution of NOS. The calcium-calmodulin–dependent NOS has been found to be identical to NADPH diaphorase and the localization of this enzyme has been studied by histochemical, immunocytochemical and in situ hybridization techniques (Bredt et al., 1991; Dawson et al., 1991; Schmidt et al., 1992; Vincent and Kimura, 1992). From all these studies, there is little doubt that the enzyme is present in only scattered interneurons—at least in hippocampal CA1 field, which is also the place where most LTP studies have been performed. This poses a serious problem for the NO hypothesis, as one would think that NOS should be present in pyramidal neurons (more specifically in the dendrites of the pyramidal neurons). The usual counterargument is that there exists another enzyme (an isozyme of NOS) that is not recognized by the antibodies, that do not exhibit NADPH-diaphorase activity, and which is coded by a gene sufficiently different from that coding for NOS that the cRNA probes do not hybridize with its mRNA. Although this remains a possibility, it does impose serious constraints on the properties of this isozyme.

Another problem related to the localization question is raised by the experiments of Madison's laboratory. This group first appeared to have replicated the Bonhoeffer experiments, which suggested that LTP is not entirely synapse specific as it can be elicited in an unstimulated set of synapses closed to a stimulated one (Bonhoeffer et al., 1989)—a result which per se fits well with the notion that a diffusible substance is involved in LTP induction. They further showed that LTP is prevented in both sets of synapses by intracellular loading of an NOS inhibitor in the neuron that receives the pairing stimulus. The paradoxical result is that tetanus-induced LTP in the neighboring set of synapses is blocked by intracellular injection of the calcium chelator EGTA, suggesting that NO would have to mediate its effect through interactions with another postsynaptic calcium-dependent process (Madison, personal communication).

NMDA Receptor Regulation by Oxidation Reduction

A further complicating factor in the interpretation of the experiments involving NO or NO-related drugs is the regulation of NMDA receptors by oxidation reduction (Izumi et al., 1992a, b; Lei et al., 1992; Manzoni et al., 1992). Thus, it appears clearly that NO as well as other oxidizing agents produces a functional blockade of NMDA receptors, possibly by favoring the formation of an S-S disulfide bond in the receptors. Conversely, reducing agents such as dTT enhance NMDA receptor function.

This makes it extremely difficult to provide simple interpretation of some experiments. In particular, the occlusion experiments between the NO-induced potentiation and tetanus-induced potentiation cannot be interpreted as evidence that the two processes use similar pathways, as it is likely that after perfusion of NO-generating compounds, NMDA receptor function is altered. Conversely, treatments with drugs that might alter the synthesis or disposition of NO might very well alter the state of oxidation reduction of NMDA receptors and thereby modify their function.

CONCLUSION

Although the idea that NO is a retrograde messenger linking postsynaptic and presynaptic events is attractive, it should be clear that the experimental evidence supporting it thus far is not extremely compelling. Some results have not been replicated in many laboratories, the specificity of the drugs used in the majority of studies is questionable, the localization of the synthesizing enzyme is still enigmatic, and the effects of NO itself remain unclear. Hopefully, many of the problems will be addressed in the near future, providing a clearer picture of the role of NO not only in LTP and LTD but more generally in various aspects of neural functions. Finally, it remains possible that the function assigned to NO in the case of LTP is actually performed by another diffusible molecule. Recently, evidence has been presented suggesting that carbon monoxide may function as a neurotransmitter and could play a role as retrograde messenger at hippocampal synapses (Verma et al., 1993), and recent data suggest that CO might also be involved in LTP (Zhuo et al., 1993; Stevens and Wang, 1993). As more molecules are proposed to play a retrograde messenger role in CNS, it might be useful to devise a set of criteria for such molecules along the lines of the criteria that were used to identify anterograde neurotransmitter substances. Thus (1) the putative retrograde messenger and/or its synthetic enzymes should be present in postsynaptic structures, (2) the messenger should be released under the appropriate conditions, (3) the messenger should have a specific receptor or target, and (4) there should exist some inactivation mechanism. Much remains to be done to determine whether NO or CO satisfy all these criteria.

REFERENCES

Baudry, M., and Lynch, G. (1993) Long-term potentiation: Biochemical mechanisms. In *Synaptic Plasticity: Molecular and Functional Aspects,* M. Baudry, R. F. Thompson, and J. L. Davis (eds.). Cambridge, Mass.: MIT Press, pp. 87–115.

Böhme, G. A., Bon, C., Stutzman, J. M., Doble, A., and Blanchard, A. (1991) Possible involvement of nitric oxide in long-term potentiation. *Eur. J. Pharmacol.* 199:379–381.

Bonhoeffer, T., Staiger, V., and Aertsen, A. (1989) Synaptic plasticity in rat hippocampal slice cultures: Local "Hebbian" conjunction of pre- and postsynaptic stimulation leads to distributed synaptic enhancement. *Proc. Nat. Acad. Sci. USA* 86:8113–8117.

Bredt, D. S., and Snyder, S. H. (1992) Nitric oxide, a novel neuronal messenger. *Neuron* 8:3–11.

Bredt, D. S., Glatt, C. E., Hwang, P. M., Fotuhi, M., Dawson, T. M., and Snyder, S. H. (1991) Nitric oxide synthase protein and mRNA are discretely localized in neuronal populations of the mammalian CNS together with NADPH diaphorase. *Neuron* 7:615–624.

Dawson, T. M., Bredt, D. S., Fotuhi, M., Hwang, P. M., and Snyder, S. H. (1991) Nitric oxide synthase and neuronal NADPH diaphorase are identical in brain and peripheral tissues. *Proc. Nat. Acad. Sci. USA* 88:7797–7801.

Gribkoff, V. K. and Lum, R. J. T. (1992) Evidence for nitric oxide synthase inhibitor-sensitive and insensitive hippocampal synaptic potentiation. *J. Neurophysiol.* 68:639–642.

Haley, J. E., Wilcox, G. L., and Chapman, P. F. (1992) The role of nitric oxide in hippocampal long-term potentiation. *Neuron* 8:211–216.

Izumi, Y., Benz, A. M., Clifford, D. B., and Zorumski, C. F. (1992a) Nitric oxide inhibitors attenuate *N*-methyl-D-aspartate excitotoxicity in rat hippocampal slices. *Neurosci. Lett.* 135:227–230.

Izumi, Y., Clifford, D. B., and Zorumski, C. F. (1992b) Inhibition of long-term potentiation by NMDA-mediated nitric oxide release. *Science* 257:1273–1276.

Lei, S. Z., Pan, Z. H., Aggarwal, S. K., Chen, H. S., Hartman, J., Sucher, N. J., and Lipton, S. A. (1992) Effect of nitric oxide production on the redox modulatory site of the NMDA receptor-channel complex. *Neuron* 8:1087–1099.

Manzoni, O., Prezeau, L., Marin, P., Deshager, S., Bockaert, J., and Fagni, L. (1992) Nitric oxide-induced blockade of NMDA receptors. *Neuron* 8:653–662.

McNaughton, B. L., Shen, J., Rao, G., Foster, T. C., and Barnes, C. A. (1994) Increased CA1 axon terminal excitability following repetitive electrical stimulation: dependence on NMDA receptor activity, nitric oxide synthase, and temperature. Submitted for publication.

Mizutani, A., Saito, H., and Abe, K. (1993) Involvement of nitric oxide in long-term potentiation in the dentate gyrus in vivo. *Brain Res.* 605:309–311.

Musleh, W., Shahi, K., and Baudry, M. (1993) Further studies concerning the role of NO in LTP induction and maintenance. *Synapse* 13:370–375.

O'Dell, T. J., Hawkins, R. D., Kandel, E. R., and Arancio, O. (1991) Tests of the roles of two diffusible substances in long-term potentiation: evidence for nitric oxide as a possible early retrograde messenger. *Proc. Natl. Acad. Sci. USA* 88:11285–11289.

Schmidt, H. H., Gagne, G. D., Nakane, M., Pollock, S., Miller, M.-F., and Murad, F. (1992) Mapping of neural nitric oxide synthase in the rat suggests frequent co-localization with NADPH diaphorase but not with soluble guanylyl cyclase, and novel paraneural functions for nitrinergic signal transduction. *J. Histochem. Cytochem.* 40: 1439–1456.

Schuman, E., and Mádison, D. V. (1991) A requirement for the intercellular messenger nitric oxide in long-term potentiation. *Science* 254:1503–1506.

Shibuki, K., and Okada, D. (1991) Endogenous nitric oxide release required for long-term synaptic depression in the cerebellum. *Nature* 349:326–328.

Snyder, S. H. (1992) Nitric oxide and neurons. *Curr. Opin. Neurobiol.* 2:323–327.

Stevens, C. F., and Wang, Y. (1993) Reversal of long-term potentiation by inhibitors of haem oxygenase. *Nature* 364:147–149.

Verma, A., Hirsch, D. J., Glatt, C. E., Ronnett, G. V., and Snyder, S. H. (1993) Carbon monoxide: A putative neural messenger. *Science* 259:381–384.

Vincent, S. R., and Kimura, H. (1992) Histochemical mapping of nitric oxide sunthase in the rat brain. *Neuroscience* 46:755–784.

Zhuo, M., Small, S. A., Kandel, E. R., and Hawkins, R. D. (1993) Nitric oxide and carbon monoxide produce activity-dependent long-term synaptic enhancement in hippocampus. *Science* 260:1946–1950.

II Expression and Maintenance Mechanisms

5 The Role of Calcium in the Induction of Long-Term Potentiation in the CA1 Region of the Hippocampus

Toshiya Manabe, David J. Perkel, Dimitri M. Kullmann, Pius Renner, David J. A. Wyllie, and Roger A. Nicoll

It is well established that the induction of long-term potentiation (LTP) of excitatory synaptic transmission in the CA1 region of the hippocampus requires an increase in postsynaptic calcium concentration. The neurotransmitter at these excitatory synapses is likely to be glutamate and one pathway for entry of calcium to the postsynaptic cell is via the *N*-methyl-D- aspartate (NMDA) subtype of glutamate receptor. It is generally agreed that NMDA receptor activation is essential for the induction of LTP in the CA1 region (Collingridge et al., 1983). The NMDA receptor is highly permeable to calcium as well as to monovalent cations and shows a characteristic voltage-dependent block of ionic currents by magnesium (Mayer et al., 1984; Nowak et al., 1984; Mayer and Westbrook, 1987). It is believed that depolarization induced by tetanic stimulation of afferent fibers relieves this voltage-dependent block of the NMDA receptor, resulting in an influx of calcium into the postsynaptic cell. The increased calcium levels can then activate subsequent calcium-dependent biochemical processes required for LTP. Previous reports have shown that injection of calcium chelators into the postsynaptic cell blocks LTP induction (Lynch et al., 1983; Malenka et al., 1988). In addition, LTP is also suppressed when extremely strong depolarization, which would decrease the driving force for calcium entry through NMDA receptors, is paired with afferent stimulation in an attempt to induce LTP (Malenka et al., 1988; Perkel et al., 1991). Furthermore, release of calcium into postsynaptic cells by a caged calcium compound can enhance synaptic transmission by itself (Malenka et al., 1988). It has recently been shown that calcium buffering 2–3 sec after tetanic stimulation by a photolabile calcium buffer fails to block LTP, indicating that only a brief rise in calcium is needed (Malenka et al., 1992).

Activation of voltage-sensitive calcium channels is known to cause a widespread influx of calcium in dendrites of CA1 pyramidal cells (Regehr et al., 1989; Jaffe et al., 1992). Why is it, then, that depolarization of the postsynaptic cell in the absence of synaptic activity is not sufficient to potentiate excitatory synaptic transmission (Malenka et al., 1989)? Two possibilities exist. First, if calcium alone potentiates synaptic transmission, the calcium entering through voltage-sensitive calcium channels may be segregated from that entering through NMDA receptors, since LTP is specific to the conditioned synapses. Recent calcium-imaging experiments support this hypothesis (Guthrie et al., 1991; Muller and Connor, 1991). Second, perhaps calcium alone is not sufficient for potentiation, although under certain conditions voltage-sensitive calcium channels may be involved in the induction of LTP. Tetanic stimulation of afferent fibers at 200 Hz in the presence of APV, an NMDA receptor antagonist, can induce a form of LTP that is sensitive to calcium channel blockers (Grover and Teyler, 1990). A similar potentiation can also be induced by blockade of potassium channels, which would give rise to calcium spikes in the postsynaptic cell (Aniksztejn and Ben-Ari, 1991). Therefore, it may be possible, by activating calcium channels, to increase sufficiently the levels of postsynaptic calcium in dendrites, and possibly in spines, to enhance synaptic transmission. However, in these two studies as well as those with caged calcium, synaptic stimulation was delivered during the presumed rise in intracellular calcium and, therefore, these experiments do not rigorously test the hypothesis that calcium alone can potentiate synaptic transmission.

In this chapter we will discuss the role of the calcium entering through NMDA receptors and voltage-sensitive calcium channels in determining synaptic strength in the CA1 region of the hippocampus.

NMDA RECEPTOR ACTIVATION ENHANCES SYNAPTIC TRANSMISSION

Brief application of NMDA enhances excitatory synaptic transmission in the CA1 region of the hippocampus (Collingridge et al., 1983; Kauer et al., 1988). Such potentiation is induced by an increase in postsynaptic calcium, as it is completely blocked by a high intracellular concentration of the calcium chelator BAPTA (Manabe et al., 1992). Although a presynaptic contribution to glutamate-induced potentiation of synaptic transmission has been implicated in cultured cells (Malgaroli and Tsien, 1992), postsynaptic modification of the non-NMDA glutamate receptor is likely to be involved in this enhancement in hippocampal slices (Manabe et al., 1992). The use of whole-cell recording techniques, providing a high signal-to-noise ratio, makes it possible to record quantal synaptic events much more clearly than with conventional intracellular recording. Spontaneously occurring miniature excitatory postsynaptic currents (mEPSCs) provide useful information for determining the site responsible for the change in synaptic efficacy. Based on studies at the vertebrate neuromuscular junction (Katz, 1966), a change in mEPSC size is usually interpreted as a change in postsynaptic sensitivity to neurotransmitter, while a change in mEPSC frequency is thought to represent a presynaptic change. NMDA application has an advantage over tetanus-induced LTP in that essentially all the synapses will be exposed to the drug and, therefore, the recorded spontaneous EPSCs are more likely to arise from potentiated synapses.

We have recorded spontaneous and miniature EPSCs in hippocampal pyramidal cells in slices using whole-cell techniques. The size of mEPSCs increases dramatically during NMDA-induced potentiation of evoked EPSCs (figures 5.1 and 5.2). Such potentiation requires a postsynaptic calcium increase, since it is blocked by postsynaptic administration of BAPTA (figure 5.2) or by strong hyperpolarization of the membrane,

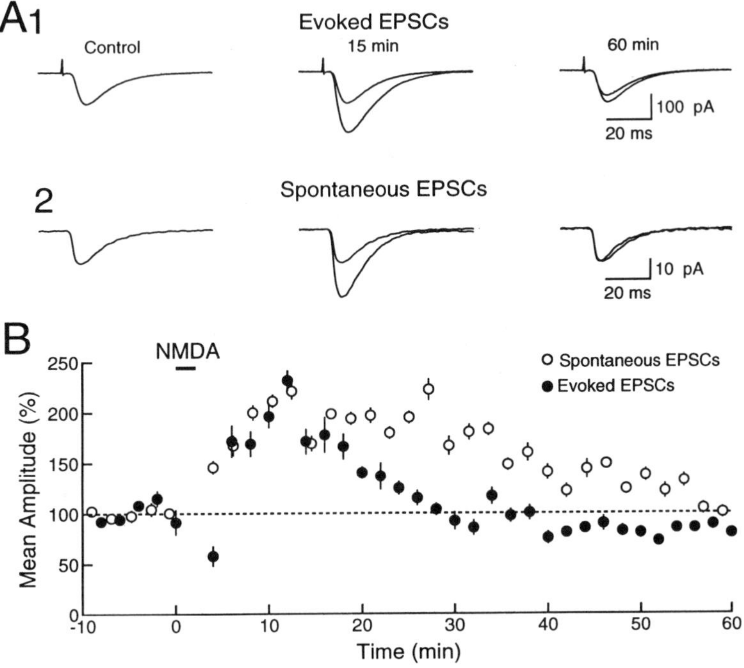

Figure 5.1 NMDA application potentiates both evoked and spontaneous EPSCs. All data are from the same cell. EPSCs were recorded in voltage clamp mode with whole-cell recording. Patch pipettes were filled with a K gluconate-based intracellular solution. (*A1*) Averaged traces of evoked EPSCs before, 15 min after, and 60 min after pressure application of 500 μM NMDA. During NMDA application the amplifier was switched to current clamp mode. (*A2*) Averaged traces of spontaneous EPSCs recorded during the same time periods as evoked EPSCs. (*B*) The time course of the potentiation of evoked and spontaneous EPSCs, partially illustrated in *A*. (From Manabe et al., 1992)

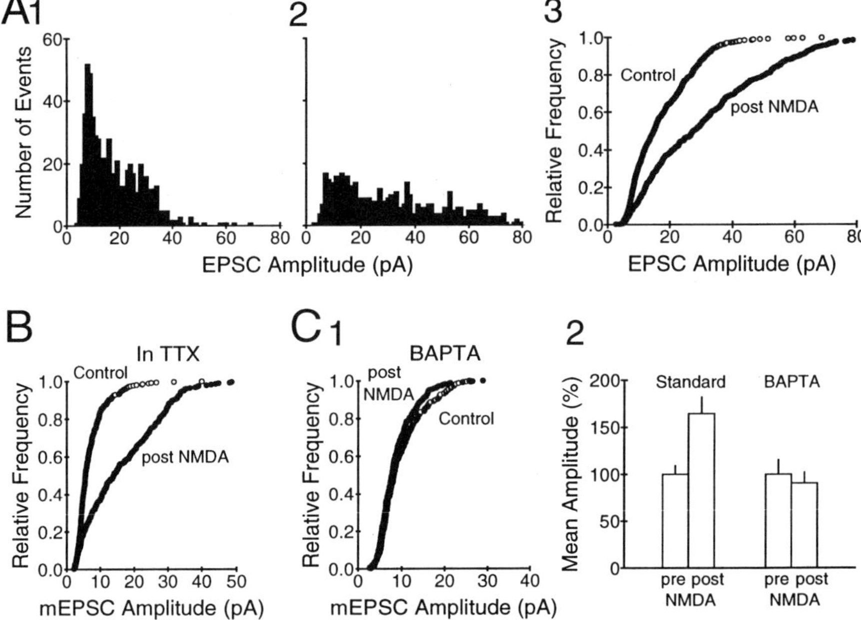

Figure 5.2 Properties of NMDA-induced increase in mEPSC amplitude. (*A*) Amplitude histograms of spontaneous EPSCs immediately before (*A1*) and 15 min after pressure application of NMDA (*A2*) in the absence of TTX. *A3* shows cumulative plots of the same data. (*B*) Cumulative plots of mEPSC amplitude distributions from an experiment which was performed entirely in the presence of TTX. NMDA (10 μM) was applied to the bath for 2 min. The curve after NMDA application comprises mEPSCs recorded between 15 min and 20 min after the application. (*C1*) Example of an experiment in which the patch pipette contained 10 mM BAPTA. After NMDA application there was a small, but insignificant decrease in the amplitude of mEPSCs. (*C2*) Summary graphs of the experiments that tested the effect of intracellular BAPTA on NMDA-induced potentiation. Although a large potentiation was observed with a standard internal solution (n = 8), intracellular BAPTA completely blocked the potentiation (n = 4). (From Manabe et al., 1992)

which blocks current flow through NMDA receptors, during NMDA application (Manabe et al., 1992). An increase in quantal size is most likely to be mediated by an increase in sensitivity of postsynaptic non-NMDA receptors to transmitter, although it could result from an increase in the amount of transmitter in each presynaptic vesicle. Our recent data on the change in postsynaptic sensitivity during NMDA-induced potentiation confirm that a modification of postsynaptic non-NMDA receptors contributes to the potentiation: responses to exogenously applied AMPA, a selective non-NMDA receptor agonist, are significantly enhanced during NMDA-induced synaptic enhancement (Manabe and Nicoll, 1992). Since an increase in mEPSC size is also observed during LTP expression, LTP is likely to be mediated, at least in part, by the same postsynaptic mechanism (Manabe et al., 1992).

ACTIVATION OF POSTSYNAPTIC CALCIUM CHANNELS POTENTIATES SYNAPTIC TRANSMISSION

Although there are some similarities between LTP and NMDA-induced potentiation, the latter is generally transient; synaptic strength returns to the baseline within 30 min after NMDA application. Under certain conditions, especially in raised extracellular calcium, the potentiation can be longer lasting (Thibault et al., 1989; Malenka, 1991; Manabe et al., 1992). However, it is still unclear whether a rise in postsynaptic calcium by itself is sufficient to induce LTP. Two recent reports have shown that strong activation of postsynaptic voltage-sensitive calcium channels induces long-lasting potentiation of synaptic transmission in the CA1 region (Grover and Teyler, 1990; Aniksztejn and Ben-Ari, 1991). In these studies synaptic stimulation accompanied the LTP-inducing procedure and thus, factors other than calcium might have contributed to the potentiation.

We have found conditions in which activation of voltage-sensitive calcium channels in the absence of afferent stimulation results in a potentiation of excitatory synaptic transmission (Perkel et al., 1991; Kullmann et al., 1992). Repeated depolarizing voltage pulses, which would activate calcium channels and lead to increased postsynaptic calcium concentration, enhance synaptic strength in the presence of APV either in conventional intracellular recording or in whole-cell recording (figure 5.3). The protocol comprises 20 depolarizing pulses (from -90 mV to ~ 0 mV for 3 sec) repeated at 0.2 Hz. The potentiation is not accompanied by any changes in series resistance and input resistance. This manipulation has the advantage that the effect of an increase in postsynaptic calcium on synaptic transmission can be examined in isolation.

This depolarizing pulse–induced potentiation is dependent on extracellular calcium concentrations. When higher calcium is used in the external solution, the potentiation is observed more frequently and is usually larger, supporting the notion that this phenomenon is induced by an influx of external calcium into the postsynaptic cell. Steady depolarization does not enhance synaptic transmission, presumably due to inactivation of calcium channels (figure 5.3), which is consistent with the previous study (Malenka et al., 1989). Although the potentiation can be induced repeatedly with conventional intracellular recording, the ability to evoke it is lost with time with whole-cell recording. In this regard, this potentiation has a similarity with LTP, which displays washout during whole-cell recording (Malinow and Tsien, 1990). Depolarizing pulse–induced potentiation is completely blocked when BAPTA is included in the internal solution during whole-cell recording. Nifedipine, a calcium channel blocker, also inhibits the potentiation significantly, confirming that the potentiation is

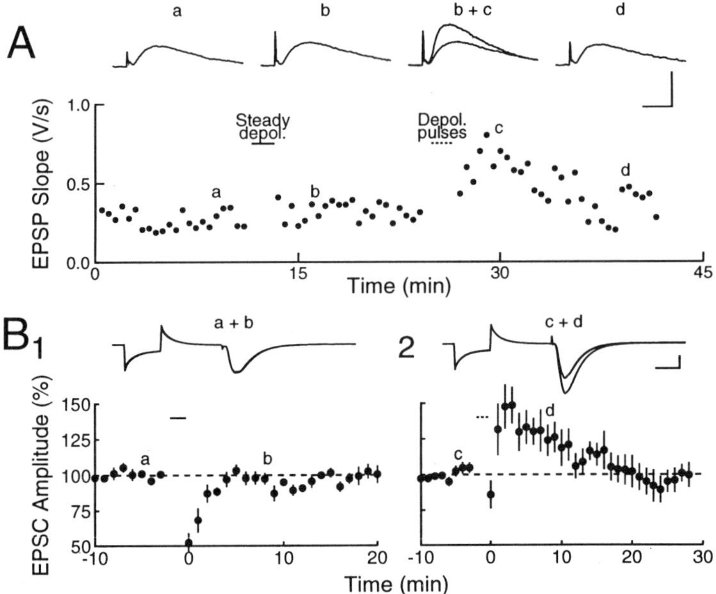

Figure 5.3 Effects of steady depolarization and repeated depolarizing pulses on synaptic transmission. *A* shows an experiment, with a conventional intracellular pipette (2.5 M CsCl), in which the cell was depolarized to + 10 mV and held at this potential for 100 sec (Steady depol.), and later depolarizing pulses were applied (Depol. pulses). Depolarizing pulses induced a transient potentiation although steady depolarization did not potentiate synaptic transmission. D-APV (25 μM) was present in this experiment. Scale bars: 5 mV, 20 msec. *B* summarizes the results from two groups of cells recorded with whole-cell recording. In *B1* the cells were depolarized to 0 mV and held at this potential for 100 sec, while in *B2* depolarizing pulses were given. D-APV (50 μM) was present in these experiments. Scale bars: 100 pA, 20 msec. (From Kullmann et al., 1992)

triggered by some mechanism(s) activated by increased calcium in the postsynaptic cell. However, nifedipine has no effect on tetanus-induced LTP of the field EPSPs, suggesting that the calcium entering through calcium channels does not significantly contribute to NMDA receptor–dependent LTP induced by high-frequency stimulation (Kullmann et al., 1992).

The depolarizing pulse–induced potentiation has some properties in common with NMDA-induced potentiation as already described above. In addition, the potentiation is also associated with the same postsynaptic sensitivity increase as that seen during NMDA-induced potentiation. The size of mEPSCs increases during this potentiation (Wyllie et al., unpublished observations) and the postsynaptic transmitter sensitivity to exogenously applied AMPA is potentiated. The increase in AMPA sensitivity displays the same time course as the change in evoked EPSC amplitude (Manabe et al., unpublished observations).

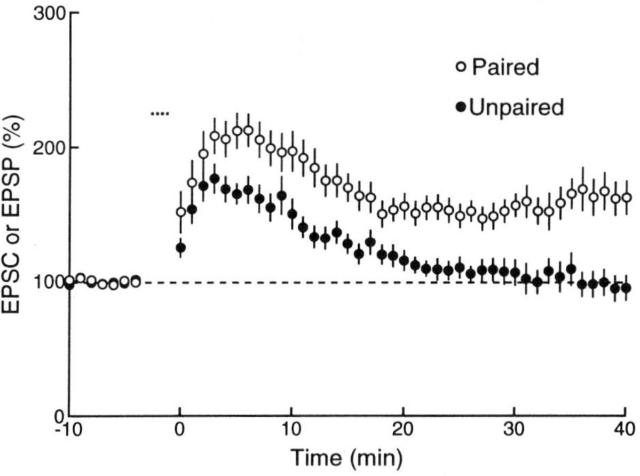

Figure 5.4 Pairing synaptic stimulation with depolarizing pulses in the presence of APV induces sustained potentiation. Two independent pathways were stimulated at 0.1 Hz in the presence of 50 μM D-APV. Stimulation of one pathway was continued at 2 Hz during depolarizing pulses (open circles), while that of the other pathway was interrupted (closed circles). Identical results were obtained in conventional intracellular recording (n = 13) and in whole-cell recording (n = 14), and the results from these two groups have been averaged together. (From Kullmann et al., 1992)

The most consistent finding with depolarizing pulse–induced potentiation is that, although it can be as large as threefold, it is always transient and synaptic strength decays to baseline values within about 30 min. This suggests that an increase in postsynaptic calcium alone may not be sufficient to induce LTP. However, when synaptic stimulation is continued during depolarizing pulses in the presence of a high concentration of APV, which prevents NMDA receptor–dependent long-lasting synaptic enhancement, this transient potentiation can be converted to a long-lasting form (figure 5.4).

It is unlikely that synaptic activation simply increases the amount of calcium entering into postsynaptic cells through calcium channels since, even if synaptic stimulation is given only when the membrane potential is depolarized to the reversal potential for the EPSC, sustained potentiation can still be induced. These results suggest that a factor(s) other than calcium may be necessary to induce LTP. One possible mechanism for inducing sustained potentiation is that synaptically released glutamate may be required for activating other types of glutamate receptors in addition to NMDA receptors. Glutamate might provide extra calcium by activating non-NMDA receptors (Iino et al., 1991) or metabotropic glutamate receptors. Another possibility includes protein kinase C activation by metabotropic glutamate receptors, analogous to LTD in the cerebellum (Linden and Connor, 1991). A number of studies have suggested that protein kinase C activation is involved in LTP (Malenka et al., 1986; Anwyl, 1989; Linden and Routtenberg, 1989; Malinow et al., 1989; Wang and Feng, 1992). Although a previous study failed to induce sustained potentiation by applying glutamate (Kauer et al., 1988), it has recently been reported that repeated ionophoresis of glutamate can induce long-lasting potentiation (Cormier et al., 1993). Furthermore, coapplication of NMDA and *trans*-ACPD, an agonist for metabotropic glutamate receptors, is reported to produce sustained synaptic potentiation (Musgrave et al. 1993). A second possibility is that some cotransmitter(s) released with glutamate may be responsible for the induction of LTP. For instance, ATP, which is known to be present in many types of synaptic vesicle (Lagercrants, 1976; Volknandt and Zimmermann, 1986), might have some effect on LTP induction. Finally, it might be possible that presynaptic activity can make the stimulated synapses receptive to some retrograde factor.

CONCLUSION

We have found that an increase in postsynaptic calcium in the absence of synaptic stimulation can enhance excitatory synaptic transmission and that NMDA-induced potentiation and depolarizing pulse–induced potentiation share many properties, suggesting that these two potentiations might be the same phenomenon. The potentiation is usually transient in both cases. However, when afferent stimulation is paired with depolarizing pulses in the presence of APV, the potentiation is converted to a sustained form, suggesting that synaptic activation, coincident with a rise in postsynaptic calcium, may be necessary for LTP induction. Thus, simultaneous presynaptic and postsynaptic activities, in addition to NMDA receptor activation, may be a crucial factor for inducing LTP.

REFERENCES

Aniksztejn, L., and Ben-Ari, Y. (1991) Novel form of long-term potentiation produced by a K^+ channel blocker in the hippocampus. *Nature* 349:67–69.

Anwyl, R. (1989) Protein kinase C and long-term potentiation in the hippocampus. *Trends Pharmacol. Sci.* 10:236–239.

Collingridge, G. L., Kehl, S. J., and McLennan, H. (1983) Excitatory amino acids in synaptic transmission in the Schaffer collateral-commissural pathway of the rat hippocampus. *J. Physiol.* 334:33–46.

Cormier, R. J., Mauk, M. D., and Kelly, P. T. (1993) Glutamate iontophoresis induces long-term potentiation in the absence of evoked presynaptic activity. *Neuron* 10:907–919.

Grover, L. M., and Teyler, T. J. (1990) Two components of long-term potentiation induced by different patterns of afferent activation. *Nature* 347:477–479.

Guthrie, P. B., Segal, M., and Kater, S. B. (1991) Independent regulation of calcium revealed by imaging dendritic spines. *Nature* 354:76–80.

Iino, M., Ozawa, S., and Tsuzuki, K. (1991) Permeation of calcium through excitatory amino acid receptor channels in cultured rat hippocampal neurones. *J. Physiol.* 424:151–166.

Jaffe, D. B., Johnston, D., Lasser-Ross, N., Lisman, J. E., Miyakawa, H., and Ross, W. N. (1992) The spread of Na^{+} spikes determines the pattern of dendritic Ca^{2+} entry into hippocampal neurons. *Nature* 357:244–246.

Katz, B. (1966) *Nerve, Muscle, and Synapse.* New York: MacGraw-Hill.

Kauer, J. A., Malenka, R. C., and Nicoll, R. A. (1988) NMDA application potentiates synaptic transmission in the hippocampus. *Nature* 334:250–252.

Kullmann, D. M., Perkel, D. J., Manabe, T., and Nicoll, R. A. (1992) Ca^{2+} entry via postsynaptic voltage-sensitive Ca^{2+} channels can transiently potentiate excitatory synaptic transmission in the hippocampus. *Neuron* 9:1175–1183.

Lagercrants, H. (1976) On the composition and function of large dense cored vesicles in sympathetic nerves. *Neuroscience* 1:81–92.

Linden, D. J., and Connor, J. A. (1991) Participation of postsynaptic PKC in cerebellar long-term potentiation in culture. *Science* 254:1656–1659.

Linden, D. J., and Routtenberg, A. (1989) Role of protein kinase C in long-term potentiation: A testable model. *Brain Res. Rev.* 14:279–296.

Lynch, G., Larson, J., Kelso, S., Barrionuevo, G., and Schottler, F. (1983) Intracellular injections of EGTA block induction of hippocampal long-term potentiation. *Nature* 305: 719–721.

Malenka, R. C. (1991) Postsynaptic factors control the duration of synaptic enhancement in area CA1 of the hippocampus. *Neuron* 6:53–60.

Malenka, R. C., Madison, D. V., and Nicoll, R. A. (1986) Potentiation of synaptic transmission in the hippocampus by phorbol esters. *Nature* 321:695–697.

Malenka, R. C., Kauer, J. A., Zucker, R. J., and Nicoll, R. A. (1988) Postsynaptic calcium is sufficient for potentiation of hippocampal synaptic transmission. *Science* 242:81–84.

Malenka, R. C., Kauer, J. A., Perkel, D. J., and Nicoll, R. A. (1989) The impact of postsynaptic calcium on synaptic transmission—its role in long-term potentiation. *Trends Neurosci.* 12:444–450.

Malenka, R. C., Lancaster, B., and Zucker, R. S. (1992) Temporal limits on the rise in postsynaptic calcium required for the induction of long-term potentiation. *Neuron* 9:121–128.

Malgaroli, A., and Tsien, R. W. (1992) Glutamate-induced long-term potentiation of the frequency of miniature synaptic currents in cultured hippocampal neurons. *Nature* 357: 134–139.

Malinow, R., and Tsien, R. W. (1990) Presynaptic enhancement shown by whole-cell recordings of long-term potentiation in hippocampal slices. *Nature* 346:177–180.

Malinow, R., Schulman, H., and Tsien, R. W. (1989) Inhibition of postsynaptic PKC or CaMKII blocks induction but not expression of LTP. *Science* 245:862–866.

Manabe, T., and Nicoll, R. A. (1992) Increase in postsynaptic non-NMDA receptor sensitivity induced by NMDA in hippocampal slices. *Soc. Neurosci. Abstr.* 18:403.

Manabe, T., Renner, P., and Nicoll, R. A. (1992) Postsynaptic contribution to long-term potentiation revealed by the analysis of miniature synaptic currents. *Nature* 355:50–55.

Mayer, M. L., and Westbrook, G. L. (1987) Permeation and block of *N*-methyl-D-aspartic acid receptor channels by divalent cations in mouse cultured central neurones. *J. Physiol.* 394:501–527.

Mayer, M. L., Westbrook, G. L., and Guthrie, P. B. (1984) Voltage-dependent block by Mg^{2+} of NMDA responses in spinal cord neurones. *Nature* 309:261–263.

Muller, W., and Connor, J. A. (1991) Dendritic spines as individual neuronal compartments for synaptic Ca^{2+} responses. *Nature* 354:73–76.

Musgrave, M. A., Ballyk, B. A., and Goh, J. W. (1993) Coactivation of metabotropic and NMDA receptors is required for LTP induction. *NeuroReport* 4:712–714.

Nowak, L., Bregestovski, P., Ascher, P., Herbet, A., and Prochiantz, A. (1984) Magnesium gates glutamate-activated channels in mouse central neurones. *Nature* 307:462–465.

Perkel, D. J., Manabe, T., and Nicoll, R. A. (1991) Role of membrane potential and calcium in the induction of long-term potentiation (LTP). *Soc. Neurosci. Abstr.* 17:2.

Regehr, W. G., Connor, J. A., and Tank, D. W. (1989) Optical imaging of calcium accumulation in hippocampal pyramidal cells during synaptic activation. *Nature* 341:533–536.

Thibault, O., Joly, M., Muller, D., Schottler, F., Dudek, S., and Lynch, G. (1989) Long-lasting physiological effects of bath applied *N*-methyl-D-aspartate. *Brain Res* 476:170–173.

Volknandt, W., and Zimmermann, H. (1986) Acetylcholine, ATP, and proteoglycan are common to synaptic vesicles isolated from the electric organs of electric eel and electric catfish as well as from rat diaphragm. *J. Neurochem.* 47:1449–1462.

Wang, J.-H. and Feng, D.-P. (1992) Postsynaptic protein kinase C essential to induction and maintenance of long-term potentiation in the hippocampal CA1 region. *Proc. Natl. Acad. Sci. USA* 89:2576–2580.

6 Mechanisms of Expression of Long-Term Potentiation: Time-dependent Reversal and Role of Protein Kinases

Dominique Muller, Kohji Fukunaga, and Eishichi Miyamoto

In many pathways of the central nervous system, activation of synapses using trains of high frequency stimulation results in a stable and long-lasting increase in the efficacy of synaptic transmission usually referred to as long-term potentiation or LTP. This form of plasticity, discovered and mostly studied in the hippocampus (Bliss and Lømo, 1973), is believed to underlie some of the mechanisms of information storage by the brain.

Important efforts have been devoted during the last years to the identification of the different events contributing to this change in synaptic efficacy. Chapter 8, by R. Malenka, reviews some of the early events contributing to the induction of this form of plasticity and the role and time course of the rise in intracellular calcium concentration that is required in the postsynaptic element (Lynch et al., 1983; Malenka et al., 1988; Colino et al., 1992; Malenka et al., 1992). While activation of NMDA receptors has been shown to represent the major route for calcium entry (Collingridge et al., 1983) in area CA1 of hippocampus, there are also indications from recent work that other pathways involving other receptor types such as metabotropic receptors (Izumi et al., 1991; Zheng and Gallagher, 1992) may also contribute to the induction phase. Recently, kinases and more specifically protein kinase C (PKC) has been implicated in these events and proposed to play a role in determining the threshold level for LTP induction (Kelso et al., 1992; Aniksztejn et al., 1992).

A second important issue that has been intensely debated during the last two years concerns the locus of the changes responsible for the long-lasting increase in synaptic activity. The major approach used for addressing this issue has been based, as described in chapter 7 by R. Malinow and colleagues, on quantal analyses. The results that have been obtained are consistent with both an increase in quantal content and quantal size of synaptic responses and therefore support the idea of changes in both presynaptic and postsynaptic properties (Malinow and Tsien, 1990; Bekkers and Stevens, 1990; Foster and McNaughton, 1990; Malinow,

1991; Manabe et al., 1992; Malgaroli and Tsien, 1992; Kullmann and Nicholl, 1992).

Finally the third important question about LTP mechanisms concerns the identification and the role played by various molecules, enzymes, or second messengers in the increase in synaptic efficacy. Among these, protein kinases are very often cited as candidates, and the present discussion will focus on the possible involvement of these enzymes, and more specifically protein kinase C (PKC) and calcium/calmodulin–dependent protein kinase II (CaM kinase II), at the different steps of the cascade of events leading to synaptic potentiation.

To address these issues, two approaches have essentially been used. One consists in analyzing the biochemical modifications in enzyme activity or substrate phosphorylation that are associated with LTP induction. This approach provides important information about events that accompany LTP induction, but the results that are obtained remain essentially correlative.

The second approach consists in testing the effects of activators or inhibitors of protein kinases on the induction of LTP or on preestablished LTP. If a kinase is critically involved in LTP, it is expected that its activation should generate and its inhibition prevent synaptic potentiation. Ideally, these two criteria should be fulfilled to convincingly establish the role of a molecule or enzyme in LTP mechanisms. It should be noted, however, that failure to observe synaptic potentiation following treatment with an activator cannot be interpreted as arguing against a role in LTP, since a conjunction of several events might be required. In contrast, failure to prevent or reverse synaptic potentiation by an inhibitor, provided that the inhibitor is active, is particularly informative and strongly argues against a critical role in LTP.

EVIDENCE FOR DIFFERENT PHASES IN LTP

Analyses of the role of protein kinases in LTP have been based on these two approaches with controversial and sometimes contradictory results. An important step in assessing the role of protein kinases in LTP is to distinguish between different phases in the cascade of events responsible for synaptic potentiation and thus the possible involvement of kinases at different points and for varying periods of time. There are now several indications supporting the existence of such phases, which probably depend on the involvement of different molecules or enzymes. As described in chapter 8 by Malenka, the most important events for triggering LTP—the rise in calcium in postsynaptic spines and its activation of a still unknown process—probably take place during a very short initial phase of a few seconds following high-frequency stimulation (Malenka et al., 1992). A very interesting result obtained in these and previous studies is that partial interference with this step invariably results in a decaying form of potentiation, with the duration of the synaptic enhancement correlating with the magnitude of the calcium influx during high-frequency stimulation (Malenka, 1991; Malenka et al., 1992). Thus the rise in calcium concentration that takes place during this initial phase affects the synaptic enhancement observed during the next 10–30 min.

This first induction period could be then followed by a second phase of about 10–20 min, which might be required for the consolidation of LTP. Evidence for this stabilization period is based mainly on results indicating the possibility to reverse LTP by specific manipulations or treatments applied during this specific time window (Krelstein et al., 1990; Fujii et al., 1991). We recently obtained interesting results supporting this interpretation. By applying fast and transient cooling shocks from 33° to 24°C at different times after LTP induction, we observed that these cooling episodes were able to fully reverse synaptic potentiation when applied 10–20 min after LTP induction. In contrast, the same treatments applied later than 20 min after stimulation no longer affected established LTP (figure 6.1; Muller and Bittar, 1992). Evidence for reversal of LTP was based on two criteria: the elimination of the potentiation generated by high-frequency stimulation applied prior to the cooling shock and the possibility to reinstate LTP after the cooling episode. These experiments clearly established that LTP could be completely abolished by rapid cooling in a time-dependent manner. These results are thus consistent with the idea that one or several temperature-sensitive events take place during the 20 min following high-frequency stimulation and that this stabilization phase might be required in order to allow consolidation of synaptic potentiation. The stable increase in synaptic efficacy that is expressed 20–30 min after high-frequency stimulation might thus involve modifications generated only after a given delay (Davies et al., 1989).

Protein kinases, but more specifically PKC and CaM kinase II have been implicated in or may contribute to these different phases.

CHANGES IN KINASE ACTIVITY WITH LTP

Among protein kinases, PKC is certainly the most cited candidate for playing a role in LTP. Among the most suggestive results supporting a role for PKC in LTP are those obtained using a biochemical approach. During the last years, several groups have reported an increased activity of PKC following LTP induction. Evidence has been obtained for translocation of PKC to the membrane as a result of high-frequency stimulation

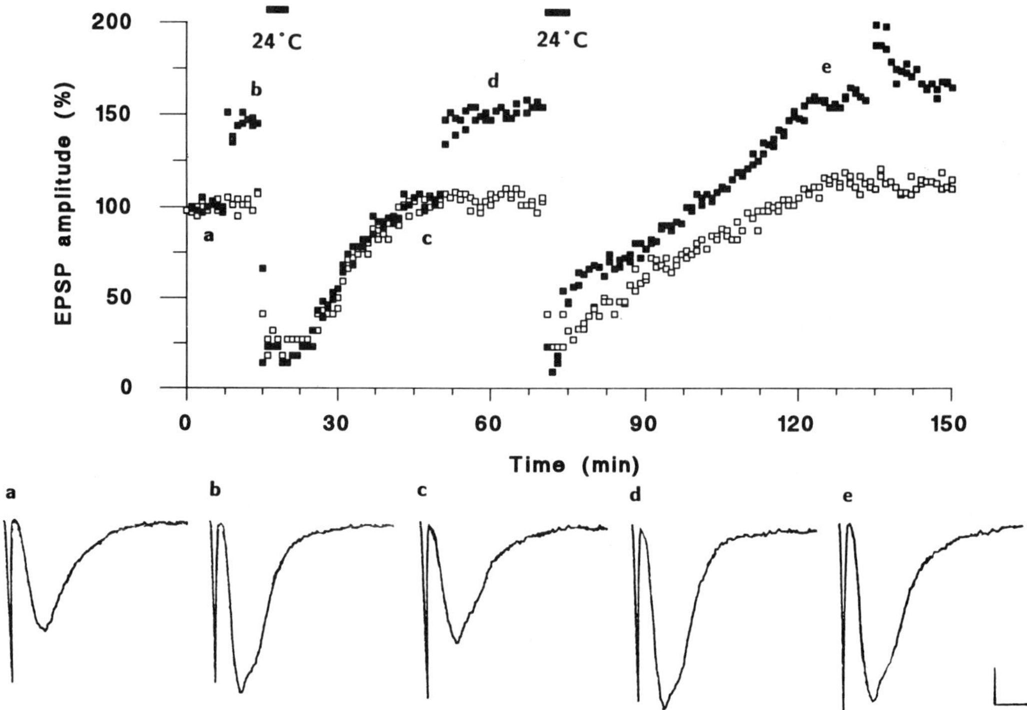

Figure 6.1 Evidence for a transient phase of 10–20 min during which LTP can be reversed. Illustration of an experiment in which brief and rapid cooling shocks from 33° to 24°C were applied at different times after high-frequency stimulation. The filled squares relate to responses recorded on the test (potentiated) pathway and open squares to control EPSPs. The first cooling shock was applied 10 min after high-frequency stimulation and resulted in a complete reversal of LTP. Synaptic potentiation, however, could be reinstated following recovery by a second application of high-frequency stimulation. In contrast, application of the second cooling shocks 30 min after reinduction of LTP no longer affected established potentiation. The field potentials were recorded on the potentiated pathway (filled squares) at the indicated times.

(Akers et al., 1986), for an increased phosphorylation activity due to PKC (Klann et al., 1991), and for an increased phosphorylation of several PKC substrates such as GAP-43, a major presynaptic protein (Lovinger et al., 1986; Gianotti et al., 1992) and possibly neurogranin (Klann et al., 1992), a postsynaptic protein. All these changes have been detected quite early after high-frequency stimulation and were shown to persist for at least 1 hr. They were NMDA receptor–dependent and could be blocked by the NMDA receptor antagonist D-AP5. Finally, they were only produced by the stimulation patterns used to generate LTP. They were thus interpreted as strongly supporting a role for PKC in the maintenance of LTP.

In the last few years, however, the same approach has also been used to test for changes in other kinases activity. Evidence has been obtained for a transient increase in casein kinase activity (Charriault-Marlangue et al.,

1991) and recently we investigated the possibility that CaM kinase II activity might also be altered following LTP induction.

This enzyme has been proposed as an interesting candidate for playing a role in LTP mechanisms on the basis of several observations. It is highly concentrated in postsynaptic membranes where it constitutes about 20–30% of the proteins of the postsynaptic density (Kennedy et al., 1983; Ouimet et al., 1984; Fukunaga et al., 1989). In addition, it has been shown by different groups that autophosphorylation of this kinase results in an enzyme that remains constitutively active (Saitoh and Schwartz, 1985; Lai et al., 1986; Miller and Kennedy, 1986; Schworer et al., 1986). It has thus been proposed that CaM kinase II might function as a molecular switch responsible for the generation of long-lasting biochemical changes (Miller and Kennedy, 1986; Lisman and Goldring, 1988).

Using a phosphorylation assay, we tested the possibility that modifications of the calcium-independent activity of CaM kinase II and thereby of its degree of autophosphorylation might occur as a result of LTP induction. This was done by measuring the phosphorylation activity present in the CA1 areas of hippocampal slices that have received either low frequency trains, which did not generate LTP, or high-frequency stimulation, which induced synaptic potentiation. The phosphorylation activity expressed in the tissue was assessed using substrates of CaM kinase II (syntide-2 and synapsin-1). Selectivity for CaM kinase II was obtained by adding inhibitors of other kinases during the assay and by verifying that the phosphorylation activity that was detected could be antagonized by the inhibitory peptide $CAMKII_{281-309}$ (Fukunaga et al., 1993).

The assay was carried out under two different conditions: in the presence of calcium and calmodulin, to measure the total activity of the enzyme, and in the presence of EGTA, to obtain the Ca^{2+}-independent activity. Several previous studies have shown that this Ca^{2+}-independent activity reflects autophosphorylation of the enzyme (Miller et al., 1988; Fukunaga et al., 1989; Hanson et al., 1989). The ratio of Ca^{2+}-independent to total activity is thus usually considered as an index of the degree of CaM kinase II autophosphorylation.

As illustrated in figure 6.2, the results of our study showed that LTP is associated with a stable increase in the activity of CaM kinase II. The increase in CaM kinase II activity that we detected was mainly due to an increase in Ca^{2+}- independent activity, but also to a small change in total activity. As a consequence, the ratio of Ca^{2+}-independent to total activity was also increased in most slices. These changes were observed in all conditions where LTP could be induced, but not when low-frequency trains which produced only transient modifications of synaptic transmission were applied. These changes in CaM kinase II activity were NMDA

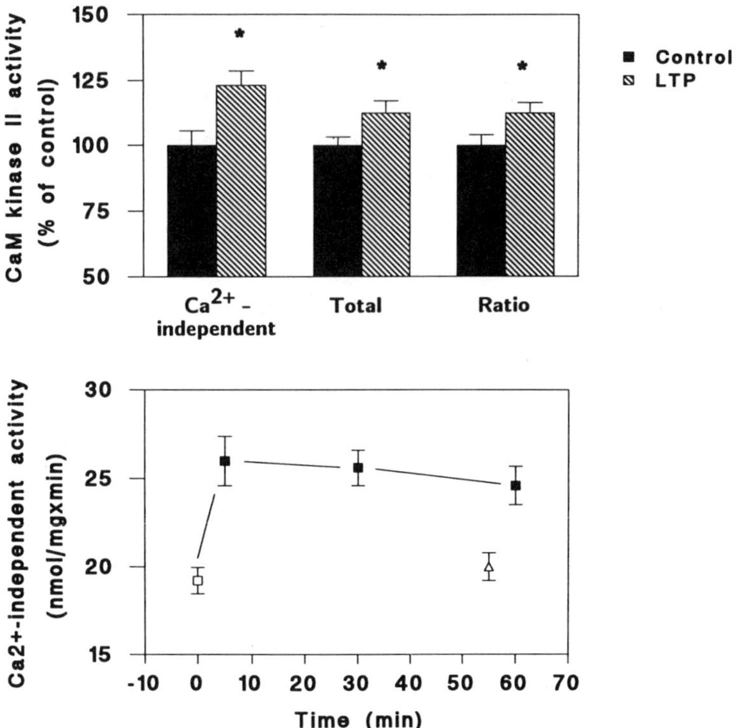

Figure 6.2 High-frequency stimulation in area CA1 results in a long-lasting increase in the activity of CaM kinase II. (*Top*) Changes in Ca^{2+}-independent and total activity, and in ratio of Ca^{2+}-independent to total activity (degree of autophosphorylation) measured in 23 slices following LTP induction. (*Bottom*) Time course of the changes in Ca^{2+}-independent CaM kinase II activity following LTP induction. Filled squares, activity measured at different times after LTP induction in potentiated slices; open square, activity in control nonstimulated slices; open triangle, activity in slices stimulated with low-frequency trains which did not induce synaptic potentiation.

dependent, since they were abolished in the presence of the NMDA receptor antagonist D-AP5. We concluded, therefore, that these changes were produced only by the specific patterns of stimulation that generate LTP (Fukunaga et al., 1993).

What then to conclude about the role of CaM kinase II in LTP? The increase in Ca^{2+}-independent activity and the increase in ratio of Ca^{2+}-independent to total activity are consistent with the idea that autophosphorylation of CaM kinase II takes place after high-frequency stimulation and that this change is stable for at least 1 hr. While these observations could be consistent with the proposal by Miller and Kennedy (1986) and Lisman and Goldring (1988) that CaM kinase II autophosphorylation might function as a molecular switch, there are, however, other results that are at variance with this interpretation. It appears from our experiments as well as from previous work of several authors (Molloy and Kennedy,

1991; Ocorr and Schulman, 1991) that the basal level of autophosphorylation of CaM kinase II is quite high and stable in hippocampal slices or even cultures and, on the other hand, that the changes produced by high-frequency stimulation are rather small (Fukunaga et al., 1993). In addition, we observed that treatment of slices with a phosphatase inhibitor produces very important changes in the degree of autophosphorylation of the kinase without major effects on synaptic transmission or the possibility to induce LTP. This would suggest, therefore, that there is no direct correlation between the ratio of autophosphorylation and synaptic potentiation.

Another intriguing result of our experiments is the observation that LTP was associated with a small increase in total CaM kinase II activity. This result was quite surprising and could have been produced by several mechanisms such as translocation of the enzyme, synthesis of new enzyme, or changes in phosphatase activity. We recently tested this last possibility and found in preliminary experiments that the activity of phosphatase IIA was indeed inhibited in an NMDA-dependent way in slices in which LTP has been induced (unpublished observations). An increase in CaM kinase II mRNA following synaptic potentiation has also been reported recently (Mackler et al., 1992). If confirmed, these observation could be of interest and indicate that subtle changes in the balance between kinase and phosphatase activity might take place following high-frequency stimulation. This could account for several of the changes in phosphorylation activity that have been found to accompany LTP induction. Additional work will be necessary to establish this further.

In summary, it appears that high-frequency stimulation in hippocampus results in long-lasting modifications in the activity of several kinases among which are protein kinase C and CaM kinase II. In addition to this or related to this, LTP seems also to be associated with modifications of the activity of phosphatases. As a result, it may be expected that the state of phosphorylation of several substrates may be altered following high frequency stimulation.

PHARMACOLOGICAL INTERACTIONS WITH KINASE ACTIVITY

If the changes in kinase activity reported above are suggestive for a role of these enzymes in LTP mechanisms, experiments based on the use of activators or inhibitors of these kinases are certainly most important to establish if and when a phosphorylation activity is required for synaptic potentiation.

With regard to PKC, the results obtained by this type of experiment are rather controversial and sometimes difficult to interpret for several reasons: the compounds used are usually not selective, they diffuse and penetrate poorly in the tissue, and thus they have been almost systematically applied at rather high concentrations. In addition, several of these inhibitors affect various cellular parameters, including regular synaptic transmission and ionic channels or receptors (Muller et al., 1988; Amador and Dani, 1991; Muller et al., 1992b).

A general finding, however, has been that inhibitors of protein kinase C prevent LTP induction when applied before high-frequency stimulation. The effect that has been invariably reported is a decaying potentiation similar to that obtained with partial antagonism of NMDA receptors. While these results have usually been interpreted as supporting a role for PKC in mechanisms of LTP expression (Lovinger et al., 1987; Reymann et al., 1988), other explanations are also possible. In analogy with the effects of NMDA receptor antagonists, the decaying potentiation observed with PKC inhibitors could simply reflect an interaction with the events controlling the rise in intracellular postsynaptic calcium concentration and thereby an effect on LTP induction rather than expression mechanisms (see chapter 8 by Malenka). Consistent with this interpretation are the recent results indicating that PKC activity may control the functioning of NMDA receptors (Chen and Huang, 1992) and that PKC is involved as a second messenger mediating the effects of metabotropic receptor activation on NMDA responses and LTP induction (Kelso et al., 1992; Aniksztejn et al., 1992). Also in favor of this interpretation are the reports indicating that PKC inhibitors are able to interfere with NMDA receptor–mediated synaptic responses. This has been observed with H7 (Muller et al., 1990; Amador et al., 1991), staurosporine (Muller et al., 1992b), and calphostin C (unpublished observation). These effects by themselves would be sufficient to account for the decaying potentiation produced by PKC inhibitors; taken together, these results favor the idea that PKC might work as a factor controlling the threshold for the induction of synaptic potentiation (Muller et al., 1990; Aniksztejn et al., 1992; Muller et al., 1992b).

Is PKC also critically involved in the mechanisms of LTP expression either during the consolidation phase or the maintenance phase? If this was the case, then one would expect to reversibly eliminate LTP by applying a PKC inhibitor after high-frequency stimulation. Alternatively, treatments with an agent able to activate PKC should generate synaptic potentiation. This possibility has been examined in many different studies with, again, variable and contradictory results. Four studies in which PKC inhibitors were applied after high-frequency stimulation reported a reversible inhibition or blockade of synaptic potentiation (Malinow et al., 1988; Reymann et al., 1990; Huang et al., 1992; Wang and Feng, 1992). There is, however, disagreement about when the inhibitors have to be applied to reverse LTP. In two studies, reversal has been described up to 1 hr after induction (Malinow et al., 1988; Wang and Feng, 1992), in another one, only when the inhibitor was applied 15–30 min after induction (Huang et al., 1992), and in the last, only when the inhibitor was applied immediately after high-frequency stimulation but not later (Reymann et al., 1990). In contrast with these results, several studies concluded that application of a PKC inhibitor after high-frequency stimulation did not affect established LTP (Denny et al., 1990; Muller et al., 1990; Reymann et al., 1990; Leahy and Vallano, 1991; Muller et al., 1992a). We tested this using H7 (Muller et al., 1990), staurosporine (Muller et al., 1992a), quercetin and calphostin C (unpublished observations), using various concentrations of these compounds, applied at different times after high-frequency stimulation. However, in none of these experiments have we been able to obtain clear evidence for reversal of LTP. In the case of staurosporine, a very potent although not selective inhibitor of PKC, indications have been obtained that the drug does penetrate into cells (Denny et al., 1990). Staurosporine does not interfere with regular transmission and, using a phosphorylation assay, we were able to demonstrate a marked inhibition of PKC in treated slices (figure 6.3). The degree of inhibition that we measured in those experiments was even probably underestimated, due to dilution and dissociation during the phosphorylation assay. However, even under those conditions, robust and stable LTP could still be induced when using coactivation of converging groups of afferents to a dendritic area (Muller et al., 1992a). Although we cannot exclude subcellular variations in the degree of PKC inhibition such as, for example, in nerve terminals, these observations make it unlikely that PKC played a crucial role for LTP expression mechanisms.

Results arguing against a role of PKC in LTP expression were also obtained when analyzing the effects of compounds able to activate PKC such as phorbol esters. Although phorbol esters do enhance synaptic responses (Malenka et al., 1986), the effect is not stable (Muller et al., 1988) and most important, no occlusion with stimulation-induced LTP has been observed (Gustafsson et al., 1988; Muller et al., 1990). It seems reasonable, therefore, to conclude that (1) a PKC-dependent phosphorylation activity is not directly responsible for the stable increase in synaptic efficacy expressed more than 20–30 min after high-frequency stimulation and (2) PKC activity, if involved, is most likely not necessary during the transient consolidation phase of 5–20 min which follows high-frequency stimulation.

There appears thus to be a dissociation between the biochemical results indicating activation of PKC following LTP induction and the pharmacological and electrophysiological data, which do not support the idea that

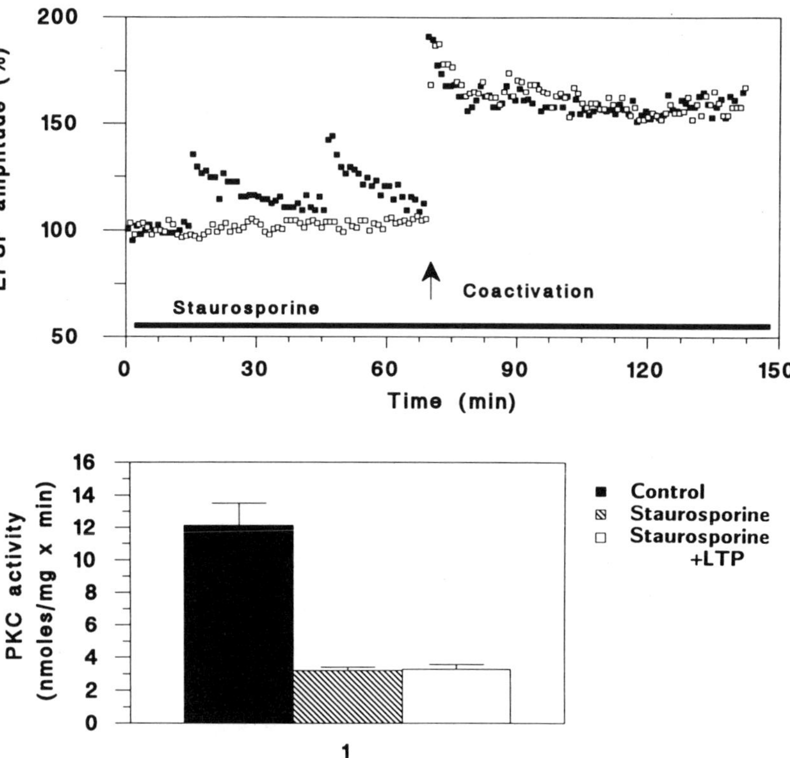

Figure 6.3 Inhibition of PKC by staurosporine (1–5 μM) did not prevent induction of robust LTP. (*Top*) Experiment in which high-frequency stimulation was applied either to a single input or to two inputs simultaneously (coactivation) in the presence of staurosporine. While single input activation resulted in a decaying LTP, coactivation still generated robust LTP. (*Bottom*) Inhibition of PKC activity measured in slices treated with similar concentrations of staurosporine using a phosphorylation assay.

PKC is directly and crucially involved in the maintenance of synaptic potentiation.

What about the role of CaM kinase II in LTP mechanisms? Pharmacological studies of the type described above have also been carried out to test for a possible involvement of CaM kinase II in LTP. Essentially, four studies have concluded that CaM kinase II activity might be important for LTP expression (Malenka et al., 1989; Malinow et al., 1989; Ito et al., 1991; Silva et al., 1992). In three cases, this interpretation was based on the observation that inhibitors of the enzyme, applied before high-frequency stimulation, resulted in a decaying form of synaptic potentiation. In the fourth study, transgenic mice were obtained in which expression of the major subunit of CaM kinase II was prevented. Electrophysiological analyses in slices prepared from these animals revealed an impossibility to obtain LTP with no other apparent changes in the properties of synaptic transmission (Silva et al., 1992). Thus, as for PKC, blockade of CaM kinase II prevents LTP induction.

It is important to note, however, that in the case of the experiments carried out with the inhibitors, application of the drugs after high-frequency stimulation did not affect established LTP. This was tested by injection of inhibitory peptides in the postsynaptic cell (Malinow et al., 1989) and by application of an antagonist in the perfusion medium (Ito et al., 1992). Again, this would suggest, as for PKC, that if the enzyme is involved at some time point for the expression of LTP, this should take place during a very short period of time immediately after high-frequency stimulation.

SUMMARY

Taken together, the results obtained so far concerning the role of protein kinases in LTP lead to the following conclusions. High-frequency stimulation of a group of afferents is accompanied by a long-lasting increase in the activity of several kinases, among which are PKC and CaM kinase II. These changes were shown in all cases to be NMDA dependent and produced exclusively by the patterns of activity required to generate LTP. In contrast with this, the results of experiments carried out with activators or inhibitors of these kinases indicate that established LTP is essentially unaffected by inhibition of one or the other enzyme and that protein kinase activity seems to be important only during a short period of time immediately after high-frequency stimulation.

Based on these conclusions, the hypothesis can be proposed that protein kinases are essentially involved in the initial events that are responsible for triggering LTP. In the case of PKC, most results, with the exception of the conflicting reports on the effects of PKC antagonists on established LTP, could be accounted for by an involvement of the enzyme in the control and modulation of NMDA receptors. Thus, PKC could act by modulating the threshold for LTP induction. In the case of CaM kinase II, however, there is so far no indication that the kinase might somehow affect the rise in calcium necessary to trigger synaptic potentiation. Accordingly, one would have to postulate a more direct implication of CaM kinase II in the triggering cascade responsible for generating synaptic potentiation. As measured in our experiments, this enzyme is very quickly activated following high-frequency stimulation and its role could be to set up conditions during the first minutes after stimulation that would be necessary for other events to take place and lead to synaptic potentiation.

According to this view, kinases would not be directly implicated as part of the mechanisms responsible for the maintenance of LTP, and the biochemical changes in enzyme activity and substrate phosphorylation that can be detected up to 1 hr after high-frequency stimulation would not be directly responsible for synaptic potentiation. This does not, however, exclude the possibility that these kinases and phosphatases could contribute even at later times to maintain a given level of phosphorylation activity which could be important for regulatory events controlling synaptic plasticity. Such a regulatory role of phosphorylation mechanisms could help explain the discrepancies reported concerning the effects of kinase inhibitors on established LTP.

In conclusion, several important questions remain yet to be answered before more precise interpretations can be proposed about the role of these kinases in LTP. Identification of presynaptic and postsynaptic substrates of these kinases, of the time course, specificity, and regulation of their phosphorylation, and understanding of the functional implication of these molecules in synaptic efficacy are certainly among the most interesting issues that will have to be addressed in the next years.

ACKNOWLEDGMENTS

This work has been supported by the Swiss National Science Foundation, FNRS 30-30980.91.

REFERENCES

Akers, R. F., Lovinger, D. M., Colley, P. A., Linden, D. J., and Routtenberg, A. (1986) Translocation of protein kinase C activity may mediate hippocampal long-term potentiation. *Science* 231:587–589.

Amador, M., and Dani, J. A. (1991) Protein kinase inhibitor, H-7, directly affects *N*-methyl-D-aspartate receptor channels. *Neurosci. Lett.* 124:251–255.

Aniksztejn, L., Otani, S., and Ben-Ari, Y. (1992) Quisqualate metabotropic receptors modulate NMDA currents and facilitate induction of long-term potentiation through protein kinase C. *Eur. J. Neurosci.* 4:500–505.

Bekkers, J. M., and Stevens, C. F. (1990) Presynaptic mechanism for long-term potentiation in the hipocampus. *Nature* 346:724–729.

Bliss, T. V. P., and Lømo, T. (1973) Long-lasting potentiation of synaptic transmission in the dentate area of the anaesthetized rabbit following stimulation of the perforant path. *J. Physiol. (Lond.)* 232:331–356.

Charriaut-Marlangue, C., Otani, S., Creuzet, C., Ben-Ari, Y., and Loeb, J. (1991) Rapid activation of hippocampal casein kinase II during long-term potentiation. *Proc. Natl. Acad. Sci. USA* 88:10232–10236.

Chen, L., and Huang, L.-Y. M. (1992) Protein kinase C reduces Mg^{2+} block of NMDA-receptor channels as a mechanism of modulation. *Nature* 356:521–523.

Colino, A., Huang, Y.-Y., and Malenka, R. C. (1992) Characterization of the integration time for the stabilization of long-term potentiation in area CA1 of the hippocampus. *J. Neurosci.* 12:180–187.

Collingridge, G. L., Kehl, S. J., and McLennan, H. (1983) Excitatory amino acids in synaptic transmission in the Schaffer collateral-commissural pathway of the rat hippocampus. *J. Physiol. (Lond.)* 334:33–46.

Davies, S. N., Lester, R. A. J., Reymann, K. G., and Collingridge, G. L. (1989) Temporally distinct pre- and post-synaptic mechanisms maintain long-term potentiation. *Nature* 338: 500–503.

Denny, J. B., Polan-Curtain, J., Rodriguez, S., Wayner, M. J., and Armstrong, D. L. (1990) Evidence that protein kinase M does not maintain long-term potentiation. *Brain Res.* 534:201–208.

Foster, T. C., and McNaughton, B. L. (1990) Long-term synaptic enhancement in hippocampal field CA1 is due to increased quantal size, not quantal content. *Hippocampus* 1:79–91.

Fujii, S., Saito, K., Miyakawa, H., Ito, K., and Kato, H. (1991) Reversal of long-term potentiation (depotentiation) induced by tetanus stimulation of the input to CA1 neurons of guinea pig hippocampal slices. *Brain Res.* 555:112–122.

Fukunaga, K., Rich, D. P., and Soderling, T. R. (1989) Generation of the Ca^{2+}-independent form of Ca^{2+}/calmodulin-dependent protein kinase II in cerebellar granule cells. *J. Biol. Chem.* 264:21830–21836.

Fukunaga, K., Stoppini, L., Miyamoto, E., and Muller, D. (1993) Long-term potentiation is associated with an increased activity of calcium/calmodulin-dependent protein kinase II. *J. Biol. Chem.* 268:7863–7867.

Gianotti, C., Nunzi, M. G., Gispen, W. H., and Corradetti, R. (1992) Phosphorylation of the presynaptic protein B-50 (GAP-43) is increased during electrically induced long-term potentiation. *Neuron* 8:843–848.

Gustafsson, B., Huang, Y. Y., and Wigström, H. (1988) Phorbol ester-induced synaptic potentiation differs from long-term potentiation in the guinea pig hippocampus in vitro. *Neurosci. Lett.* 85:77–81.

Hanson, P. I., Kapiloff, M. S., Lou, L. L., Rosenfeld, M. G., and Schulman, H. (1989) Expression of a multifunctional Ca^{2+}/calmodulin-dependent protein kinase and mutational analysis of its autoregulation. *Neuron* 3:59–70.

Huang, Y.-Y., Colley, P. A., and Routtenberg, A. (1992) Postsynaptic then presynaptic protein kinase C activity may be necessary for long-term potentiation. *Neuroscience* 49: 819–827.

Ito, I., Hidaka, H., and Sugiyama, H. (1991) Effects of KN-62, a specific inhibitor of calcium/calmodulin-dependent protein kinase II, on long-term potentiation in the rat hippocampus. *Neurosci. Lett.* 121:119–121.

Izumi, Y., Clifford, D. B., and Zorumski, C. F. (1991) 2-Amino-3-phosphonopropionate blocks the induction and maintenance of long-term potentiation in rat hippocampal slices. *Neurosci. Lett.* 122:187–190.

Kelso, S. R., Nelson, T. E., and Leonard, J. P. (1992) Protein kinase C-mediated enhancement of NMDA currents by metabotropic glutamate receptors in *Xenopus* oocytes. *J. Physiol. (Lond.)* 449:705–718.

Kennedy, M. B., Bennett, M. K., and Erondu, N. E. (1983) Biochemical and immunochemical evidence that the major postsynaptic density protein is a subunit of a calmodulin-dependent protein kinase. *Proc. Natl. Acad. Sci. USA* 80:7357–7361.

Klann, E., Chen, S.-J., and Sweatt, J. D. (1991) Persistent protein kinase activation in the maintenance phase of long-term potentiation. *J. Biol. Chem.* 266:24253–24256.

Klann, E., Chen, S.-J., and Sweatt, J. D. (1992) Increased phosphorylation of a 17-kDa protein kinase C substrate (P17) in long-term potentiation. *J. Neurochem.* 58:1576–1579.

Krelstein, M. S., Thomas, M. P., and Horowitz, J. M. (1990) Thermal effects on long-term potentiation in the hamster hippocampus. *Brain Res.* 520:115–122.

Kullmann, D. M., and Nicoll, R. A. (1992) Long-term potentiation is associated with increases in quantal content and quantal amplitude. *Nature* 357:240–244.

Lai, Y., Nairn, A. C., and Greengaard, P. (1986) Autophosphorylation reversibly regulates the Ca^{2+}/calmodulin-dependence of Ca^{2+}/calmodulin-dependent protein kinase II. *Proc. Natl. Acad. Sci. USA* 83:4253–4257.

Leahy, J. C., and Vallano, M. L. (1991) Differential effects of isoquinolinesulfonamide protein kinase inhibitors on CA1 responses in hippocampal slices. *Neuroscience* 44:361–370.

Lisman, J. E., and Goldring, M. A. (1988) Feasibility of long-term storage of graded information by the Ca^{2+}/calmodulin-dependent protein kinase molecules of the post-synaptic density. *Proc. Natl. Acad. Sci. USA* 85:5320–5324.

Lovinger, D. M., Colley, P. A., Akers, R. F., Nelson, R. B., and Routtenberg, A. (1986) Direct relation of long-term synaptic potentiation to phosphorylation of membrane protein F1, a substrate for membrane protein kinase C. *Brain Res.* 399:205–211.

Lovinger, D. M., Wong, K. L., Murakami, K., and Routtenberg, A. (1987) Protein kinase C inhibitors eliminate hippocampal long-term potentiation. *Brain Res.* 436:177–183.

Lynch, G., Larson, J., Kelso, S., Barrionuevo, G., and Schottler, F. (1983) Intracellular injections of EGTA block induction of hippocampal long-term potentiation. *Nature* 305: 719–721.

Mackler, S. A., Brooks, B. P., and Eberwine J. H. (1992) Stimulus-induced coordinate changes in mRNA abundance in single postsynaptic hippocampal CA1 neurones. *Neuron* 9:539–548.

Malenka, R. C. (1991) Postsynaptic factors control the duration of synaptic enhancement in area CA1 of the hippocampus. *Neuron* 6:53-60.

Malenka, R. C., Madison, D. V., and Nicoll, R. A. (1986) Potentiation of synaptic transmission in the hippocampus by phorbol esters. *Nature* 321:175–177.

Malenka, R. C., Kauer, J. A., Zucker, R. J., and Nicoll, R. A. (1988) Postsynaptic calcium is sufficient for potentiation of hippocampal synaptic transmission. *Science* 242:81–84.

Malenka, R. C., Kauer, J. A., Perkel, D. J., Mauk, M. D., Kelly, P. T., Nicoll, R. A., and Waxman, I. (1989) An essential role for postsynaptic calmodulin and protein kinase activity in long-term potentiation *Nature* 340:554–556.

Malenka, R. C., Lancaster, B., and Zucker, R. S. (1992) Temporal limits on the rise in postsynaptic calcium required for the induction of long-term potentiation. *Neuron* 9:121–128.

Malgaroli, A., and Tsien, R. W. (1992) Glutamate-induced long-term potentiation of the frequency of miniature synaptic currents in cultured hippocampal neurons. *Nature* 357: 134–139.

Malinow, R. (1991) Transmission between pairs of hippocampal slice neurons: Quantal levels, oscillations, and LTP. *Science* 252:722–724.

Malinow, R., and Tsien, R. W. (1990) Presynaptic enhancement shown by whole-cell recordings of long-term potentiation in hippocampal slices. *Nature* 346:177–180.

Malinow, R., Madison, D. V., and Tsien, R. W. (1988) Persistent protein kinase activity underlying long-term potentiation. *Nature* 335:820–824.

Malinow, R., Schulman, H., and Tsien, R. W. (1989) Inhibition of postsynaptic PKC or CaMKII blocks induction but not expression of LTP. *Science* 245:862–866.

Manabe, T., Renner, P., and Nicoll, R. A. (1992) Postsynaptic contribution to long-term potentiation revealed by the analysis of miniature synaptic currents. *Nature* 355:50–55.

Miller, S. G., and Kennedy, M. B. (1986) Regulation of brain type II Ca/calmodulin-dependent protein kinase by autophosphorylation: A Ca^{2+}-triggered molecular switch. *Cell* 44:861–870.

Miller, S. G., Patton, B. L., and Kennedy, M. B. (1988) Sequences of autophosphorylation sites in neuronal type II CaM kinase that control Ca^{2+}-independent activity. *Neuron* 1: 593–604.

Molloy, S. S., and Kennedy, M. B. (1991) Autophosphorylation of type II Ca^{2+}/calmodulin-dependent protein kinase in cultures of postnatal rat hippocampal slices. *Proc. Natl. Acad. Sci. USA* 88:4756–4760.

Muller, D., and Bittar, P. (1992) LTP expression mechanisms: Evidence for a time window during which LTP can be reversed. *Soc. Neurosci. Abstr.* 18:1498.

Muller, D., Tumbull, J., Baudry, M., and Lynch, G. (1988) Phorbol ester-induced synaptic facilitatlon is different than long-term potentiation. *Proc. Natl. Acad. Sci. USA* 85:6997–7000.

Muller, D., Buchs, P.-A., Dunant, Y., and Lynch, G. (1990) Protein kinase C activity is not responsible for the expression of long-term potentiation in hippocampus. *Proc. Natl. Acad. Sci. USA* 87:4073–4077.

Muller, D., Bittar, P., and Boddeke, H. (1992a) Induction of stable long-term potentiation in the presence of the protein kinase C antagonist staurosporine. *Neurosci. Lett.* 135:18–22.

Muller, D. Buchs, P.-A., Stoppini, L., and Boddeke, H. (1992b) Long-term potentiation, protein kinase C and glutamate receptors. *Mol. Neurobiol.* 5:277–288.

Ocorr, K. A., and Schulman, H. (1991) Activation of multifunctional Ca^{2+}/calmodulin-dependent kinase in intact hippocampal slices. *Neuron* 6:907–914.

Ouimet, C. C., McGuiness, T. L., and Greengaard, P. (1984) Immunocytochemical localization of calcium/calmodulin-dependent protein kinase II in rat brain. *Proc. Natl. Acad. Sci. USA* 81:5604–5608.

Reymann, K. G., Frey, U., Jork, R., and Matthies, H. (1988) Polymyxin B, an inhibitor of protein kinase C, prevents the maintenance of synaptic long-term potentiation in hippocampal CA1 neurons. *Brain Res.* 440:305–314.

Reymann, K. G., Davies, S. N., Matthies, H., Kase, H., and Collingridge, G. L. (1990) Activation of a K-252b-sensitive protein kinase is necessary for a post-synaptic phase of long-term potentiation in area CA1 of rat hippocampus. *Eur. J. Neurosci.* 2:481–486.

Saitoh, T., and Schwartz, J. H. (1985) Phosphorylation-dependent subcellular translocation of a Ca^{2+}/calmodulin-dependent protein kinase produces an autonomous enzyme in Aplysia neurons. *J. Cell Biol.* 100:835–842.

Schworer, C. M., Colbran, R., and Soderling, T. R. (1986) Reversible generation of a Ca^{2+}-independent form of Ca^{2+}(calmodulin)-dependent protein kinase II by an autophosphorylation mechanisms. *J. Biol. Chem.* 261:8581–8584.

Silva, A. J., Stevens, C. F., Tonegawa, S., and Wang, Y. (1992) Deficient hippocampal long-term potentiation in calcium-calmodulin kinase II mutant mice. *Science* 257:201–206.

Wang, J., and Feng, D.-P. (1992) Postsynaptic protein kinase C essential to induction and maintenance of long-term potentiation in the hippocampal CA1 region. *Proc. Natl. Acad. Sci. USA* 89:2576–2580.

Zheng, F., and Gallagher, J. P. (1992) Metabotropic glutamate receptors are required for the induction of long-term potentiation. *Neuron* 9:163–172.

7 Two Different Forms of Quantal Changes during Long-Term Potentiation and Their Impact on Neuronal Output

Roberto Malinow, Dezhi Liao, Nikolai Otmakhov, and Aneil M. Shirke

In this chapter we investigate several issues regarding quantal transmission in the CNS. Briefly, we ask whether such transmission can be observed, what happens to quantal levels with long-term potentiation (LTP), and whether the quantal variability seen at individual synapses makes a contribution to the trial-to-trial variability for a threshold stimulus.

Recent work in hippocampal slices has resolved quantal levels in elicited transmission (Larkman et al., 1991; Malinow, 1991). This allows direct measurement of quantal amplitude, the response to a single quantum of transmitter (a measure expected to change with postsynaptic modifications), and quantal content, the average number of quanta released per impulse (an index of presynaptic efficacy) (del Castillo and Katz, 1954). We measured quantal amplitude and quantal content of elicited transmission before and after induction of LTP. We find changes in both quantal parameters consistent with modifications in presynaptic and postsynaptic function.

We studied synaptic transmission in rat hippocampal slices by minimal stimulation of Schaffer collateral/commissural afferents while recording excitatory postsynaptic currents (EPSCs) from CA1 neurons under whole-cell voltage clamp (figure 7.1; Malinow and Tsien, 1990). We analyzed postsynaptic responses by constructing peak EPSC amplitude distribution density estimates (figure 7.1A; Silverman, 1986; Malinow, 1991) for epochs of 280–820 trials. This technique allows one to determine the amplitude distribution of responses without requiring binning of data. We also constructed binned amplitude distribution histograms (figure 7.1B–D), which allow simple statistical tests. These amplitude distributions often contain peaks that are regularly spaced (figures 7.1–7.3, 7.5, 7.6, 7.8, 7.9) and do not depend on binning (figure 7.1). The peaks could represent quantal levels of transmission or a statistical artifact due to finite sampling from a continuous unimodal distribution (Clements, 1991). To test between these two possibilities, several experimental, analytical, and statistical manipulations were performed.

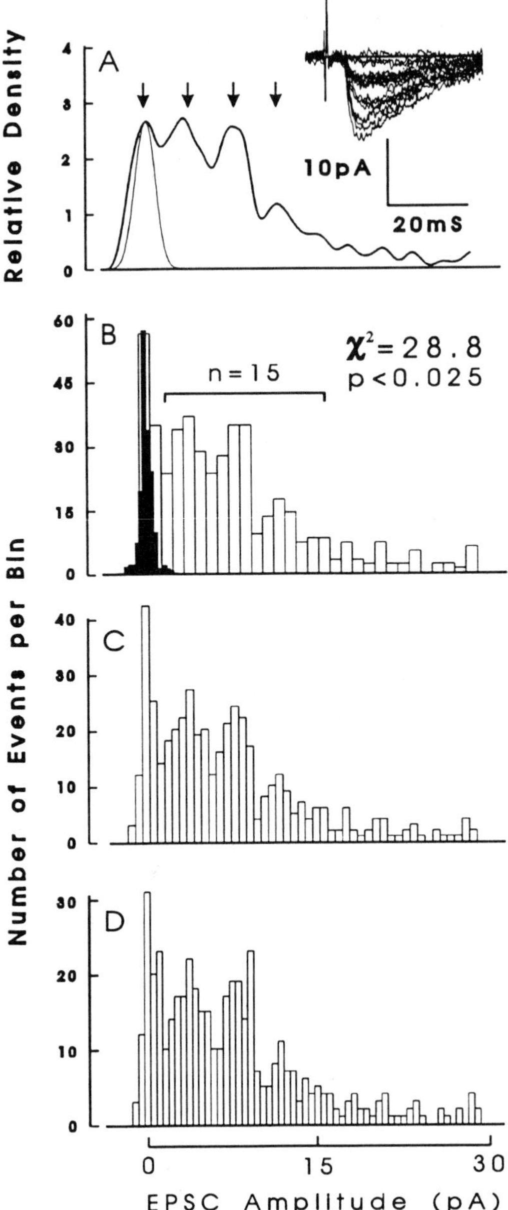

Figure 7.1 Amplitude distribution histograms from 450 consecutive sweeps without (*A*) or with (*B, C, D*) binning show evenly spaced peaks. Changing binning from 1/4 (*A*), 1/6 (*B*), or 1/8 (*C*) of the estimated quantal amplitude does not change location of amplitude peaks. The likelihood that this distribution (*B*) results from a unimodal distribution is <.025 (calculated for the 15 bins indicated; see experimental procedures). *Inset*: 18 consecutive sweeps displaying quantal levels.

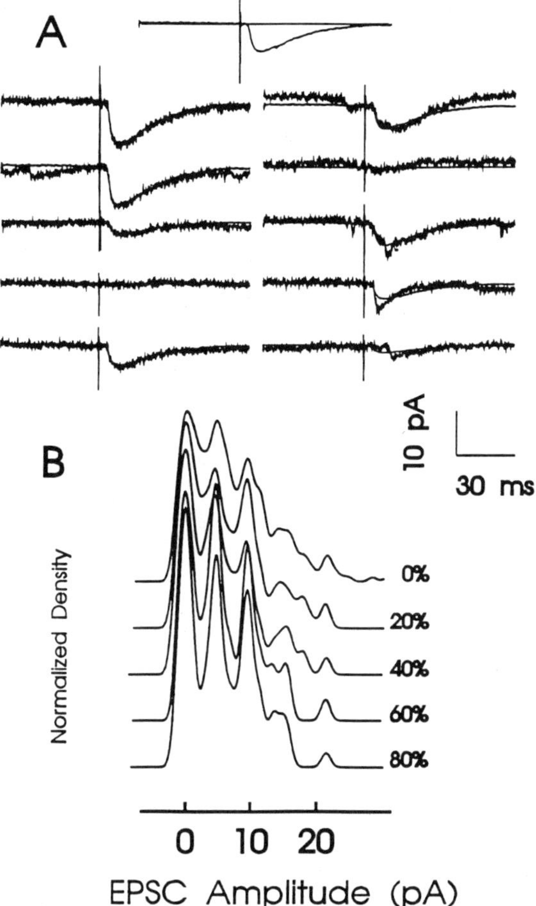

Figure 7.2 Noise-free EPSC amplitudes are restricted to discrete levels. (*A*) Representative sweeps that fit well (*left column*) and fit poorly (*right column*) a scaled ensemble average (superimposed on each sweep) of all 280 sweeps in epoch (template, top trace). (*B*) Amplitude distribution density estimate of responses after removing 0%, 20%, 40%, 60%, and 80% of sweeps in epoch based on how well sweeps fit the scaled template (see Experimental Procedures).

We tested statistically the possibility that the observed amplitude histograms were from a unimodal population of responses (i.e., a nonquantal distribution). We generated a unimodal distribution by smoothing the binned histograms, and determined the probability that the observed data could be fit by this expected distribution by computing a χ^2 value of the difference (see Experimental Procedures). Setting the degrees of freedom as the number of bins minus one yields the probability that the data came from this unimodal distribution (less than 0.025 in this case, figure 7.1B).

If the peaks in an amplitude distribution correspond to restricted levels of transmission (rather than a finite sampling artifact from a continuous distribution) then removal of noisy responses should produce an amplitude distribution with more prominent peaks that are at the same locations as the peaks in the parent amplitude distribution. Computing an ensemble average of all sweeps in an epoch produces a noise-free EPSC that we call an *epoch template* (figure 7.2). The relative amount of noise during an individual response was estimated by comparing the response to a scaled

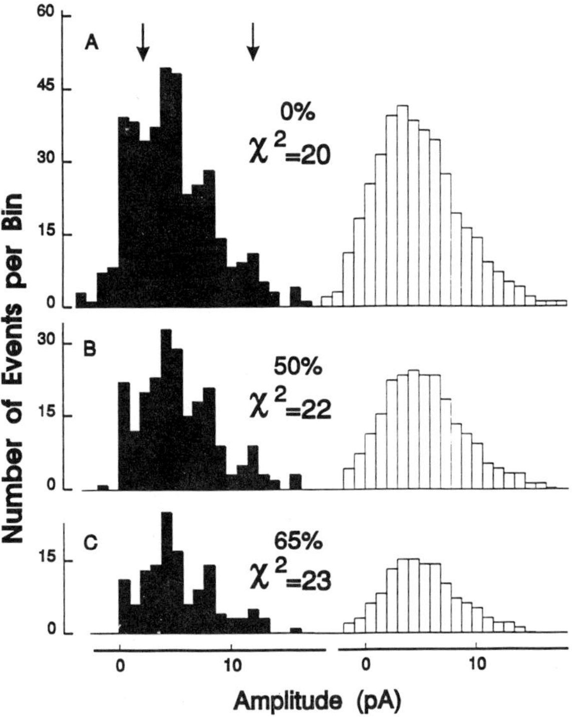

Figure 7.3 EPSC amplitude distributions are significantly different from a unimodal (nonquantal) distribution and remain so after template-based selection. (*A, left*) Amplitude distribution histogram of 400 consecutive sweeps (observed); (*right*) 5-bin smoothing of left histogram yields a unimodal distribution (expected). Computing a χ^2 of the difference over the bins between arrows. Same calculation on the amplitude distribution of sweeps remaining after (*B*) 50% and (*C*) 65% were removed by template-based selection.

epoch template (see Experimental Procedures). Some sweeps fit the scaled template very well (figure 7.2A, left) whereas other sweeps do not (figure 7.2A, right). A family of amplitude distributions was constructed by removing 20%, 40%, 60%, and 80% of the sweeps in an epoch. Sweeps were removed based on their deviation from the scaled epoch template; this process we call *template-based selection.* As more sweeps are removed, peaks become more prominent and their locations remain invariant. This finding supports further the view that transmission is restricted to distinct levels.

Does template-based selection randomly remove sweeps? To answer this question we studied the effect of template-based selection on the χ^2 value computed between an observed and unimodal distribution. For a homogeneous (random) removal of sweeps from an epoch, the χ^2 value will decrease in proportion to the percent of sweeps removed (see figure 8.4 legend). However, with template-based selection the χ^2value does not decrease (figures 7.3, 7.4), showing that this procedure does not randomly remove traces.

Having observed that transmission appears to be restricted to particular amplitudes, we tested if these levels behave as one would expect for quantal release of transmitter. We performed experimental manipulations to change either quantal content or quantal amplitude: delivery of paired pulses should show a shift in the second response distribution toward

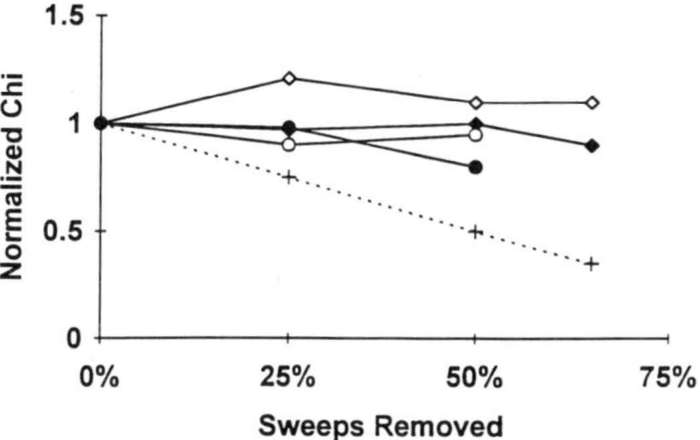

Figure 7.4 Effect of template-based selection on the difference between an observed and unimodal distribution. A χ^2 of the difference between the observed and unimodal distribution was computed for experiments A2U2 (circles) and A1Y4 (diamonds) for epochs before (open) and during (filled) LTP. For these experiments, the χ^2 test rejected the unimodal model, indicating systematic differences between observed and unimodal model. Amplitude distribution histograms were generated after removing 25%, 50%, and 65% of sweeps with template-based selection. A χ^2 from resulting histograms was recalculated and expressed as a fraction of the χ^2 from the histogram of all sweeps (0% removal). Homogeneous removal of events from a histogram where there is a systematic difference between x_i and s_i results in linear decay of χ^2 (dotted line). This is because if χ^2(original) = $\sum (x_i - s_i)^2/s_i$, then upon homogeneous removal of $P \cdot 100\%$ events from each bin, $\chi^2\text{(new)} = \sum_i (P \cdot x_i - P \cdot s_i)^2/P \cdot s_i = P \cdot \chi^2\text{(original)}$.

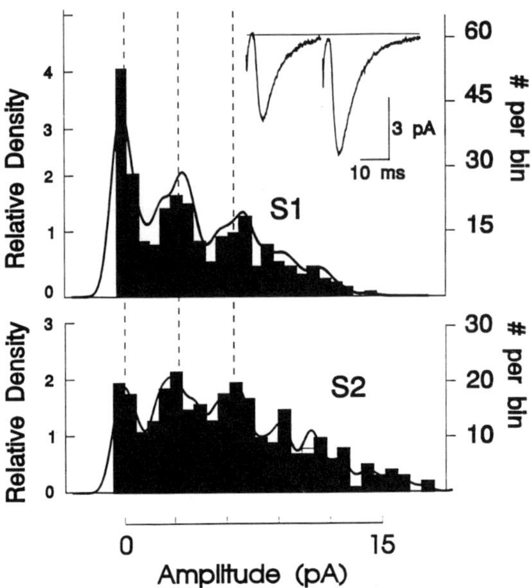

Figure 7.5 Amplitude distribution histograms of EPSCs show quantal levels that change as predicted for paired-pulse facilitation. Amplitude distributions of first (S1) and second (S2) responses separated by 50 msec for 300 consecutive sweeps. Note distributions have peaks at same locations with different relative frequencies as would be expected for increased release of quanta with paired pulse facilitation. *Inset*: Average of 100 sweeps delivered paired stimuli.

more multiquantal events; increasing the synaptic current driving force by postsynaptic hyperpolarization should increase quantal amplitude. With delivery of pairs of pulses, amplitude distributions of the first and second responses show peaks at similar EPSC amplitudes but a shift in their distribution (figures 7.5, 7.8E; n = 5). Changing the postsynaptic membrane potential from −50 mV to −90 mV enhances the mean amplitude of the EPSC with a comparable increase in the spacing between amplitude peaks (figure 7.6; n = 3). These results support the view that the amplitude distribution peaks correspond to quantal levels of transmitter released.

To induce LTP the postsynaptic membrane was depolarized to approximately −10 mV while afferent stimuli were continued for 100–150 trials (Malinow and Tsien, 1990). After this conditioning period, the membrane potential was returned to the baseline level and responses to test stimuli were recorded. There was a large persistent increase in transmission following this pairing protocol (figure 7.7B). From a group of 50 experiments where LTP was attempted, we selected 15 based on the following criteria: (1) baseline periods showed a stable amplitude of transmission (figure 7.7C); (2) transmission increased at least 50% after pairing; (3) amplitude distributions generated from epochs with at least 300 trials ob-

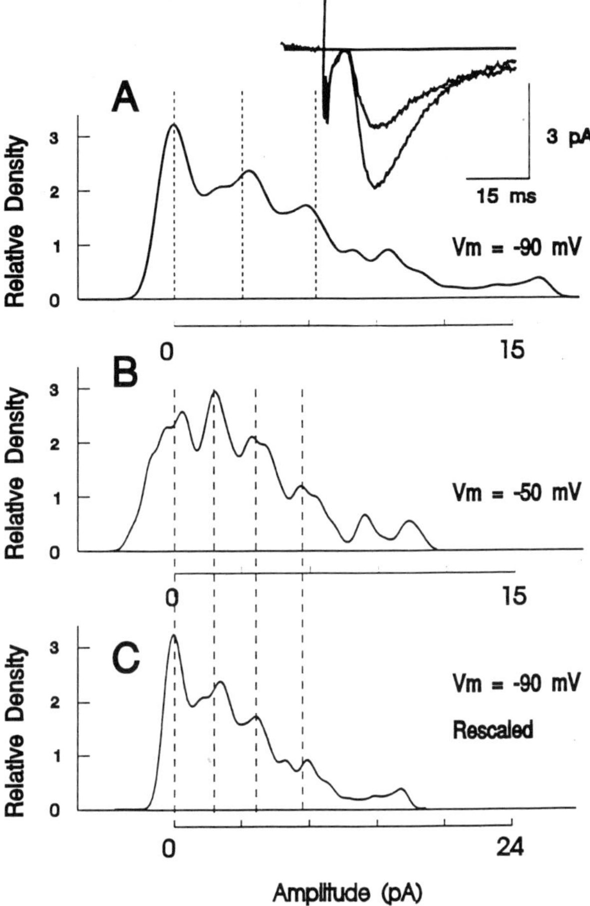

Figure 7.6 Effect of changes in membrane potential on amplitude distribution density estimates. Amplitude distributions for (*A*) 300 consecutive EPSCs obtained at −90 mV membrane potential and (*B*) 300 obtained at −50 mV. (*C*) Data obtained at −90 mV scaled down 1.6-fold, the decrease in mean amplitude observed at −50 mV. Note the superposition of amplitude peaks upon rescaling of data. The increased noise at −50 mV artificially increases the amplitude of the central peaks (due to spillover of noise from neighboring peaks) and can account for the apparent change in amplitude distribution (noise: 0.5 pA at Vm = −90 mV; 0.8 pA at Vm = −50 mV). The increased quantal amplitude at Vm = −90 mV despite the decreased background noise argues that the peaks of the signal amplitude distribution are not a statistical sampling artifact related to background noise (Clements, 1991). *Inset*: Average of 100 sweeps at −90 mV and −50 mV.

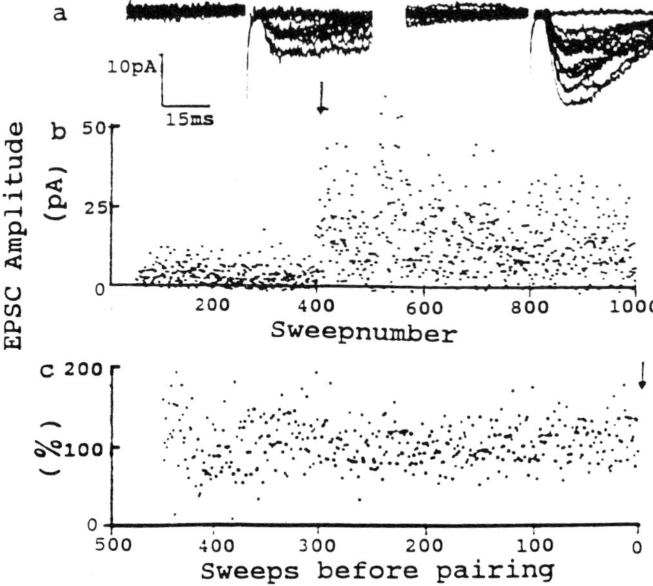

Figure 7.7 LTP of stable synaptic transmission. (*a*) Twelve of 15 consecutive minimally elicited EPSCs obtained before (*left*) and after (*right*) pairing protocol (see table 8.1 legend) to induce LTP (H1G4). (*b*) Amplitude of elicited EPSCs plotted versus time for an individual experiment (H1G4); pairing at arrow. (*c*) Ensemble average of transmission during the baseline period for all 15 experiments used in the analysis. Different experiments were time-registered with respect to pairing (arrow) and normalized with respect to the average of 50 EPSCs immediately before pairing.

tained before and after pairing showed clear evenly spaced peaks (figures 7.1–7.3, 7.5, 7.6, 7.8, 7.9); (4) the amplitude distribution of the noise matched the zero amplitude peak in the signal (figure 7.8).

For each experiment the quantal amplitude (q) for an epoch was calculated as the average distance between peaks in the amplitude density estimate. A maximum likelihood estimate (MLE) with no constraint on exact number of gaussian components coincided well with the spacings obtained by visual inspection of the data (table 7.1). The signal-to-noise ratio from the MLE (computed as the quantal amplitude divided by the standard deviation about each gaussian component) was 2.9 ± 0.2 (greater than 2.5 for every case). Quantal content (m, the average number of quanta released for each trial) was computed as $m = M/q$, where M is the mean transmission for that epoch. Figure 7.8A shows the amplitude distribution of 400 consecutive sweeps obtained before and 500 consecutive sweeps obtained after induction of LTP. We noted an increase in the spacing between peaks as well as a shift in the distribution. In figure 7.8B we superimposed the density estimate obtained before pairing with the density estimate obtained after pairing, scaling the amplitude axis of the latter to offset the calculated increase in spacing between peaks. Such superimposed density estimates reveal the increase in quantal amplitude (33/20 in this case) as well as the enhanced quantal content (upward and downward arrowheads). After rescaling for change in quantal amplitude, all 15 experiments showed close agreement of the underlying peaks with a clear shift in amplitude distribution to greater number of quanta.

For these 15 experiments, before LTP q was 2.7 $\pm$ 1.4 pA (mean $\pm$ SD), and m was 2.2 $\pm$ 1.2 quanta. After pairing, q increased to 4.2 $\pm$ 2.4 pA, and m to 4.1 $\pm$ 1.0 quanta (table 7.1). In all experiments we noted a marked decrease in the proportion of synaptic failures, as measured by the height of the zero amplitude peak, consistent with previous results (Malinow and Tsien, 1990; Malinow, 1991). On the whole, there was a small increase in the background noise following pairing (1.8 to 2.1 pA for experimentally measured noise and 0.95 to 1.5 pA for MLE noise). This may be due to an increased amplitude and/or frequency of spontaneous miniature EPSCs.

We tested further if the location of peaks in the amplitude distributions for these LTP experiments was due to chance. We generated density estimates after template-based selection (as in figure 7.2B). As shown in figure 7.8C, D, the location of the peaks was constant upon removal of noisy traces and in general, the peaks became more prominent. Similarly, peak locations were constant with paired pulse facilitation (figure 7.8E, F). Figure 7.9 displays the amplitude distribution of the data from figure 7.7. This is our clearest example to date demonstrating (1) evenly spaced amplitude peaks before and after LTP (before: $\chi^2 = 21.4$, $P < .01$; after: $\chi^2 = 51$ $P < .001$); (2) an increase in quantal content (arrows, 2.6-fold); and (3) an increase in quantal amplitude (2.1-fold).

In most individual experiments we noted an increase in both m and q during LTP, suggesting presynaptic and postsynaptic modifications underlying the potentiated transmission. On average, the increase in m accounted for 61% of the potentiated transmission (table 7.1). However, the percent increase in these parameters varied greatly, from 0% to 150% increase in quantal amplitude, and from 20% to 305% enhanced quantal content (table 7.1). By plotting the percent change in m versus initial m, we noted a strong negative correlation; low initial levels of m led to large increases and large initial levels increased little with LTP (figure 7.10A). To compare quantal amplitudes among different experiments, we normalized the measured quantal amplitude by an EPSC shape index to account for variation in electrotonic distance and access resistance across different experiments (See Experimental Procedures). For such normalized quantal amplitudes, plotting the change versus its initial value displayed the same negative correlation: the change in q was large in experiments for which q

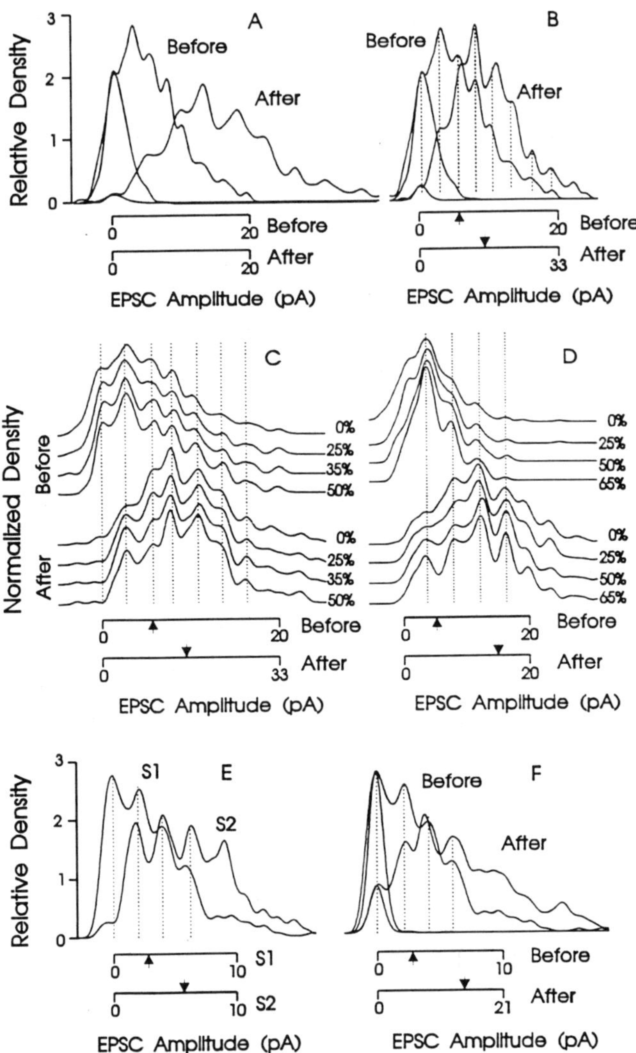

Figure 7.8 Amplitude distribution density estimates of EPSCs elicited before and after induction of LTP show clear peaks and an increase in both quantal amplitude and quantal content. (*A*) EPSC amplitude distributions of 400 consecutive EPSCs before and 500 consecutive EPSCs after pairing protocol. An estimate of the noise distribution using the same gaussian kernel as that used for the signal yields a single peak that matches the first peak in the signal amplitude distribution. (*B*) Amplitude distributions from the same data with the EPSC amplitude axis (x-axis) for the after-pairing data scaled by the calculated increase in quantal amplitude (experiment A1Y4). (*C*), (*D*) Family of density estimates obtained by template-based selection (see figure 8.2 and Methods). Amplitude axis for data obtained after LTP has been scaled by calculated increase in quantal amplitude. Location of peaks remains invariant upon removal of noisy sweeps. Note increase in quantal amplitude and quantal content in *C* (experiment A1Y4) with change only in quantal content in *D* (experiment A2U2). (*E*) Amplitude distribution of responses to first and second stimuli (S1, S2) in paired-pulse stimulation protocol (H1G4). (*F*) Amplitude distribution of S1 and noise, before and after induction of LTP, with rescaling of after-potentiation amplitude axis (experiment H1G4). Epochs used for density estimate are

was initially small and the increase was small when q was initially large (figure 7.10B). As the validity of this normalization procedure has not been fully investigated with electrotonic models, we only take this result as suggestive. A similar plot without normalization gave no significant correlation (not shown).

These results presented support to the quantal hypothesis of transmission at CNS synapses (Kuno, 1971; Jack et al., 1981; Walmsley et al., 1987; Redman, 1990; Korn and Faber, 1987; Edwards et al., 1989; Foster and McNaughton, 1991; Larkman et al., 1991; Malinow, 1991). This view is suggested by the following findings: (a) a simple statistical analysis gives clear indication that the observed responses are very unlikely to have come from a unimodal (nonquantal) distribution; (b) an objective removal of noisy traces improves the separation between quantal levels; and (c) the quantal levels change predictably with experimental manipulations. Our values for quantal amplitude are comparable to spontaneous miniature EPSC amplitudes reported for these cells (Manabe et al., 1992; Raastad et al., 1992). Not all of our experiments showed clear quantal levels; this may reflect the excitation of multiple synapses with different quantal amplitudes by a single or multiple presynaptic fiber(s).

The increased signal resolution of the whole-cell patch clamp technique (even without template-based selection) allows direct measurement of quantal levels. This avoids assumptions about release statistics that are often required to deconvolve quantal levels from a relatively large level of background noise. By looking at quantal parameters in elicited rather than spontaneous (Manabe et al., 1992) transmission, we are able to measure the contribution made by changes in presynaptic and postsynaptic quantal parameters to potentiated transmission.

To study the changes in these quantal parameters with LTP, we took care to choose experiments for which basal transmission was stable, thus decreasing the possibility of artifacts arising from drift in transmission (See figure 7.7B,C). We also chose only those experiments that showed a clear potentiation triggered by the pairing protocol. In 15 such experiments for which amplitude distribution density estimates showed clear peaks, we saw an increase in both quantal amplitude and quantal content during LTP, with a slightly predominant effect on the latter.

An increase in quantal content is generally considered to result from a presynaptic change in the release of quanta. However, this could be due to

given in table 7.1. Dotted vertical lines indicate visual estimate of quantal levels. Quantal content is indicated by upward (before) and downward (after) arrowheads; values are read with respect to dotted lines. Note the superposition of quantal levels and the shift in distribution toward greater number of quanta after the pairing procedure.

Table 7.1 Quantal parameters before and after LTP induction

Experiment	Epoch	χ^2 ($P <$)	q (pA)	qMLE (pA)	m (quanta)	$\Delta m/(\Delta q + \Delta m)$ (%)
A1E2						
Before	1–300		5.4	5.2	3.0	
After	750–1350	0.05	9.0	8.1	3.6	20
A1F2						
Before	1–300	0.001	2.6	1.9	1.8	
After	900–1500	0.05	3.3	3.5	4.2	83
A1W4						
Before	1–300		2.1	2.0	2.6	
After	400–950		4.0	3.5	3.4	26
A1X2						
Before	1–350	0.01	1.8	1.6	1.7	
After	550–1050	0.05	3.0	3.0	3.2	57
A1X4						
Before	1–350		2.4	2.7	0.1	
After	500–1300	0.1	2.4	2.6	3.2	100
A1Y4						
Before	1–400	0.01	2.7	2.8	2.0	
After	550–1050	0.01	4.4	4.4	3.5	54
A2B2						
Before	1–350	0.1	2.4	2.4	2.9	
After	1050–1500	0.001	2.6	2.2	4.5	87
A2K4						
Before	1–350	0.05	1.7	2.1	3.7	
After	1500–2000	0.01	2.2	2.9	5.4	61
A2O4						
Before	1–350		3.8	4.7	1.4	
After	460–840		4.3	4.7	3.3	91
A2U2						
Before	1–400	0.05	3.7	3.8	1.3	
After	550–1200	0.05	3.8	4.0	3.7	99
H1G4						
Before	50–350	0.01	2.2	2.3	1.4	
After	400–1000	0.001	4.5	4.7	3.6	60
H1P4						
Before	1–470	0.01	1.0	1.0	5.2	
After	480–1300	0.01	2.5	2.9	6.1	10
R50C3						
Before	1–420	0.001	1.7	1.4	1.4	
After	420–720	0.05	3.4	3.6	5.7	75
R50L3						
Before	1–280	0.05	5.8	4.6	1.3	
After	300–600	0.1	11.0	11.0	2.5	51

Table 7.1 (cont.)

Experiment	Epoch	χ^2 ($P <$)	q (pA)	qMLE (pA)	m (quanta)	$\Delta m/(\Delta q + \Delta m)$ (%)
R63I2						
Before	1–600	0.001	1.8	1.9	3.2	
After	800–1200		3.1	3.1	5.0	44
Average						
Before			2.7 ± 1.3	2.7 ± 1.2	2.2 ± 1.3	
After			4.2 ± 2.5	4.3 ± 2.2	4.1 ± 1.0	61 ± 27

Quantal parameters for 15 experiments satisfying acceptance criteria. For experimental series A, pairing occurred during 100–150 trials (between Before and After epochs) delivered at same frequency as test stimuli, every 1.6 sec, accounting for the break between the before and after epochs. For series H and R, test stimuli were delivered every 4 sec; pairing was delivered at 2 Hz, during which time no trials are counted. Series H and R63I2 were obtained using perforated patch technique (Horn and Marty, 1988), by inclusion of nystatin (167 μg/ml) in internal solution. To minimize problems of nonstationarity, epochs used to determine quantal parameters were chosen immediately before and after pairing, except for a few cases where clear peaks were not seen immediately after pairing but in a later epoch. No attempt was made to compare quantal parameters at different times after pairing. q refers to quantal amplitude, computed as the average spacing between quantal peaks determined by visual inspection of amplitude distribution estimates. m refers to quantal content computed as $m = M/q$, where M = mean transmission for epoch. $\Delta m/(\Delta q + \Delta m)$ refers to the percent increase in transmission accounted by increase in quantal content computed as $(\Delta m/m)/[(\Delta q/q) + (\Delta m/m)]$. χ^2 was calculated as in experimental methods.

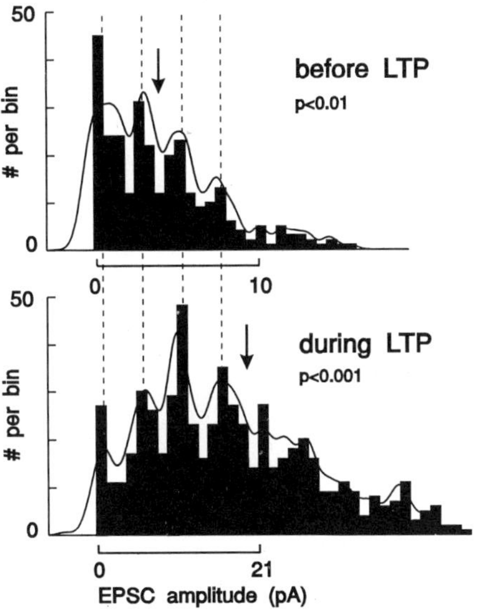

Figure 7.9 Clearest example of quantal levels and quantal changes with LTP. (*Top*) Amplitude distribution histogram and superimposed density estimate of 300 consecutive sweeps before LTP. (*Bottom*) Same for 600 sweeps during LTP. Note change in amplitude axis. Arrows denote quantal content with respect to dotted lines (H1G4).

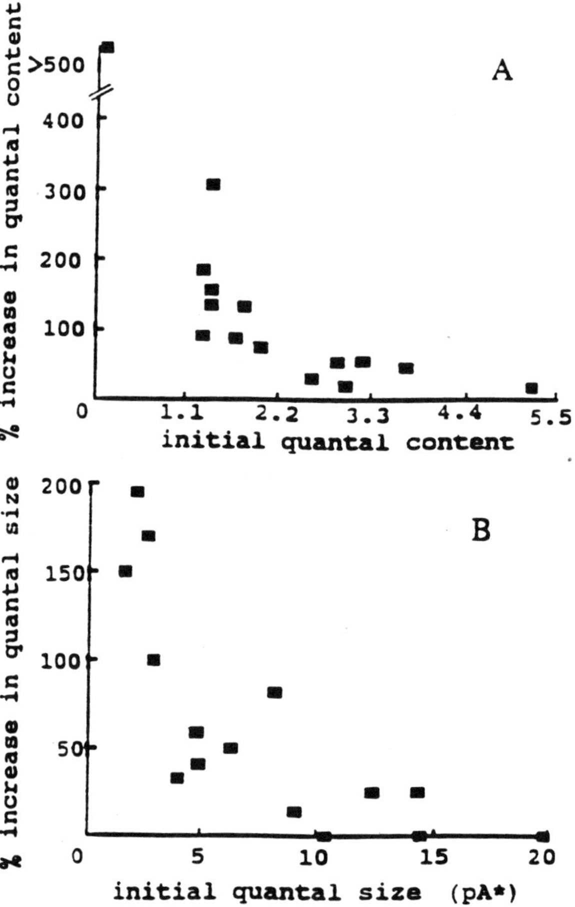

Figure 7.10 The percent increase in quantal amplitude or quantal content during LTP is inversely related to its initial value. (*A*) Plot of the percent increase in quantal content versus the initial quantal content. (*B*) Plot of the percent increase in quantal amplitude, versus the initial quantal amplitude, measured in normalized current (*) (see Experimental Procedures).

recruitment of previously silent synapses. The latter possibility would be most easily reconciled with LTP induction hypotheses if silent synapses contain only NMDA receptors before potentiation (and thus are effectively "silent" at hyperpolarized membrane potentials) with addition of AMPA receptors after activation of NMDA receptors during LTP induction. If there is such recruitment of new synapses, however, it is important to keep in mind that after LTP, quantal amplitude peaks are still resolvable and that we generally also see an increase in quantal amplitude. This requires that the quantal amplitude at newly recruited synapses be the same as (or an integer multiple of) the new quantal amplitude at previously existing synapses. This could be satisfied if the final quantal amplitude were determined by the degree of NMDA-receptor activation during LTP induction, which could be similar in our induction protocols for all (active and silent) synapses. A more parsimonious model for the changes with LTP is that we are exciting a number of functioning synapses during an experiment and there is increased release of quanta and receptor function at these functioning synapses. This model still requires that all synapses increase their quantal size equally, so that quantal levels are resolvable after LTP. This could be satisfied if the increase in quantal amplitude were determined by the degree of NMDA-receptor activation. The simplest model is that in our experiments we are exciting a single synapse, and LTP produces increased release and receptor function at this synapse. Since multiquantal levels are observed, this latter possibility requires that all receptors at a synapse are not saturated by a single vesicle of transmitter. The apparent lack of increase in peak width for multiquantal peaks (all histograms in this study; Larkman et al., 1991; Malinow, 1991) implies very low quantal variance and has been used to argue that receptors are in fact saturated by a single vesicle. The increase in quantal amplitude could reflect an increased number or efficacy of postsynaptic receptors or an increase in transmitter per vesicle.

LTP in cultured hippocampal neurons can be completely explained by increases in quantal content (Bekkers and Stevens, 1990; Malgaroli and Tsien, 1992). Our finding that LTP in slices also shows increased quantal amplitude suggests that a critical factor responsible for increasing quantal amplitude is missing in cultured cells (e.g., α-calcium/calmodulin protein kinase II is very low at ages when cultures are normally made; Hanson and Schulman, 1992) or that quantal amplitude is maximally increased during the culture process.

The initial values of quantal parameters showed a large variation across experiments. This suggests a heterogeneity in synaptic function, consistent with the various synaptic morphologies identified by electron microscopy (Harris and Stevens, 1989). The increase in quantal parameters following LTP induction showed a strong negative correlation with their initial values. This suggests that prior induction of LTP (possibly in the behaving animal) occluded further increases during our experiments. Interestingly, the initial values of the two parameters, quantal amplitude and quantal content, were not correlated. This lack of correlation suggests that physiological mechanisms may independently modify presynaptic or postsynaptic efficacy. It will be interesting to determine if different experimental manipulations can predispose activation of presynaptic or postsynaptic expression mechanisms for LTP.

Different studies have presented evidence suggesting presynaptic or postsynaptic modifications during LTP. Our study provides evidence that the change of a presynaptic or postsynaptic quantal parameter is directly related to its initial value. It is possible that experimental conditions that inadvertently modify initial quantal values (such as a high Ca^{2+} to Mg^{2+} ratio in perfusate) will affect the mechanism of LTP that is observed.

We thus have evidence for two different forms of plasticity that appear to be independently regulated. We may ask: Why are there two kinds of plasticity? They both can provide an increase in the mean transmission. However, these two forms of plasticity will have a different effect on the trial-to-trial variability of individual synapses. That is, an increase in quantal content will decrease the trial-to-trial variability at that synapse (measured as the coefficient of variance [CV]; see below). An increase in quantal amplitude will not change the variability of the synapse. We have investigated whether such changes in variability at individual synapses produces an effect on the input-output characteristics of these neurons.

It is generally believed that the input-output relations of single neurons are important determinants of nervous system function. A critical parameter with influence on these relations is the signal to noise ratio (S/N). In fact, computer simulation studies suggest that there may be an optimal S/N in the communication between components that produces the best performance of a network (Little and Shaw, 1975; Hinton et al., 1984; Buhman and Schulten, 1987; Amit and Treves, 1989; Gibson et al., 1991). In this study we measure the magnitude and the modulation of the S/N for synaptic transmission onto a central neuron, and assess the effects of long-term potentiation. We consider the relevant signal to be an elicited synaptic input sufficient to trigger a postsynaptic action potential; the noise we measure is the trial-to-trial deviation in the postsynaptic response to this input.

One possible source of noise is the probabilistic release of transmitter at activated afferent inputs (Katz, 1969). The impact of such transmission on the input-output relations of neurons has received considerable theoretical (Taylor, 1972; Shaw and Vasudevan, 1974; Hinton et al., 1984; Burnod and Korn, 1989; Bressloff and Taylor, 1990; Gibson et al., 1991), but little experimental evaluation. One view holds that the generation of an action potential requires the summed input of 100–400 synapses (McNaughton et al., 1981; Andersen, 1987). The summed effects of many synapses will be expected to show little trial-to-trial deviation about an average response, an application of the law of large numbers (figure 7.11). Thus, this view predicts that changes in the statistics of elicited transmitter release would have little effect on the S/N of a threshold input and that other possible sources of noise, such as changes over time in membrane excitability or membrane potential, would determine the S/N.

An alternate view is that a small number of synaptic events may be enough to elicit a postsynaptic action potential. This view is suggested by

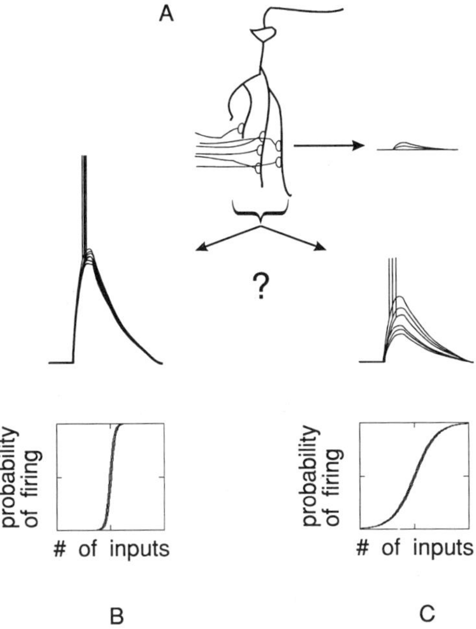

Figure 7.11 Diagram of two possible scenarios regarding the contribution of synapse variability to a threshold input. (*A*) A typical neuron with several afferent synapses, each of which responds in a quantal manner. If a large number of synapses must be coactivated to elicit an action potential, the variability of postsynaptic responses will be low and the input-output curve of the neuron will be relatively steep (*B*). If fewer synapses must be coactive, trial-to-trial variability will be high and the input-output curve of the neuron will be more shallow (*C*).

some data (Miles and Wong, 1986), and is supported by the recent finding that neuronal input resistance and membrane time constant are much larger than previously thought (Raastad et al., 1992; Spruston and Johnston, 1992). If this is the case then the variability at each synapse could contribute significantly to the variability at a threshold input (Korn and Faber, 1987; figure 7.11). Thus, this alternate view predicts that changes in the statistics of elicited transmitter release could have significant effects on the S/N at a threshold input.

In this study we measure the S/N for an elicited synaptic input that produces a postsynaptic action potential in a hippocampal slice neuron. At this level of input, we note a significant trial-to-trial variability in the postsynaptic response. The S/N is governed by synaptic mechanisms as it can be modulated by pharmacological and physiological manipulations affecting transmitter release. We also find that two different cell populations (CA1 and fascia dentata [FD] neurons) have different S/N at threshold.

We have studied the predicted and observed effects of long-term potentiation (LTP) on S/N at threshold. At the level of a single synapse, presynaptic or postsynaptic potentiating modifications will have different effects on trial-to-trial variability. For a threshold input, we find an increase in the S/N after LTP, consistent with presynaptic modifications. If elicited responses are returned to the threshold level by recruiting fewer fibers, the trial-to-trial variability also returns to that observed before potentiation. We thus observe a maintained input-output S/N at a threshold input strength despite the large gain changes observed with LTP. We hypothesize that this invariance results from the presynaptic changes observed during LTP (Bekkers and Stevens, 1990; Malinow and Tsien, 1990; Malinow, 1991; Liao et al., 1992; Kullman and Nicoll, 1992) and may be important for the performance of a plastic neuronal network.

HOW LARGE IS VARIABILITY AT THRESHOLD?

Synaptic transmission was studied in the CA1 and FD regions of hippocampal slices. After placement of stimulation electrodes in afferent regions (see Experimental Procedures) a tight "gigaseal" recording was obtained from a postsynaptic neuron and maintained in the cell-attached patch configuration. Synaptic transmission was elicited by passing brief current pulses through the stimulating electrode. If the stimulus current was sufficient, all-or-none action potentials with a 8–50 msec delay were clearly observed through the cell-attached pipette (figure 7.12A) (Fenwick et al., 1982; Lynch and Barry, 1989). A threshold stimulus was set so that about 50% of trials produced an action potential in the recorded cell. This configuration was maintained for about five minutes to ensure stable recording conditions. In several control experiments intrapipette potential was temporarily reduced from −60 mV to 0 mV. No effect on the probability of action potential generation was observed with this manipulation. This implies minimal disruption of the membrane patch under these conditions.

After setting the threshold level of afferent stimulation in the cell-attached patch configuration, we obtained a whole-cell recording from the same cell. Current clamp recordings revealed a large variability in excitatory postsynaptic potential (EPSP) amplitude from trial to trial, with action potentials triggered by the larger responses (figure 7.12B). The synaptic depolarization required to generate an action potential was 12.4 $\pm$ 2.8 mV ($N = 5$) from rest and quite constant within any given cell (figure 7.12B). To measure the trial-to-trial variability in peak EPSP response, action potential generation was prevented by passing a small (0.1–0.2 nA) d.c. hyperpolarizing current through the recording pipette (figure 7.12C). To quantify this variability, we computed the coefficient of variation (CV) in postsynaptic responses by calculating the mean peak response amplitude

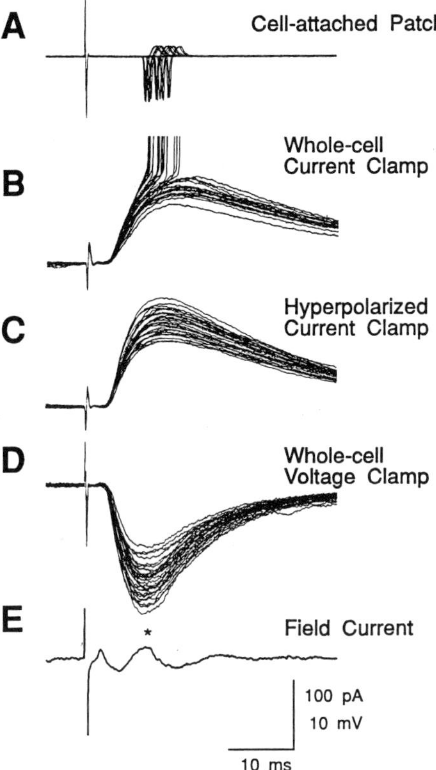

Figure 7.12 Synaptic transmission onto a CA1 neuron at a threshold stimulus. (*A*) Fifty superimposed postsynaptic currents recorded in a cell-attached patch configuration under voltage clamp. About 50% of the traces fail to generate a postsynaptic action potential. (*B*) The same cell, after rupture of the membrane patch, while recording in whole-cell current clamp (Vm = −65 mV). Note that threshold for action potential generation is quite constant (CV = 1.5 ± 0.35%; N = 5). (*C*) The full size EPSPs variability as resolved when the cell was hyperpolarized (Vm = −80 mV) to prevent action potential generation. (*D*) Synaptic responses from the same cell in whole-cell voltage clamp. (*E*) Field currents recorded after the cell was disrupted by gentle positive pressure. Upward deflection (*) is presumably due to synchronous action potentials generated (population spike) in nearby cells.

(M) and the standard deviation about this mean (σ) for an epoch of 50 consecutive trials (CV = σ/M × 100%). This measure normalizes variability and thus allows comparison for different sized mean responses. The CV may also be interpreted as the inverse of the S/N. Under these conditions the average trial-to-trial variability of EPSPs was 13.7 ± 3% and the mean amplitude of response was 11.4 ± 2.5 mV (n = 11). Converting to voltage-clamp revealed EPSCs (204 ± 53 pA; N = 19) that also showed a similar trial-to-trial variability (CV = 15.8 ± 3.4%; N = 19). Paired comparison of the variability of EPSPs and EPSCs in the same cells revealed the latter to be slightly but significantly larger ($P < .005$; N = 11; paired t-test). The difference may be attributed to nonlinear summation of postsynaptic potentials (McNaughton et al., 1981; Langmoen and Andersen, 1983). After recording from a single neuron, the cell was discarded by gentle positive pressure through the pipette. Recordings through the pipette now reflected population currents, which revealed population spikes demonstrating that some proportion of nearby neurons were also firing with this threshold stimulus.

HOW MANY QUANTA ARE REQUIRED TO FIRE A POSTSYNAPTIC ACTION POTENTIAL?

If the variability for a threshold stimulus is due to the stochastic release of transmitter, then this suggests that few quanta are required to produce a threshold depolarization. We tested this by comparing the amplitude of spontaneous miniature EPSCs elicited by local application of 500 mM sucrose with the amplitude of a threshold response. The amplitude of miniature EPSCs varied considerably (~2–30 pA), as previously described (Bekkers et al., 1990; Manabe et al., 1992; McBain and Dingledine, 1992; Raastad et al., 1992). The average amplitude was 7.9 ± 1.2 pA (n = 9) with coefficient of variance, c_q = 63 ± 14%. The average threshold EPSC current was 204 pA with inhibition intact and 125 pA without inhibition (see below). Thus, the number of synchronously active quanta required to reach threshold would be about 16 in the absence of inhibition and about 26 in the presence of inhibition.

In several experiments, threshold and miniature EPSCs were measured in the same cell (figure 7.13). As described for the above experiments, a threshold stimulus level was set while in the cell-attached patch configuration and the elicited synaptic currents were measured after conversion to whole-cell configuration (figure 7.13A,B). The average number of miniature EPSCs required to generate a threshold current in these experiments for which inhibition was blocked was 14 (mean threshold EPSC: 107 ± 11

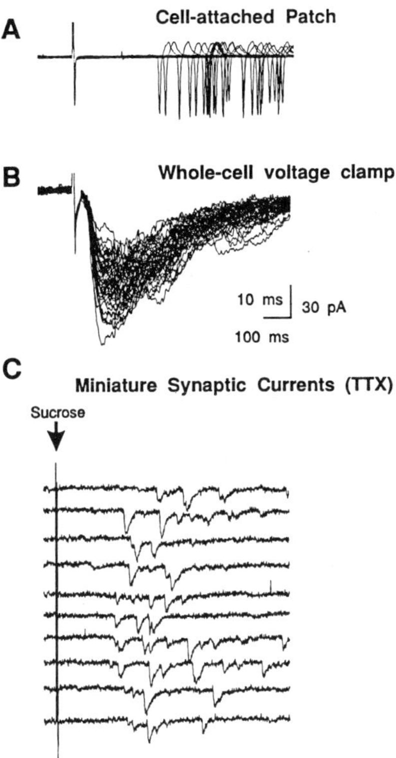

Figure 7.13 Comparison of threshold EPSC to quantal amplitudes. (*A, B*) 50 consecutive responses recorded in cell-attached patch (*A*) and whole-cell (*B*) configuration at a threshold stimulus. (*C*) Quantal responses recorded in the presence of 1 μM tetrodotoxin elicited by local pressure application of 500 mM sucrose. Note different time scales in *B* and *C*.

pA; mean miniature EPSC: 7.6 $\pm$ 0.4 pA; n = 3). Thus, since few quanta are required to reach threshold, this supports the view that the trial-to-trial variability observed at threshold is of synaptic origin.

DO CHANGES IN THE STATISTICS OF TRANSMITTER RELEASE AFFECT VARIABILITY AT THRESHOLD?

To answer this question we studied excitatory transmission in the absence of inhibition and performed several manipulations expected to specifically affect the statistics of transmitter release: (a) paired pulse facilitation, (b) changes in the ratio of extracellular concentrations of calcium to magnesium, and (c) pharmacological blockade and activation of presynaptic receptors that control transmitter release. If the variability we see at a threshold stimulus is at least partly due to probabilistic release of transmitter, manipulations that increase release probability should decrease variability (Malinow and Tsien, 1990).

If pairs of stimuli are delivered less than ~ 100 msec apart, the second response is enhanced, due to increased probability of excitatory transmitter release (del Castillo and Katz, 1954; Zucker, 1989). We investigated the effect of such facilitation on the variability of responses. In the second response we saw increased transmission as well as a marked reduction in trial-to-trial variability (figure 7.14A–D; mean: from 74 pA to 117 pA; $P < .005$; CV: from 26% to 20%; $P < .005$; $N = 9$; paired t-test). Another method of increasing release probability is to raise the ratio of extracellular calcium to magnesium concentration. Elevating calcium (with a concomitant decrease in magnesium to keep divalent cation concentration and fiber excitability constant) resulted in an increase in transmission and a decrease in trial-to-trial variability in both the first (I) and second (II) responses of paired stimuli (figure 7.14A–D; mean(I): from 74 pA to 110 pA; $P < .005$; CV(I): 26% to 19%; $P < .005$; mean(II): from 117 pA to 154 pA; $P < .005$; CV(II): from 20% to 14%; $P < .005$; $N = 9$; paired t-test). Fiber excitability, as measured by the amplitude of extracellularly measured fiber volley, showed no change during these different manipulations (figure 7.14E,F).

We also used drugs known to have effects specifically on transmitter release mechanisms. Adenosine agonists at low concentrations have been shown to inhibit release of transmitter (Scholz and Miller, 1992) with no effect on postsynaptic input resistance (Siggins and Schubert, 1981), membrane potential or presynaptic fiber potential (Nicoll et al., 1990). Moreover all postsynaptic effects on K^+ conductance observed at higher adenosine concentrations should be blocked by the Cs^+ in our intrapipette solution (Trussell and Jackson, 1985). We found that bath application of 0.3 μM 2-chloro-adenosine significantly decreased transmission and concomitantly increased variability (mean: from 90 pA to 50 pA; $P < .005$; CV: 23% to 33%; $P < .05$; $N = 4$; paired t-test). Endogenously released adenosine may provide a constant inhibition of excitatory transmission (Nicoll et al., 1990). We blocked adenosine receptors with 25–100 μM IBMX (Nicoll et al., 1990) or 10 μM SPT (Gustafsson et al., 1988) and found a significant increase of synaptic transmission and a decreased variability (mean: from 92 pA to 131 pA; $P < .005$; CV: 23% to 20%; $P < .025$; $N = 6$; paired t-test).

Does stochastic release account for all of the variability observed at threshold? Could some of the observed variability be an artifact of the experimental conditions? It is possible that a variable number of fibers are stimulated from trial to trial, since extracellular electrical stimulation in a slice may not reliably excite the same number of fibers every time. If we could decrease the observed variability in responses to zero by manipulating synaptic release, this would strongly argue that *all* of the observed variability is due to probabilistic transmitter release. This is expe-

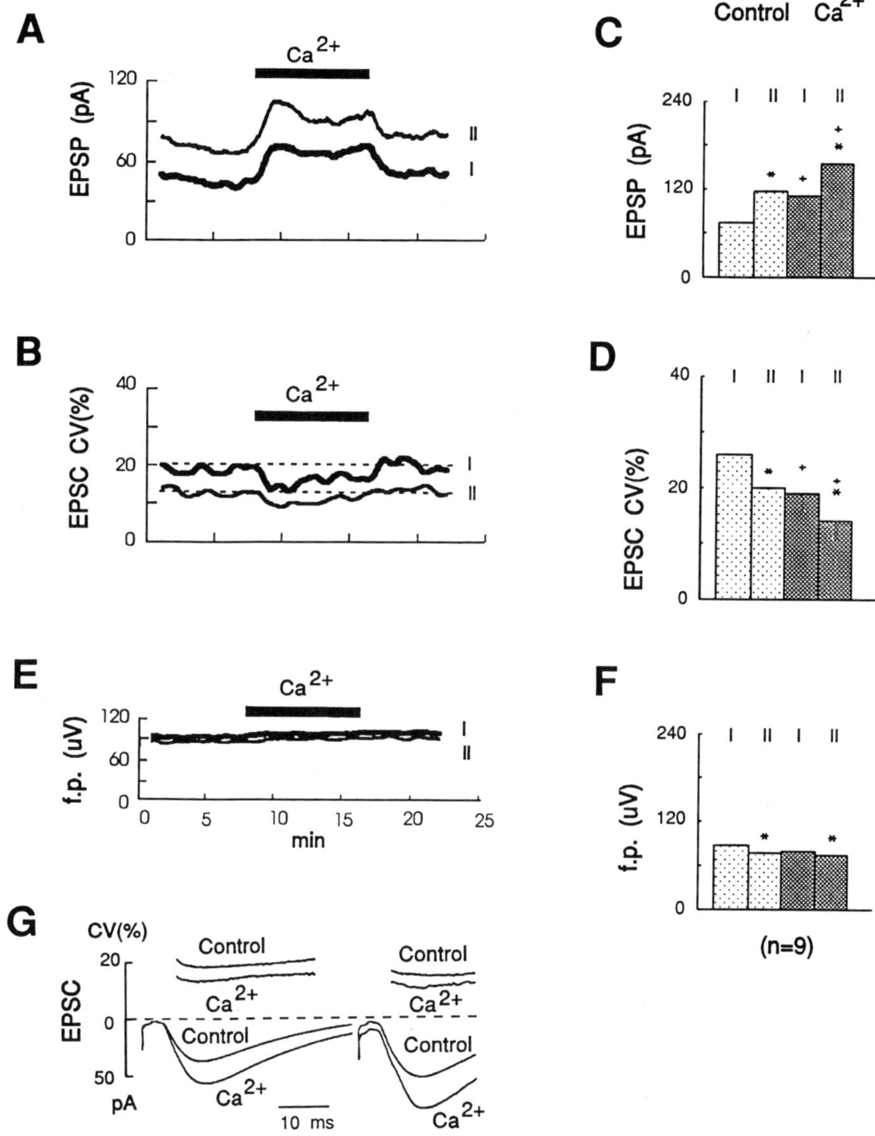

Figure 7.14 Paired pulse facilitation and high Ca^{2+}/low Mg^{2+} solution increases amplitude and decreases variability of synaptic responses, but has no effect on axonal excitability. (*A, B, C*) Fifty-point running average of the EPSC amplitude (*A*) EPSC coefficient of variation (CV) (*B*) and fiber potential amplitude (*C*) of the first (I) and second (II) elicited event. Events are elicited 45 msec apart every 2 sec and the stimulus strength was set (in a patch configuration) to fire an action potential ~50% of trials on the second elicited event. Bathing solution was changed from 2 mM Ca^{2+}/2 mM Mg^{2+} to 3 mM Ca^{2+}/1 mM Mg^{2+} where shown. (*D, E, F*) Averages of data collected from several experiments similar to *A, B,* and *C*. Differences in mean and CV under these manipulations (paired-pulse facilitation and change in Ca/Mg) were significant at the 0.01 level (paired t-test). (*G*) Time course of mean amplitude (below) and CV (above) of 50 consecutive responses to paired pulses in control and 3 mM Ca^{2+}/1 mM Mg^{2+} solutions.

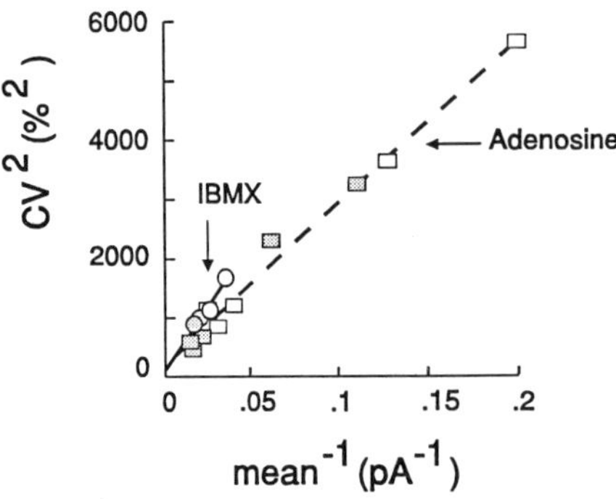

Figure 7.15 Linear relationship between transmission amplitude and intertrial variability for different release efficacies. CV^2 is plotted against $mean^{-1}$ computed over 50 consecutive responses elicited in the presence of 0, 0.3, 1, and 3 μM 2-chloro-adenosine and 50 μM IBMX. Open and filled circles represent the response to the first and second, respectively, of a pair of pulses delivered 45 msec apart. The y-intercepts (17 and 8) represents the contribution to total CV^2 of variable fiber stimulation. Stimulus strength was set to threshold level in the cell-attached patch configuration.

rimentally difficult because it is not possible to achieve arbitrarily high levels of transmitter release. We can, however, extrapolate to an arbitrarily high release condition if we can find a linear relation between the observed variability and the mean response for different release efficacies. We found that for manipulations affecting release, plotting CV^2 versus (mean response)$^{-1}$ generally produced a linear relation (figure 7.15; 8/10 experiments had coefficient of correlation >0.9; see experimental methods for why these parameters may be expected to be related linearly). With this plot we can estimate the relative contributions to the observed variability made by synaptic variability and fiber stimulation variability. Intuitively, one may reason that as release efficacy is increased, synaptic variability should decrease, whereas fiber stimulation variability should be unchanged. Thus, at the projected maximal release (y-intercept) the variability should be primarily due to stimulus variability. For these plots the y-intercept value was on average $7 \pm 14\%$ of the threshold value. We also found that plotting CV^2 versus (mean response)$^{-1}$ for experiments in which the stimulus strength was varied also produced linear plots (figure 7.16C; coefficient of correlation >0.9 for 6/6 experiments) with a small intercept (y-intercept $= 6 \pm 18\%$ of threshold value, $n = 6$; see experimental methods for why this may also be expected). These results argue that the variability due to fiber excitation makes a negligible contribution to the observed fluctuation of responses at a threshold stimulus.

With all of these experiments, manipulations that decreased transmitter release increased the observed variability in postsynaptic responses; increases in transmitter release produced a decrease in variability. These

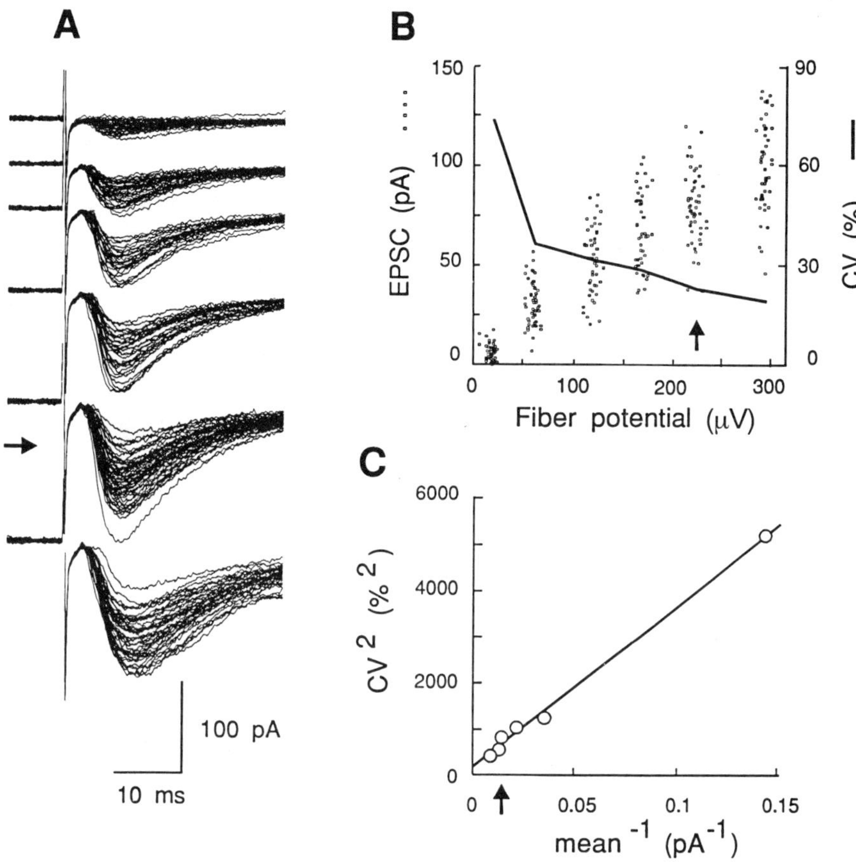

Figure 7.16 The effect of increasing stimulus strength on mean and variance of EPSC and fiber potentials. (*A*) 30 superimposed EPSCs recorded at successively higher stimulus strengths. The response to the lowest stimulus strength is uppermost. (*B*) Plot of EPSC amplitude (squares) and CV (solid line) versus extracellularly recorded fiber potentials for different stimulus strengths. Note greater variability in the EPSC amplitude than in the fiber potential amplitude. (*C*) Plot of CV^2 versus $mean^{-1}$ revealing linear relation. The y-intercept (18) corresponds to the contribution of fiber potential excitation variability (see text). Arrows in *A*, *B*, and *C* indicate responses to threshold stimuli.

results support the view that the observed variability at a threshold input is controlled by the statistics of transmitter release.

PHYSIOLOGICAL MANIPULATION OF SYNAPTIC S/N

Previous theoretical studies suggest that the S/N between neurons controls the performance of a neuronal network. We were interested in determining if different physiological manipulations changed the S/N at a threshold stimulus. In this way, physiological changes, such as changes in afferent firing frequency or different levels of inhibition, could modify the function of a group of neurons. Furthermore, neurons in different brain regions that perform different functions may have different S/N. We investigated these possibilities in the hippocampal slice.

We tested the effect of stimulus frequency on synaptic variability, recording combined excitatory and inhibitory postsynaptic currents. Synaptic variability was significantly larger at a stimulus frequency of 0.5 Hz as compared to 0.05 Hz stimulation, with little or no difference in EPSC amplitudes (figure 7.17B; mean = 173 pA at 0.5 Hz, mean = 168 pA at 0.05 Hz; $P < .05$; N = 8; paired t-test; CV = 10% at 0.05 Hz, 15% at 0.5 Hz; $P < .0005$; N = 8; paired t-test). The change in variability while maintaining the same peak response may be due to a combination of effects on excitatory and inhibitory transmission. A decreased release at excitatory synapses, as noted previously for increased stimulus frequency, could decrease peak EPSC amplitude (Larkman et al., 1991), and increase the trial-to-trial variability, as observed in all of the above pharmacological manipulations. The increased stimulus frequency could also produce a decrease in inhibition (McCarren and Alger, 1985; Deisz and Prince, 1989; Davies et al., 1990) that would tend to increase peak EPSC amplitude. The combined result could be no change in amplitude with an increased variability.

We tested the effect of synaptic inhibition on threshold by comparing experiments in the presence and absence of picrotoxin, a GABA-A receptor blocker. The mean EPSC amplitude required to drive a cell to generate an action potential was considerably lower in the presence of picrotoxin (figure 7.17C; with PTX: 125 ± 30 pA, N = 26; without PTX: 204 ± 53 pA, N = 19; $P < .0005$; t-test). Accordingly, the synaptic variability at this lowered level of elicited response was considerably higher (with PTX: CV = 21 ±2.7%, N = 26; without PTX: CV = 15.8 ± 3.4%, N = 19, $P < .0005$, t-test). Since the experiments were carried out at a 0.5 Hz stimulation frequency where inhibition may already be reduced, the effect of blocked inhibition may be even more pronounced at lower stimulus frequencies.

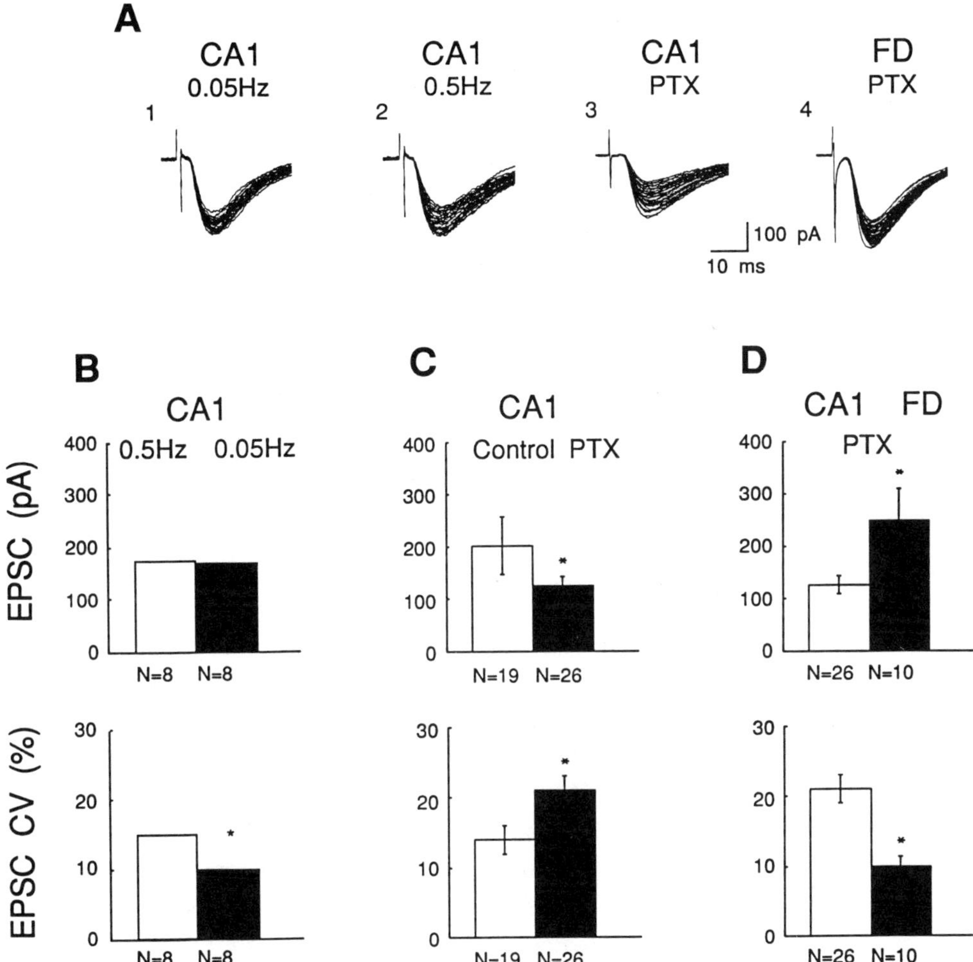

Figure 7.17 The effects of pharmacologic and physiologic manipulations on variability at a threshold stimulus. (*A*) 50 consecutive evoked responses at a threshold stimulus. (Al, A2, A3) are responses recorded in the CA1 region to stimulation every 20 sec (A1) every 2 sec (A2, same cell as A1), in the presence of picrotoxin (A3, different cell), and from fascia dentata in picrotoxin (A4). (*B, C, D*) Collected results from several experiments showing mean EPSC amplitude (*upper*) and CV (*lower*) under similar conditions as in *A*. Error bars indicate standard deviation among different experiments; N indicates number of experiments; (*) indicates $P < .05$, paired t-test.

We compared the threshold level and synaptic variability in two different hippocampal synaptic regions, CA1 and FD. To investigate those factors arising from just excitatory inputs, we performed the experiments in the presence of picrotoxin. The amplitude of synaptic currents necessary to generate a postsynaptic action potential was considerably higher in FD cells (figure 7.17D; FD: 248 $\pm$ 62 pA, N = 10; CA1: 125 $\pm$ 30 pA, N = 26; $P < .0005$; t-test). There was also a lower synaptic variability (FD: 10 $\pm$ 1.4%, N = 10; CA1: 21 $\pm$ 2.7%, N = 26; $P < .0005$; t-test). The reason for this difference in threshold current is not explained by differences in input resistance (R_{in}(CA1) = 330 $\pm$ 36 MΩ, N = 26; R_{in}(FD) = 708 $\pm$ 180 MΩ, N = 10; $P < .0005$; t-test) but may be due to different membrane excitability in cells from the CA1 and FD regions (Andersen et al., 1980, 1987; McNaughton et al., 1981; Staley et al., 1992).

EFFECT OF LTP ON SYNAPTIC VARIABILITY AT THRESHOLD

In the context of this study, we were interested in determining the effect of LTP on synaptic variability at threshold. If the efficacy at individual synapses is increased, the number of synapses required to drive a cell to threshold will be proportionally reduced. Decreasing the number of activated synapses, as was shown above, will increase the amount of trial-to-trial variability. We asked: Do the mechanisms underlying LTP compensate for this anticipated increase in variability?

We studied transmission between Shaffer collaterals/commissural fibers and CA1 neurons in the presence of picrotoxin. The synaptic stimulus level was set to threshold while recording from a CA1 neuron in the cell-attached patch configuration. Postsynaptic responses were subsequently monitored with whole-cell voltage clamp and baseline synaptic response amplitude and variability was measured (figure 7.18A,C). LTP, induced by pairing continued presynaptic activity with postsynaptic depolarization to −10 mV, produced an increase in mean transmission as well as a decrease in synaptic variability. Reducing the stimulus strength so that the mean postsynaptic response was approximately the same as before LTP (figure 7.18A,B, arrows) produced an increase in synaptic variability (figure 7.18C). Interestingly, the increased variability resulting from the reduced stimulus strength almost exactly matched the decreased variability produced by LTP (figure 7.18C). Thus, when measured at the same postsynaptic threshold level, the synaptic variability before LTP virtually matched that after LTP. A comparable reduction in stimulus strength of the control (nonpotentiated) pathway produced an increase in variability. Thus, if LTP had not decreased variability, upon reduction of stimulus strength, the variability at a threshold stimulus would have increased. This

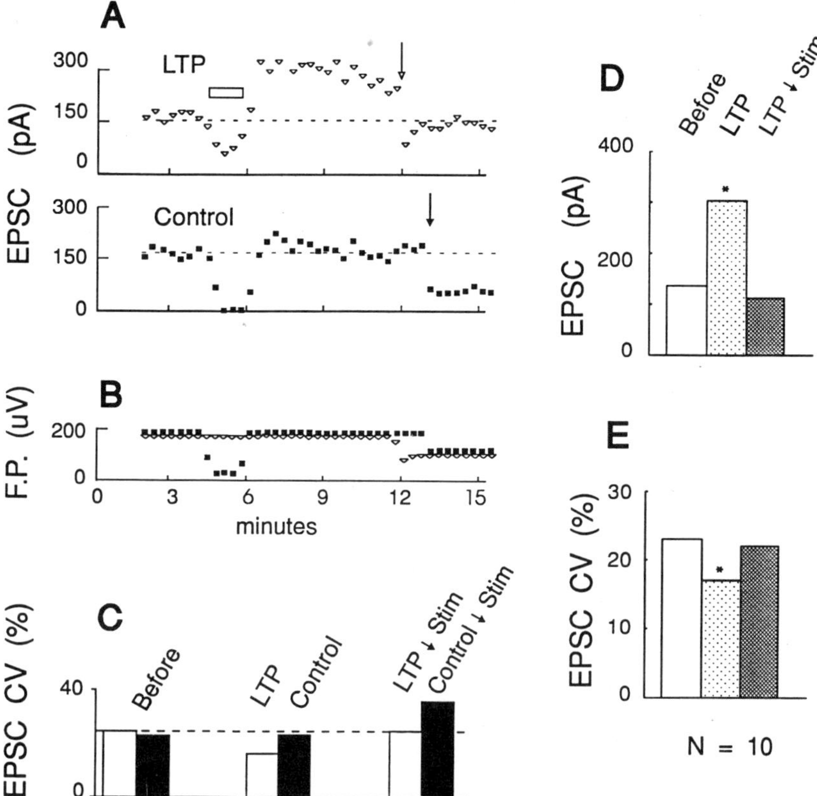

Figure 7.18 The effect of LTP on synaptic variability at threshold. (*A*) Plot of EPSC amplitude (20-sweep mean) versus time for transmission at threshold stimulus delivered through a test pathway (*upper*) and control pathway (*lower*). Dashed line indicates threshold response level. Pairing of postsynaptic depolarization and presynaptic activity of the test pathway is indicated by the bar. Stimulus strength was decreased at arrows. (*B*) Plot of fiber potential amplitude (20-sweep mean) versus time for the two pathways. (*C*) Coefficient of variance of EPSC amplitude calculated over 50 consecutive sweeps for test (open bars) and control (filled bars) pathways before pairing procedure (Before), after pairing procedure (LTP, Control), and after subsequent reduction of stimulus strengths (LTP ↓ Stim, Control ↓ Stim). (*D*) Compiled results from 10 experiments showing mean EPSC (*D*) and mean EPSC CV (*E*) before induction of LTP (open bars), during LTP (stippled bars), and after reduction of stimulus strength during LTP (dark shaded bars). Note that CV decreases during LTP and returns to pre-LTP levels after reduction of stimulus strength during LTP. (*) indicates $P < .05$, paired t-test.

maintenance of synaptic variability at threshold, in spite of decreased number of active synapses, was a general finding (N = 10). After an average of 2.2-fold potentiation and reduction of stimulus strength to match the mean postsynaptic response (136 pA to 112; $P > .05$; paired figure 7.18D) the CV also coincided (23 to 22; $P > .05$; paired figure 7.18E).

In this part of the study we have investigated the nature of the input sufficient to elicit a postsynaptic action potential. The input stimulus level was set while recording from a postsynaptic cell in the cell-attached patch configuration. By setting the threshold stimulus level before violating the cell membrane, little if any perturbation of the membrane potential, input resistance or cytosolic milieu of the postsynaptic cell is likely. Such effects could affect the generation of an action potential, and thus preclude an accurate setting of a threshold input. Field measurements revealed population spikes indicating that the input was sufficient to fire surrounding cells.

We were surprised to find a relatively large amount of trial-to-trial variability in the amplitude of postsynaptic responses at this threshold stimulus once a whole-cell recording was obtained. Several lines of inquiry lead us to conclude that this variability is of synaptic origin: (1) the observed variability can be modulated experimentally by modifying synaptic release efficacy (paired-pulse facilitation, extracellular calcium concentration, and adenosine agonists and antagonists) and by changing stimulus strength or frequency; (2) the variability extrapolates to nearly zero as transmitter release is enhanced, and (3) the variability predicted from unitary transmission can fully account for the observed variability at threshold (see experimental methods). None of these findings would be expected if the variability we observed were due to the variable stimulation of afferent fibers.

We have estimated the number of synchronously active quanta required to fire an action potential. As few as 5 large excitatory quanta in the absence of inhibitory action could drive a postsynaptic cell, although on average 15–30 would be required, depending on the level of inhibition. The apparent requirement for very few quanta to activate a postsynaptic neuron suggests that the spontaneous simultaneous firing of a small set of afferents may occasionally (particularly in the absence of inhibitory transmission) drive a postsynaptic cell to fire an action potential. These ideas are consistent with the view emerging from whole-cell patch recordings that neurons are more compact than previously thought. This may enable even small synaptically induced currents to produce a prominent deviation of membrane potential (Spruston and Johnston, 1992). For instance, it has been recently shown that in olfactory neurons even a single membrane channel opening may generate an action potential (Lynch and Barry, 1989).

Although in our experimental conditions the predominant factor contributing to input-output variability is the probabilistic nature of elicited transmitter release, other possible sources of noise could play a larger role in different brain regions or under in vivo conditions. Changes in the membrane threshold for firing could be prominent if the neurons display endogenous pacemaker activity. Spontaneous release of transmitter and spontaneous afferent activity are both likely to be greater at warmer temperatures and in intact preparations, and may thus cause fluctuations in membrane potential and input conductance that will affect the generation of action potentials (Burnod and Korn, 1989; Ropert et al., 1990; Bernander et al., 1991; Ferster and Jagadeesh, 1992).

Theoretical and computer simulation studies have been used to investigate the relation between the performance of a network and the input output characteristics of its components. Performance has been measured with respect to input pattern recognition, associative learning, and memory storage capacity and accessibility (Little and Shaw, 1975; Buhman and Schulten, 1987; Bressloff and Taylor, 1990; Gibson et al., 1991). These studies find that some level of noise enhances performance. In particular, it has been suggested that modulation of the level of noise could aid in solving certain problems. For instance, the performance of a network that seeks to find a global minimum for a surface may be facilitated if initially the components have a low S/N in their transmission, incrementally increasing the S/N over a number of trials (Hinton et al., 1984; see also Kirkpatrick et al., 1983). This may allow the network to find initially the general area of a global minimum without getting "trapped" in a local minimum. Subsequently, with lower noise, the exact location of the minimum can be located. We have investigated whether the S/N in our system could be modulated.

We found that manipulations mimicking different physiological conditions affected the S/N of transmission at threshold. A decrease in afferent frequency produced an increased S/N, suggesting a relation between the rate at which serial information is processed and the performance of the processing. We also observed a decreased threshold as well as a diminished S/N upon blocking inhibitory transmission. Inhibitory transmission in hippocampus may be decreased by activation of cholinergic system (Krnjević and Ropert, 1982; Rovira et al., 1983; Pitler, Alger, 1992) particularly active during the theta rhythm—a pattern of activity present during orientation reflex, attention (see ref. in Lopes da Silva et al., 1990) and learning (Eichenbaum et al., 1992; Mecklinger et al., 1992).

One of the primary goals of this study was to determine the effect of LTP on the S/N for a threshold stimulus. Previous studies using a minimal stimulus show that trial-to-trial variability decreases with LTP (Bekkers and Stevens, 1990; Malinow and Tsien, 1990; Malinow, 1991). In this study, we used a threshold stimulus that recruits a larger and more heterogeneous population of synapses. Despite this, in every case we still noted an increase in the S/N with LTP (decrease in CV). Does this finding make implications for higher order processes?

A learning process is often characterized by a diminishing input requirement to reach a criterion. For example, recognition of a previously encountered predator requires presentation of only a very limited subset of features (e.g., its scent). Does the fidelity of recognition change when only a subset of the inputs is used? At the level of single cells we have asked: What happens to the S/N of transmission if only a subset of inputs are recruited after LTP? Reducing the number of activated synapses after LTP so that the postsynaptic response was as large as before LTP reverted the S/N to that observed before LTP. Thus, we consistently saw a maintained S/N after potentiated transmission was reduced to the threshold level. Indeed, this was because LTP increased the S/N as much as reducing the stimulus decreased the S/N. Thus, although learning is clearly a higher order process than those studied in the hippocampal slice, we can hypothesize that biological noise is important in the processing of information and will depend on the intrinsic noise of the components used and on the number of components used in parallel. After learning, the number of parallel components required to perform a task may be less, thus increasing the level of noise. This may be offset by decreasing the intrinsic noise of the components.

We can consider the predicted effect of different types of synaptic modifications on the S/N for a threshold input. For a simple case in which synapses are homogeneous and release with Poisson statistics, a threshold response T, and its CV^2 are given by,

$$T = m^*q^*S$$

$$CV_T^2 = \frac{1}{S} * CV_S^2 = \frac{1}{S} * \frac{1}{m}$$

where m is the quantal content, q is the quantal size, and S is the number of synapses activated. From this one can calculate directly the predicted effects of presynaptic or postsynaptic modifications. Stipulating that the threshold response, T, stays constant, a presynaptic (m) or postsynaptic (q) enhancement will require a proportionate decrease in S. Thus, for a postsynaptic change, the decrease in S will produce an increased CV_T^2; for a presynaptic modification, a decrease in S will be offset by the increase in m, leaving CV_T^2 constant. For more complex nonhomogeneous populations, additional mechanisms may be used to ensure a maintained CV_T^2. For instance, a change in quantal amplitude in which small values of q are increased more than larger q will tend to decrease the quantal variance of the population, thus decreasing CV^2. This differential increase in q is in fact what we see (e.g., figure 8.10). From these results we hypothesize that the maintenance of the S/N across individual nodes in a neuronal network in the face of large plastic changes is important to the proper function of the organism. Thus, whereas postsynaptic integration of information may be important to appropriately trigger synaptic gain changes (e.g., associative requirements), the actual modification may need to be, at least in part, presynaptic, so as to preserve input-output fidelity.

EXPERIMENTAL PROCEDURES

Brain Slice Preparation and Recording

Hippocampal slices, prepared from 2 to 4-week-old rats by standard methods, were submerged and superfused as previously described (Malinow and Tsien, 1990). Synaptic transmission was elicited at 0.5 Hz (unless otherwise stated) with a bipolar electrode and recorded with a patch electrode (2.5–4.0 MΩ tip resistance, (Axopatch 1-D, Axon Instruments) at room temperature. In some experiments an additional field recording electrode was placed in stratum radiatum. The stimulating electrode was placed in stratum radiatum for recordings from CA1 region, stratum moleculare for recordings from FD. After obtaining a cell-attached patch recording (1–5 GΩ), a threshold stimulus strength was set by passing sufficient current through the stimulating electrode to elicit an action potential from the recorded cell in ~50% of trials. For whole-cell voltage-clamp recordings series resistance was monitored and holding potential was kept constant at a level between −55 and −70 mV. Miniature EPSCs were elicited in a bath containing 1 μM tetrodotoxin by local application of 500 mM sucrose through a patch pipette directed toward the s. radiatum within ~100 μm of the recording pipette. Internal solution contained in mM: Cs-gluconate, 100; EGTA, 0.2; $MgCl_2$, 5; ATP, 2; GTP, 0.3; HEPES, 40 (pH 7.2 with CsOH). Superfusate contained in mM: NaCl, 119; KCl, 2.5; $MgCl_2$, 1.3; $CaCl_2$, 2.5; $NaHCO_3$, 26.2; glucose, 11; gassed with 95% O_2/5% CO_2 (pH 7.4). For experiments in which picrotoxin (100 μM, Sigma) in superfusate was used to block inhibitory transmission, a cut between CA3 and CA1 regions prevented epileptiform activity. All drugs were made in stock solutions at 1000 times the final concentration in water.

EPSC Amplitude Distributions

Individual EPSC amplitudes were determined by averaging the current over a fixed 2–5 msec window encompassing the peak amplitude and subtracting a baseline estimate immediately before the EPSC. The same windows placed in the baseline period gave an estimate of the baseline noise. Density estimates of EPSC amplitudes in an epoch were generated by (1) computing EPSC and noise amplitudes for each sweep of an epoch; (2) computing the noise s.d. for that epoch; (3) choosing a gaussian kernel with SD $\sim 1/2$ SD of the baseline noise; (4) placing a gaussian kernel on the x-axis (amplitude axis), centered about each EPSC amplitude value; and (5) summing all gaussians. Binned histograms were generated by choosing bin widths $\sim 1/4$ estimated quantal amplitude.

Comparison Between Binned Amplitude Distribution Histograms and a Unimodal Distribution

Unimodal distributions were generated by calculating a 5-bin smoothed histogram from the parent distribution (examples shown in figure 8.3). An interval over which to compute the χ^2 of the difference between the parent and smooth histograms was chosen. The interval began 3 bins after zero amplitude and ended with the farthest bin on the right that contained at least 5 events. Intervals contained an average of 16 bins. The χ^2 of the difference was calculated as

$$\chi^2 = \sum (x_i - s_i)^2/s_i,$$

where x_i = number of events in the ith bin of the parent histogram; s_i = number of events in the ith bin of the smoothed histogram; degrees of freedom = (number of bins in the interval) -1.

One aspect of this analysis makes this test more strict than necessary. To avoid edge effects resulting from the smoothing protocol, the first peak in the distribution was not included in the χ^2 calculation. Since this is generally the most clearly discernible peak, its exclusion decreases the calculated statistical difference.

Maximum Likelihood Estimates for Components Underlying Parent Distribution

Each parent distribution was assumed to be a mixture of a small number of gaussian components with equal variance. The number of gaussian components that best described the parent distribution was determined by constructing a sequence of unconstrained MLEs from the parent distribution (excluding extreme outliers), for increasing numbers of gaussian components. The best MLE in such a sequence was chosen by finding the point at which the log likelihood ceased to change significantly (about 2.5 log units, Titterington et al., 1985).

Template-based Selection

For each epoch an average of all traces was computed (the template). A window encompassing the onset and peak of the EPSC was chosen. For each trace, the template was scaled so that (within the chosen window) the template area matched the trace area. The deviation from the ensemble average was computed as

$\sum$(EPSC − scaled template)^2/(scaled template + .5 pA) over the chosen window.

Dividing by (scaled template +.5 pA) maintained the general shape of the amplitude distribution after selection with no effect on the location of peaks and valleys. An offset (arbitrarily chosen as 0.5 pA) is required so that the error measurements for zero amplitude events (for which the scaled template is zero) are not infinite.

Quantal Amplitude Normalization

To make comparisons across different experiments (figure 8.10B), quantal amplitudes were normalized by an EPSC shape index. Since a second pulse was often delivered before a complete decay of the EPSC, normalization by total integrated current (charge) was not possible. Thus, the quantal amplitude for an epoch was normalized by a factor consisting of the average synaptic current over the first 40 msec divided by the average synaptic current for the first 3.5 msec, computed from an ensemble average of all sweeps in that epoch. From other experiments, in which the series resistance changes during the recording period, we have found this factor to normalize for changes in series resistance. There was no significant difference in this normalization factor before (3.0 ± 1.1, mean ± S.D.) and after (2.8 ± 0.9) pairing.

Selection of LTP Experiments

Of the 35 experiments that were rejected, approximately 10 did not show a 50% increase in transmission; 20 did not show clear, evenly spaced amplitude distribution peaks (several had peaks before but not after pairing); 5 had an unstable baseline (greater than 30% change in EPSC amplitude between first half and second half of baseline period).

Equations Modeling Synapse and Fiber Variability

Let the synaptic response resulting from the activity of a single fiber (S_i) and the number of fibers excited (F) be independent random variables over time that contribute to the ensemble synaptic response (R). Then

$$R = \sum_{i}^{F} S_i \tag{1}$$

For the special case where synapses are homogeneous and independent,

$$\langle R \rangle = \sum_{i}^{\langle F \rangle} S_i = \langle F \rangle^* \langle S \rangle \tag{2}$$

$$\sigma_R^2 = \langle \sigma_{R|F}^2 \rangle + \sigma_{\langle R|F \rangle}^2 = \langle F \rangle^* \sigma_S^2 + \langle S \rangle^{2*} \sigma_F^2 \tag{3}$$

$$CV_R^2 = \frac{1}{\langle F \rangle} * CV_S^2 + CV_F^2 \tag{4}$$

If release satisfies Poisson statistics, $CV_S^2 = q^*(1 + CV_q^2)/\langle S \rangle$, where q and CV_q^2 are quantal mean amplitude and CV^2, respectively (Bekkers and Stevens, 1990). Thus

$$CV_R^2 = \frac{1}{\langle F \rangle} * \frac{q^*(1 + CV_q^2)}{\langle S \rangle} + CV_F^2 = \frac{q^*(1 + CV_q^2)}{\langle R \rangle} + CV_F^2 \tag{5}$$

and plotting CV_R^2 versus $1/\langle R \rangle$ produces a line with y-intercept $= CV_F^2$. If CV_F^2 is small, eq. (4) predicts that CV_R^2 is inversely related to the average number of fibers excited, $\langle F \rangle$. Since $R \propto F$, CV_R^2 will be inversely related to R and plotting CV_R^2 versus $1/\langle R \rangle$ for experiments in which F is changed should also produce a line with slope $q^*(1 + CV_q^2)$ and y-intercept CV_F^2.

Correlating Unitary and Threshold Transmission

Given the mean value of unitary transmission, $\langle S \rangle$ and variability, CV_S one can compute the predicted variability CV_R at threshold response $\langle R \rangle$ from the simple relation

$$CV_R = CV_S * \sqrt{\frac{\langle S \rangle}{\langle R \rangle}} \tag{6}$$

Computing for several values of unitary transmission (Malinow, 1991) yields a predicted variability at threshold of 10–20%, which coincides with the measured value of 15.8%.

REFERENCES

Amit, D. J., and Treves, A. (1989) Associative memory neural network with low temporal spiking rates. *Proc. Natl. Acad. Sci. USA* 86:7871–7875.

Andersen, P. (1987) Properties of hippocampal synapses of importance for integration and memory. In *Synaptic Function*, G. M. Edelman, W. E. Gall, and W. M. Cowan (eds.). New York: Wiley, pp. 403–429.

Andersen, P., Silfvenius, H., Sundberg, S. H., and Sveen, O. (1980) A comparison of distal and proximal dendritic synapses on CA1 pyramids in guinea-pig hippocampal slices in vitro. *J. Physiol.* 307:273–299.

Andersen, P., Storm, J., and Wheal, H. V. (1987) Thresholds of action potentials evoked by synapses on the dendrites of pyramidal cells in the rat hippocampus in vitro. *J. Physiol.* 383:509–526.

Bekkers, J. M., and Stevens, C. F. (1990) Presynaptic mechanism for long-term potentiation in the hippocampus. *Nature* 346:724–729.

Bekkers, J. M., Richerson, G. B., and Stevens, C. F. (1990) Origin of variability in quantal size in cultured hippocampal neurons and hippocampal slices. *Proc. Natl. Acad. Sci. USA* 87:5359–5362.

Bernander, Ö, Douglas, R. J., Martin, K. A. C., and Koch, C. (1991) Synaptic background activity influences spatiotemporal integration of single pyramidal cells. *Proc. Natl. Acad. Sci. USA* 88:11569–11573.

Bressloff, P. C., and Taylor, J. G. (1990) Random iterative networks. *Phys. Rev.* A 41:1126–1137.

Buhmann, J., and Schulten, K. (1987) Influence of noise on the function of a "physiological" neural network. *J. Biol. Cybern.* 56:313–327.

Burnod, Y., and Korn, H. (1989) Consequences of stochastic release of neurotransmitters for network computation in the central nervous system. *Proc. Natl. Acad. Sci. USA* 86:352–356.

Clements, J. (1991) Quantal synaptic transmission? *Nature* 353:396.

Davies, C. H., Davies, S. N., and Collingridge, G. L. (1990) Paired-pulse depression of monosynaptic GABA-mediated inhibitory postsynaptic responses in rat hippocampus. *J. Physiol.* 424:513–531.

Deisz, R. A., and Prince, D. A. (1989) Frequency-dependent depression of inhibition in guinea-pig neocortex in vitro by GABA-B receptor feed-back on GABA release. *J. Physiol.* 412:513–541.

del Castillo, J., and Katz, B. (1954) Quantal components of the end-plate potential. *J. Physiol.* 124:560–573.

Edwards, F. A., Konnerth, A., and Sakmann, B. (1989) Quantal analysis of inhibitory synaptic transmission in the dentate gyrus of rat hippocampal slices: A patch-clamp study. *J. Physiol. (Lond.)* 430:213–249.

Eichenbaum, H., Otto, T., and Cohen, N. J. (1992) The hippocampus—What does it do? Behavioral and Neural Biology 57, 2–36.

Fenwick, E. M., Marty, A., and Neher, E. (1982) A patch-clamp study of bovine chromaffin cells and of their sensitivity to acetylcholine. *J. Physiol.* 331:577–597.

Ferster, D., and Jagadeesh, B. (1992) EPSP-IPSP interactions in cat visual cortex studied with in vivo whole-cell patch recording. *J. Neurosci.* 12:1262–1274.

Foster, T., and McNaughton, B. (1991) Long-term enhancement of CA1 synaptic transmission is due to increased quantal size, not quantal content. *Hippocampus* 1:79.

Gibson, W. G., Robinson, J., and Bennett, M. R. (1991) Probabilistic secretion of quanta in the central nervous system: Granule cell synaptic control of pattern separation and activity regulation. *Phil. Trans. R. Soc. Lond. B.* 332:199–220.

Gustafsson, B., Huang, Y.-Y., and Wigström, H. (1988) Phorbol ester-induced synaptic potentiation differs from long-term potentiation in the guinea pig hippocampus in vitro. *Neurosci. Lett.* 85:77–81.

Hanson, P., and Schulman, H. (1992) Neuronal Ca^{2+}/calmodulin-dependent protein kinases. *Annu. Rev. Biochem.* 61:559–601.

Harris, K., and Stevens, J. (1989) Dendritic spines of CA1 pyramidal cells in the rat hippocampus: Serial electron microscopy with reference to their biophysical characteristics. *J. Neurosci.* 9:2982.

Hinton, G. E., Sejnowski, T. J., and Ackley, D. H. (1984) Boltzmann machines: Constraint satisfaction networks that learn. Pittsburgh: Carnegie-Mellon University.

Horn, R., and Marty, A. (1988) Muscarinic activation of ionic currents measured by a new whole-cell recording method. *J. Gen. Physiol.* 92:145.

Jack, J., Redman, S., and Wong, K. (1981) The components of synaptic potentials evoked in cat spinal motoneurones by impulses in single group Ia afferents. *J. Physiol. (Lond.)* 321:65–96.

Katz, B. (1969) *The Release of Neural Transmitter Substance.* Liverpool: Liverpool University.

Kirkpatrick, S., Gelatt, C. D., and Vecchi, M. P. (1983) Optimization by simulated annealing. *Science* 220:671–680.

Korn, H., and Faber, D. S. (1987) Regulation and significance of probabilistic release mechanisms at central synapses. In *Synaptic Function,* G. M. Edelman, W. E. Gall, and W. M. Cowan (eds.) New York: Wiley, pp. 57–107.

Krnjević, K., and Ropert, N. (1982) Electrophysiological and pharmacological characteristics of facilitation of hippocampal spikes by stimulation of the medial septum. *Neuroscience* 7:2165–2183.

Kullman, D. M., and Nicoll, R. A. (1992) Long-term potentiation is associated with increases in quantal content and quantal amplitude. *Nature* 357:240–244.

Kuno, M. (1971) Quantum aspects of central and ganglionic synaptic transmission in vertebrates. *Physiol. Rev.* 51:647–678.

Langmoen, J. A., and Andersen, P. (1983) Summation of excitatory postsynaptic potentials in hippocampal pyramidal cells. *J. Neurophysiol.* 50:1320–1329.

Larkman, A., Stratford, K., and Jack, J. (1991) Quantal analysis of excitatory synaptic action and depression in hippocampal slices. *Nature* 350:344–347.

Liao, D., Jones, A., and Malinow, R. (1992) Direct measurement of quantal changes underlying long-term potentiation in CA1 hippocampus. *Neuron* 9:1089–1097.

Little, W. A., and Shaw, G. L. (1975) A statistical theory of short and long term memory. *Behav. Biol.* 14:115–133.

Lopes da Silva, F. H., Witter, M. P., Boeijinga, P. H., and Lohman, A. H. M. (1990) Anatomic organization and physiology of the limbic cortex. Phys. Rev. 70, 453–509.

Lynch, J. W., and Barry, P. H. (1989) Action potentials initiated by single channels opening in a small neuron (rat olfactory receptor). *Biophys. J.* 55:755–768.

Malinow, R. (1991) Transmission between pairs of hippocampal slice neurons: Quantal levels, oscillations, and LTP. *Science* 252:722–724.

Malinow, R., and Tsien, R. W. (1990) Presynaptic enhancement shown by whole-cell recordings of long-term potentiation in hippocampal slices. *Nature* 346:177–180.

Malgaroli, A., and Tsien, R. W. (1992) Glutamate-induced long-term potentiation of the frequency of miniature synaptic currents in cultured hippocampal neurons. *Nature* 357: 134–139.

Manabe, T., Renner, P., and Nicoll, R. A. (1992) Postsynaptic contribution to long-term potentiation revealed by the analysis of miniature synaptic currents. *Nature* 355:50–55.

McBain, C., and Dingledine, R. (1992) Dual-component miniature excitatory synaptic currents in rat hippocampal CA3 pyramidal neurons. *J. Neurophysiol.* 68:16–27.

McCarren, M., and Alger, B. E. (1985) Use-dependent depression of IPSPs in rat hippocampal pyramidal cells in vitro. *J. Neurophysiol.* 53:557–571.

McNaughton, B. L., Barnes, C. A., and Andersen, P. (1981) Synaptic efficacy and EPSP summation in granule cells of rat fascia dentata studied in vitro. *J. Neurophysiol.* 46:952–966.

Mecklinger, A., Kramer, A. F., and Strayer, D. L. (1992) Event related potentials and EEG components in a semantic memory search. *Psychophysiology* 29:104–119.

Miles, R., and Wong, R. K. S. (1986) Excitatory synaptic interactions between CA3 neurones in the guinea-pig hippocampus. *J. Physiol.* 373:397–418.

Nicoll, R. A., Malenka, R. C., and Kauer, J. A. (1990) Functional comparison of neurotransmitter receptor subtypes in mammalian central nervous system. *Physiol. Rev.* 70:513–565.

Pitler, T. A., and Alger, B. E. (1992) Cholinergic excitation of GABAergic interneurons in the rat hippocampal slice. J. Physiol. 450, 127–142.

Raastad, M., Storm, J. F., and Andersen, P. (1992) Putative single quantum and single fibre excitatory postsynaptic currents show similar amplitude range and variability in rat hippocampal slices. *Eur. J. Neurosci.* 4:113–117.

Redman, S. (1990) Quantal analysis of synaptic potentials in neurons of the central nervous system. *Physiol. Rev.* 70:165–198

Ropert, N., Miles, R., and Korn, H. (1990) Characteristics of miniature inhibitory postsynaptic currents in CA1 pyramidal neurones of rat hippocampus. *J. Physiol.* 428:707–722.

Rovira, O., Ben-Ari, Y., Cherubini, E. (1983) Dual cholinergic modulation of hippocampal somatic and dendritic field potentials by the septo-hippocampal pathway. *Exp. Brain Res.* 49:151.

Scholz, K. P., and Miller, R. J. (1992) Inhibition of quantal transmitter release in the absence of calcium influx by a G protein-linked adenosine receptor at hippocampal synapses. *Neuron* 8:1139–1150.

Shaw, G. L., and Vasudevan, R. (1974) Persistent states of neural networks and the random nature of synaptic transmission. *Math. Biosci.* 21:207–218.

Siggins, G. R., and Schubert, P. (1981) Adenosine depression of hippocampal neurons in vitro: An intracellular study of dose-dependent actions on synaptic and membrane potentials. *Neurosci. Lett.* 23:55–60.

Silverman, V. (1986) *Density Estimate for Statistics and Data Analysis.* New York: Chapman and Hall.

Spruston, N., and Johnston, D. (1992) Perforated patch-clamp analysis of the passive membrane properties of three classes of hippocampal neurons. *J. Neurophysiol.* 67:508–529.

Staley, K. J., Otis, T. S., and Mody, I. (1992) Membrane properties of dentate gyrus granule cells: Comparison of sharp microelectrode and whole-cell recordings. *J. Neurophysiol.* 67:1346–1358.

Taylor, J. G. (1972) Spontaneous behavior in neural networks. *J. Theor. Biol.* 36:513–528.

Titterington, D., Smith, A., and Makov, V. (1985) *Statistical Analysis of Finite Mixture Distributions.* New York: Wiley.

Trussell, L. O., and Jackson, M. B. (1985) Adenosine-activated potassium conductance in cultured striatal neurons. *Proc. Natl. Acad. Sci. USA* 82:4857–4861.

Walmsley, B., Edwards, F. R., and Tracey, D. (1987) The probabilistic nature of synaptic transmission at a mammalian excitatory central synapses. *J. Neurosci.* 7:1037.

Zucker, R. S. (1989) Short-term plasticity. *Annu. Rev. Neurosci.* 12:13–31.

8 Multiple Forms of NMDA Receptor–dependent Synaptic Plasticity in the Hippocampus

Robert C. Malenka

Over the last decade, there has been an explosion of research on the cellular mechanisms responsible for long-term potentiation (LTP). The most active areas of experimentation have focused on one of two questions: (1) is the site of LTP expression presynaptic and/or postsynaptic? and (2) what are the biochemical mechanisms responsible for the induction and maintenance of LTP? Although conceptually simple, these questions have generated enormous quantities of data that, to the novice and expert alike, are frequently contradictory and often confusing.

Why has it been so difficult to reach any sort of consensus with regard to such seemingly simple but important issues (see Baudry and Davis, 1991)? There are innumerable answers to this question. It is now well accepted that LTP is not a unitary phenomenon but encompasses different forms of long-lasting changes in synaptic efficacy with profoundly or subtly different underlying cellular mechanisms. However, this cannot explain the many discrepancies in experimental results obtained from the same synaptic preparation. One aspect of LTP research which can be accepted as a confounding variable is the often insurmountable technical problem of using electrophysiological or biochemical techniques to study processes that occur in cellular compartments impossible to access directly. Considering that most mechanistic experiments are aimed at elucidating the events occurring in individual dendritic spines or presynaptic terminals located in the dense neuropil of a mammalian CNS structure, it can be considered quite remarkable that anything is known about some of the mechanisms responsible for LTP.

As the contents of this book indicate, much of the work on the cellular mechanisms underlying LTP has been done on the NMDA receptor–dependent form of LTP observed at the synapse between Schaffer collateral/commissural fibers and hippocampal CA1 pyramidal cells. In this chapter, evidence will be presented that the complexity of the phenomenon of LTP itself, even at this extensively studied synapse, has been underestimated and that even at single synaptic connections, there exists the capacity not for a single plastic event (i.e., LTP) but rather a repertoire of NMDA receptor–dependent forms of synaptic plasticity. This more extensive "Hebbian" synaptic phenomenology contributes additional complexity to the study of LTP but nevertheless may provide some additional clues to the reasons for some of the inconsistency and confusion in the literature. Of equal or greater importance, these additional NMDA receptor–dependent synaptic mechanisms have been predicted by some neural network theories and provide increased flexibility to neural circuits involved in information processing and storage.

FACTORS CONTROLLING THE STABILIZATION OF LTP: SHORT-TERM VERSUS LONG-TERM POTENTIATION

Over the last decade, several investigators have noted that certain patterns of physiological stimulation elicit a synaptic enhancement that lasts longer than presynaptic facilitatory phenomena such as post-tetanic potentiation (Zucker, 1989), yet is transient and decays back to control levels over a few to tens of minutes (McNaughton, 1982; Racine and Milgram, 1983; Larson et al., 1986; Sastry et al., 1986; Anwyl et al., 1989). Interest in this decremental short-term potentiation (STP) and its relationship to LTP was increased by two pharmacological observations. First, contrary to theoretical predictions iontophoretic or pressure application of NMDA onto CA1 cells caused STP but not LTP (Collingridge et al., 1983; Kauer et al., 1988; Asztely et al., 1991). Second, application of LTP-inducing stimuli in the presence of inhibitors of protein kinase activity causes a decaying synaptic enhancement like STP (for examples see Malenka et al., 1989; Malinow et al., 1988, 1989). In fact, application of inhibitors of any number of biochemical processes (for reviews see Madison et al., 1991; Baudry and Davis, 1991) also result in an inhibition of LTP and the generation of STP following tetanic stimulation. Such results have often been used to conclude that LTP can be "induced" but not "maintained" and that the inhibited biochemical process must be required for the maintenance of LTP.

One way to clarify the relationship between STP and LTP is to examine more closely the physiological factors responsible for controlling the duration of synaptic enhancement and the variables critical for converting STP to LTP. Like LTP, STP is dependent on NMDA receptor activation (Larson and Lynch, 1988; Anwyl et al., 1989; Malenka, 1991) and is blocked by loading CA1 cells with Ca^{2+} chelators, suggesting a requirement for changes in postsynaptic Ca^{2+} levels (Lynch et al., 1983; Malenka et al., 1988, 1992). Figure 8.1 shows that the degree of NMDA receptor activa-

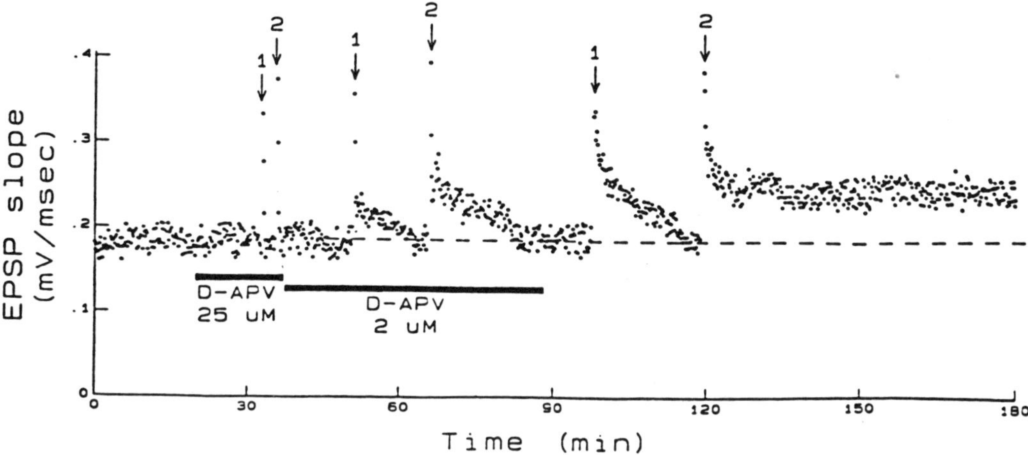

Figure 8.1 The degree of NMDA receptor activation governs the duration of synaptic enhancement elicited by high-frequency stimulation. A graph of the initial slope of a field EPSP recorded in CA1 stratum radiatum vs. time taken from an experiment in which the same single tetanus (1 or 2) was applied during varying degrees of NMDA receptor blockade using D-APV. Only post-tetanic potentiation (PTP) was elicited in the presence of 25 μM D-APV; in the presence of 2 μM D-APV the same tetani caused short-term potentiation (STP); when D-APV was washed out of the slice, the longer single tetanus (2) generated stable LTP. (Reprinted with permission from Malenka, 1991)

tion is important for determining whether STP or LTP is generated. In the presence of 25 μM D-APV, a tetanus (50 Hz for 0.5 or 1 sec) causes post-tetanic potentiation lasting only 30–60 sec. When NMDA receptors are only weakly antagonized with 2 μM D-APV, these same stimuli generate STP. After complete wash-out of the D-APV, the longer of the two tetani induced LTP. Essentially identical results have been found for LTP in the dentate gyrus (Hanse and Gustafsson, 1992).

How does the degree of NMDA receptor activation control whether STP or LTP is generated? The most obvious possibility is that the magnitude of the increase in postsynaptic Ca^{2+} during the induction protocol is critical for the stabilization of LTP (i.e., whether STP or LTP is generated). To test this hypothesis, experimental manipulations that should control the magnitude of Ca^{2+} influx through the NMDA ionophore were performed. One manipulation took advantage of the long time course of the NMDA receptor–mediated portion of the EPSC (Hestrin et al., 1990) and the ability to induce LTP by pairing synaptic stimulation with postsynaptic depolarization (Gustafsson et al., 1987). Applying a depolarizing current pulse to the postsynaptic cell with some delay following synaptic stimulation resulted in STP, whereas in the same cell, depolarizing the cell prior to

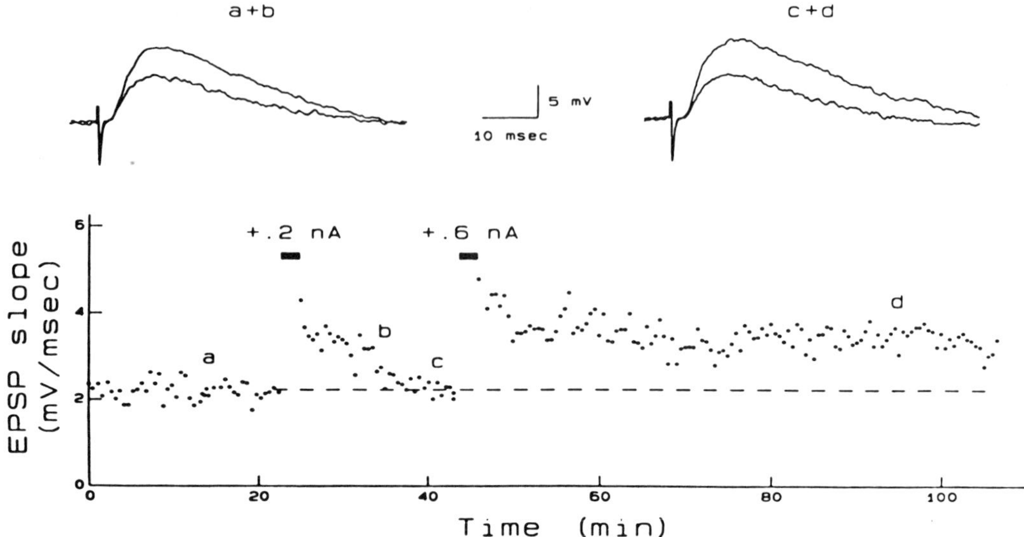

Figure 8.2 The level of postsynaptic depolarization during synaptic activation affects the duration of synaptic potentiation. Graph of EPSP slope vs. time from an experiment in which, during 15 afferent stimuli, the cell was depolarized, first with +0.2 nA, which resulted in STP; then with +0.6 nA, which caused stable LTP. Each point represents the average of three successive EPSP slope measurements. Sample raw data traces were obtained at the times indicated by the letters. The cell was held at −80 mV. (Reprinted with permission from Malenka, 1991)

the peak NMDA conductance to increase Ca^{2+} entry resulted in LTP (Malenka, 1991). A second manipulation took advantage of the well-documented voltage-dependence of NMDA responses (Mayer et al., 1987). Figure 8.2 shows that pairing synaptic stimulation with modest depolarization caused STP, whereas pairing the same number of stimuli with a larger depolarization caused LTP.

Recently, using the photolabile Ca^{2+} buffer diazo-4 it has been possible to influence more directly the level and duration of postsynaptic calcium rises generated by a LTP-inducing tetanus (Malenka et al., 1992). The affinity of diazo-4 for Ca^{2+} increases approximately 1600-fold (from 89 μM to 55 nM) following flash photolysis (Adams et al., 1989), permitting almost instantaneous buffering of intracellular Ca^{2+} levels. Initial experiments demonstrated that photolysis of diazo-4 blocked the generation of both LTP and STP. By adjusting the timing between the LTP-inducing tetanus and flash photolysis of diazo-4 it was determined that a tetanus-induced rise in postsynaptic Ca^{2+} lasting at most $2-2\frac{1}{2}$ sec was sufficient to generate LTP. By using a lower concentration of diazo-4 in the recording electrode and a lower intensity light flash, it was possible to dampen the rise in postsynaptic Ca^{2+} induced by the tetanus rather than

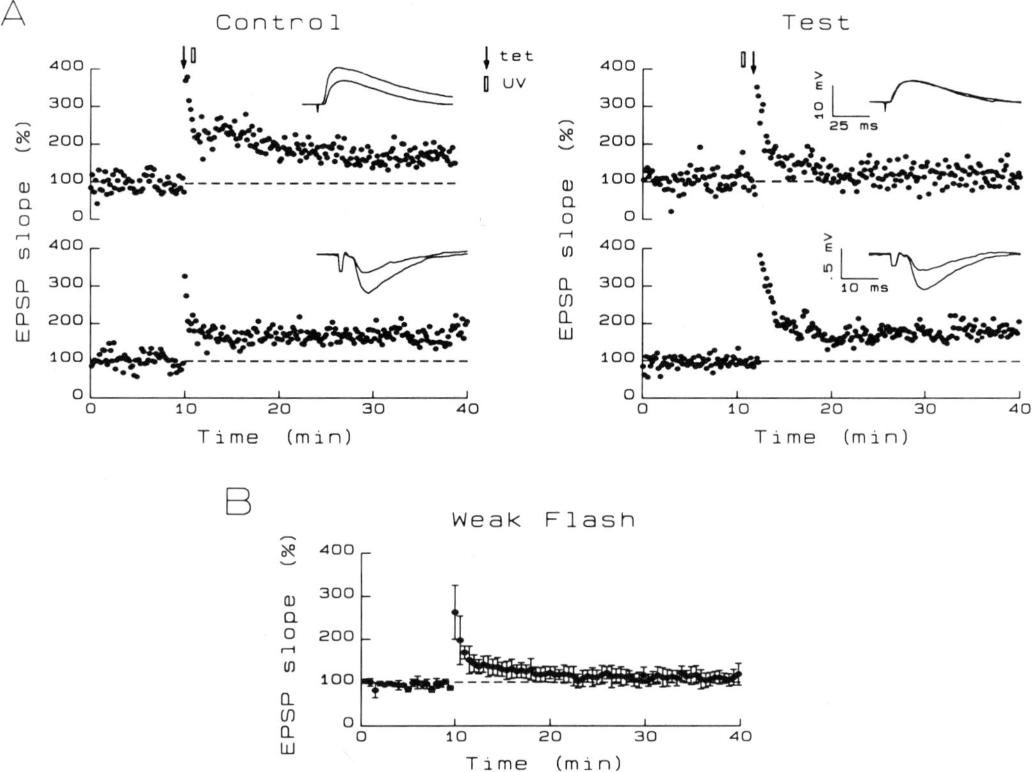

Figure 8.3 Partial diazo-4 photolysis results in the generation of STP. (*A*) Experiment in which the whole-cell recording electrode was filled with 1.0 mM diazo-4 (rather than the normal 2.5 mM) and a 30 J flash (rather than the normal 100 J flash) was used for photolysis. LTP was first induced in the control pathway and following partial diazo-4 photolysis, the test tetanus induced STP in the cell and LTP in the field EPSP. (*B*) Summary of experiments (n = 7) in which diazo-4 (1.0 mM) was photolyzed with a 30 J flash prior to application of the test tetanus. (Reprinted with permission from Malenka et al., 1992)

limiting its duration. Consistent with the previous results (Malenka, 1991), this manipulation caused the generation of STP (figure 8.3).

Together these experiments strongly suggest that the magnitude of the rise in postsynaptic calcium during a LTP induction protocol helps determine whether STP or LTP is generated (Malenka, 1991). Specifically, it can be proposed that stable LTP requires a critical "threshold" level of postsynaptic Ca^{2+} and that if smaller increases in Ca^{2+} are generated, STP (but not LTP) results. Two reports from other laboratories are consistent with this hypothesis. Hanse and Gustafsson (1992) performed very similar experments to those of Malenka (1991) in the dentate gyrus and concluded that "the stability of LTP in the dentate gyrus is critically related to the amount of postsynaptic calcium influx during the tetanus." In addition, Aniksztejn and colleagues (1992) found that *trans*-ACPD, a metabotropic glutamate receptor agonist, enhanced NMDA currents in CA1 cells and when applied with a weak tetanus incapable of inducing LTP by itself, generated LTP. In agreement with the aforementioned studies, they also concluded that the level of NMDA receptor activation during a tetanus is important for the stabilization of LTP.

One important implication of these results is that caution must be used when interpreting any experiment in which a manipulation was performed prior to or during the LTP induction protocol. For example, a pharmacological or biochemical manipulation that prevents the induction of stable LTP and results in STP does not necessarily indicate that the inhibited biochemical process was required for LTP maintenance or stabilization. Any direct or indirect effect on NMDA receptor function or the magnitude of the postsynaptic Ca^{2+} change would have significant effects on the ability to generate LTP. This point is particularly important given that NMDA receptor function appears to be strongly modulated by kinase activity (Gerber et al., 1989; MacDonald et al., 1989; Chen and Huang, 1991, 1992; Aniksztejn et al., 1992; Kelso et al., 1992; Yamazaki et al., 1992).

A critical issue is the relationship between STP and LTP in terms of underlying biochemical mechanisms. Three scenarios can be envisioned. STP may simply reflect a "weak induction" of the biochemical processes responsible for LTP, suggesting that the stabilization of LTP requires activation of a "threshold" concentration of some biochemical process(es). Alternatively, the biochemical processes responsible for STP may be a necessary step toward the stabilization of LTP but may not be sufficient alone to generate stable LTP. This would suggest that higher levels of Ca^{2+} in the postsynaptic cell activate processes not activated by the Ca^{2+} increase responsible for STP. Finally, STP may use mechanisms not required for LTP, indicating that it is a completely independent phenomenon which can be activated by the same physiological stimuli that induce LTP.

Experiments attempting to address these possibilities present a confusing picture. As mentioned above, application to slice preparations of inhibitors of any number of enzymes including protein kinases prevent the stabilization of LTP (resulting in STP) and when tested, do not appear to block STP induced by NMDA application (Kauer et al., 1988; McGuinness et al., 1991). One interpretation of these results is that STP does not require any of these biochemical processes. However, since for all these experiments the degree of enzyme inhibition within the slice is not known, these results cannot rule out the possibility that partial activation of the "inhibited" enzyme is sufficient to generate STP. One solution to this problem may be the use of gene targeting or "knock-outs" in transgenic animals (Silva et al., 1992), although even with this state-of-the-art technology one must be aware of potential complications due to compensatory mechanisms as the nervous system matures.

One physiological approach to the question of the relationship between STP and LTP involved examining whether saturating LTP "occludes" the generation of STP. In one study (Kauer et al., 1988), STP induced by NMDA iontophoresis or by a tetanus was blocked soon after LTP induction but could be generated 50–100 min later. This result suggested that the processes responsible for STP are not activated for the duration of LTP but decay after LTP induction and can be reactivated in synapses still exhibiting LTP. Furthermore, since STP was induced (i.e., was not "occluded") at a time when LTP had not decayed significantly, there is likely some important difference between the expression mechanisms responsible for STP and LTP. In support of this difference, it has been found that a transient synaptic enhancement like STP may be generated in CA1 cells by activating voltage-dependent Ca^{2+} channels with long depolarizing current pulses in the presence of D-APV and that this STP is not affected by prior induction of LTP (Kullmann et al., 1992). However, studies from another laboratory (Asztely et al., 1991; Gustafsson et al., 1989) yield a different conclusion, since it was found that STP induced by application of NMDA or by a short tetanus was occluded for more than 1–2 hr following saturation of LTP. This result would suggest STP and LTP do share some common underlying mechanisms. There appears to be no easy resolution to this contradiction in experimental results unless there are different forms of STP which are generated by distinct processes.

This brief discussion of the biochemical and physiological relationship between STP and LTP can be viewed as an alternate way of addressing the question of what is important for the stabilization or maintenance of LTP. Physiological manipulations suggest that whatever the processes necessary for LTP stabilization are, they decay significantly within 1–2 min following LTP induction (Arai et al., 1990; Colino et al., 1992). Biochemical approaches to this question have become increasingly popular and sophisticated (see Baudry and Davis, 1991) and offer perhaps the best hope for definitive answers, yet there is always a problem of demonstrating causation rather than correlation when assaying for differences between control and "potentiated" tissue. Nevertheless, by combining electrophysiological, biochemical, and molecular techniques with appropriate caution and rigor as many investigators in the field are presently doing, eventually it should be possible to unravel the complicated series of events that are responsible for converting STP to LTP and thereby the stabilization of LTP.

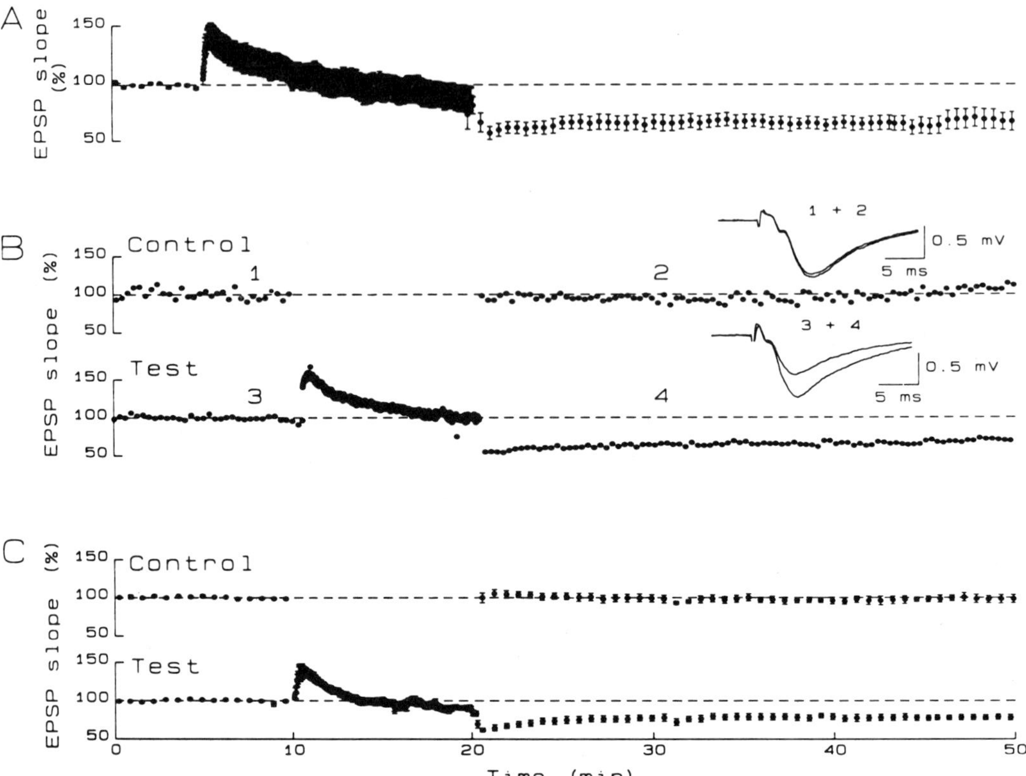

Figure 8.4 Prolonged 1 Hz stimulation produces synapse-specific LTD. (*A*) Summary of seven experiments in which 1 Hz stimulation was applied for 15 min, resulting in a decrease in EPSP initial slope of 36 ± 7%. (*B*) Experiment in which two independent fiber bundles synapsing on the same population of postsynaptic cells were alternately stimulated but only the test pathway received stimulation at 1 Hz for 10 min. The test EPSP was depressed by 37% and the control EPSP showed no change. *Insets*: superimposed data traces taken at the times indicated by the numbers above each graph. (*C*) Summary of five experiments like that shown in *B*. Synaptic strength was depressed by 21 ± 3% in the test pathway but was unaffected in the control pathway. (Reprinted with permission from Mulkey and Malenka, 1992)

HOMOSYNAPTIC LONG-TERM DEPRESSION

While most investigators interested in the mechanisms responsible for synaptic plasticity have concentrated on LTP, much less experimental work has been performed on the mechanisms underlying long-lasting decreases in synaptic efficacy or long-term depression (LTD). This is somewhat surprising given that the usefulness of LTP as a synaptic mechanism for encoding information may be quite limited if mechanisms for decreasing synaptic efficacy did not also exist. Various forms of LTD have been described in several brain regions including the cerebellum, cerebral cortex, dentate gyrus, and CA1 region of the hippocampus. However, only for the cerebellar form are the underlying mechanisms beginning to be reasonably delineated (Ito, 1989; Linden et al., 1991; Konnerth et al., 1992).

In hippocampal CA1 pyramidal cells, several different types of LTD have been described, each generated by different experimental protocols (Dunwiddie and Lynch, 1978; Stanton and Sejnowski, 1989; Pockett et al., 1990; Abraham and Wickens, 1991; Dudek and Bear, 1992). One form of LTD can be generated by repetitive low-frequency afferent stimulation (Dudek and Bear, 1992; Mulkey and Malenka, 1992) and will be the focus of the remainder of this discussion. Figure 8.4 shows that the depression of synaptic transmission caused by low-frequency stimulation occurs only at the activated synapses, indicating that this protocol induces a synapse-specific or homosynaptic form of LTP. This finding also makes it difficult to attribute this form of LTD to some sort of cell injury or cell death. Although prolonged (5–15 min) stimulation is most effective for inducing LTD, a physiological relevance to the phenomenon is suggested by the finding that repetitive bursts of shorter periods of stimulation (10–30 sec) can also be quite effective (Mulkey and Malenka, 1992).

What are the initial steps responsible for this form of LTD? D-APV was found to completely and reversibly block LTD, forcing the somewhat surprising conclusion that like STP and LTP, LTD also requires activation of NMDA receptors (Dudek and Bear, 1992; Mulkey and Malenka, 1992). A further similarity between LTD and LTP or STP is that LTD can be completely blocked by buffering changes in postsynaptic Ca^{2+} with the Ca^{2+} chelator BAPTA or by strongly hyperpolarizing CA1 cells during the induction protocol (Mulkey and Malenka, 1992). The change in postsynaptic Ca^{2+} level that seems to be required for LTD is likely to be due to influx through the NMDA channel since nifedipine, an L-type voltage-dependent Ca^{2+} channel antagonist, has no effect on LTD at concentrations that have been found to block the synaptic enhancement induced by repetitive activation of postsynaptic Ca^{2+} channels (Grover and Teyler, 1990; Huang and Malenka, 1992).

The properties of this form of LTD appear to distinguish it from other heterosynaptic or NMDA receptor-independent forms that have been reported to occur in CA1 cells (Dunwiddie and Lynch, 1978; Stanton and Sejnowski, 1989; Pockett et al., 1990; Wickens and Abraham, 1991). However, they are consistent with the proposal that LTD induction depends on a rise in postsynaptic Ca^{2+} that is less than that required for LTP (Lisman, 1989; Brocher et al., 1992; Hirsch and Crepel, 1992). Figure 8.5 shows an experiment that tested this hypothesis. A tetanus (20 Hz, 30 sec) was applied first in the presence of normal extracellular Ca^{2+} (2.5 mM) and caused LTP. The driving force for Ca^{2+} entry was then decreased by lowering the concentration of extracellular Ca^{2+} (to 0.5 mM) and the same tetanus was applied resulting in LTD. Lowering extracellular Ca^{2+} should not have an effect on the degree of activation of postsynaptic receptors at individual synapses since vesicular release of transmitter appears to be all-or-none (Korn and Faber, 1991; Perkel and Nicoll, 1992; Raastad et al., 1992).

While entry of Ca^{2+} via the NMDA receptor ionophore is required for homosynaptic LTD in CA1 cells, other forms of LTD also require Ca^{2+} changes due either to voltage-dependent Ca^{2+} channels or to release from intracellular stores (Wickens and Abraham, 1991; Brocher et al., 1992; Hirsch and Crepel, 1992). A critical issue is whether these different forms of LTD actually are due to the same or distinct biochemical mechanisms. That is, does Ca^{2+} accumulation due to these different sources activate the same or different biochemical processes? Examination of the synaptic enhancement caused by activation of voltage-dependent Ca^{2+} channels (Grover and Teyler, 1990; Aniksztejn and Ben-Ari, 1991) revealed that it may involve biochemical mechanisms that are at least partially the same as those activated during NMDA receptor–dependent LTP (Charriaut-Marlangue et al., 1991; Huang and Malenka, 1993; Powell et al., 1992; but see Kullman et al., 1992). This suggests that postsynaptic Ca^{2+} changes due to voltage-dependent Ca^{2+} channels can activate at least some of the same processes as those activated by Ca^{2+} entering via the NMDA receptor channel. Much more experimental work will be necessary to determine whether this explains why quite different experimental protocols can evoke LTD.

How can NMDA receptor–dependent changes in postsynaptic Ca^{2+} levels mediate LTD, STP as well as LTP? Evidence has already been presented that the magnitude of the Ca^{2+} change in the postsynaptic cell may be a critical variable controlling which form of synaptic modification is generated. The feasibility of such a mechanism is apparent given that Ca^{2+} and calmodulin-dependent enzymes have affinities for Ca^{2+}/calmodulin that can vary by two to three orders of magnitude (Cohen and Klee, 1988).

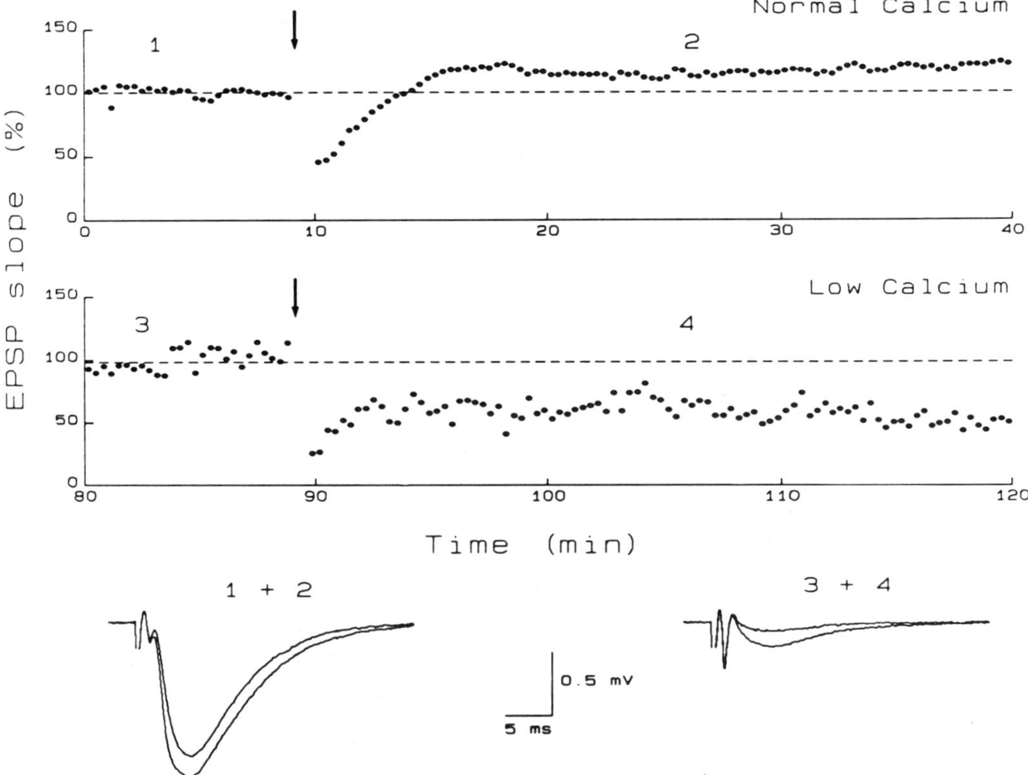

Figure 8.5 Lowering extracellular Ca^{2+} results in the induction of LTD by a tetanus that previously induced LTP. A 20 Hz, 30 sec tetanus (arrow, top graph) in normal extracellular Ca^{2+} (2.5 mM) induced LTP. Extracellular Ca^{2+} was then lowered to 0.5 mM and the same tetanus (arrow, bottom graph) induced LTD. Values in the bottom graph were normalized according to the baseline EPSP slope after it had stabilized following complete exchange of the solutions. Sample traces were taken at the times indicated by the numbers above the graphs. (Reprinted with permission from Mulkey and Malenka, 1992)

Taking advantage of these differences, one theoretical model has demonstrated how the quantitative level of rise in postsynaptic Ca^{2+} may be encoded in the relative activities of Ca^{2+}/calmodulin–dependent protein kinase II (CaM kinase II) and phosphatase I (Lisman, 1989).

However, LTD was never observed when using diazo-4 (Malenka et al., 1992) or weak tetanic stimulation (Malenka, 1991), manipulations that would be expected occasionally to result in a change in Ca^{2+} of the correct magnitude to elicit LTD. In addition, it has been difficult to elicit homosynaptic LTD in the hippocampus without using some prolonged (>10 sec) patterned afferent stimulation. Taken together, these results suggest that temporal features of the Ca^{2+} signal may also be quite important. Consistent with this hypothesis, recent work on the biochemical properties of CaM kinase II (Meyer et al., 1992; Schulman et al., 1992) suggest that the temporal pattern or frequency of Ca^{2+} transients may significantly influence the duration and degree of activity of Ca^{2+}-dependent processes. A final possibility is that a change in postsynaptic Ca^{2+} level is a necessary but not sufficient trigger for these different forms of synaptic plasticity and that each form also requires some additional distinct factors. This hypothesis has some experimental evidence to support it (Kauer et al., 1988; Kullmann et al., 1992), although the identification of any such putative factor has remained elusive (Malenka, 1991).

ADJUSTING THE THRESHOLD FOR LTP INDUCTION

A common assumption underlying virtually all studies of LTP has been that the threshold for LTP induction is relatively fixed and does not vary from preparation to preparation or during the course of an experiment. However, at least one sort of neural network model (Bienenstock et al, 1982; Bear et al., 1987) absolutely depends on the ability to adjust continually the threshold for synaptic modifications according to the history of postsynaptic cell activity. Such a mechanism may be particularly important during development, when synaptic connections are actively being made and pruned. These theoretical predictions led to a test of whether the history of synaptic activity could influence the threshold for the induction of LTP in CA1 cells (Huang et al., 1992). To complete this study, it was necessary to design experiments in which synapses were repetitively activated without inducing LTP. Figure 8.6A shows the results of one such experiment in which repetitive application to a test pathway of a weak tetanus, capable of eliciting STP but not LTP, inhibited the generation of LTP by a subsequent stronger tetanus which was capable of eliciting LTP in a control pathway in the same slice. A summary of these experiments indicates that the inhibition of LTP caused by the prior synaptic activity was synapse specific; LTP was inhibited in the test, but not the control pathway.

Further experimental work demonstrated several interesting features of this inhibition of LTP by prior synaptic activity. First, the inhibition of LTP was not permanent. A tetanus that was incapable of inducing LTP 10 min after application of repetitive weak tetani could induce LTP when given 50–80 min later (figure 8.6D). Second, NMDA receptor activation was required during the repetitive synaptic activity to inhibit LTP. Thus, applying D-APV during the inhibitory protocol prevented the subsequent inhibition of LTP, and iontophoretic application of NMDA could mimic the effects of the repetitive synaptic activity. Finally, the inhibition of LTP was not absolute but appeared to reflect an increase in the threshold amount of synaptic activation required to elicit LTP. Consistent with this

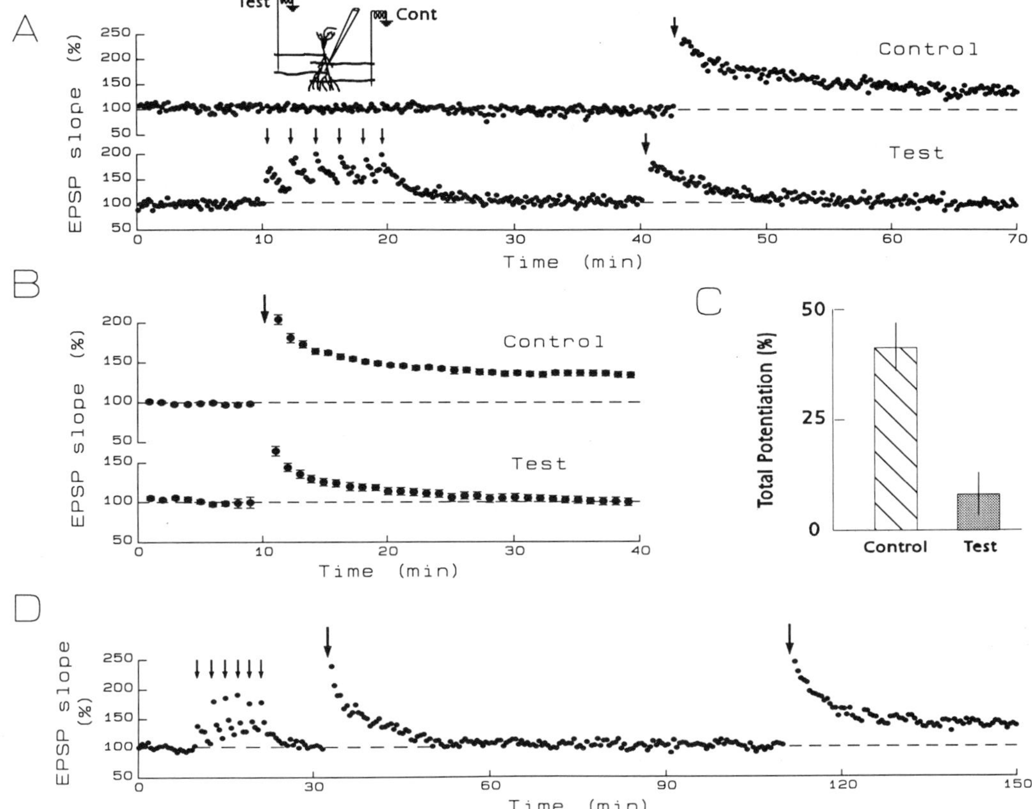

Figure 8.6 Prior synaptic activity inhibits the induction of LTP. (*A*) Plot of a typical experiment in which field EPSPs were recorded from a single site (*inset*) in response to two independent inputs (control and test). The small arrows indicate the times at which a weak tetanus (30 Hz, 0.15 sec) was applied to the test pathway. The large arrows indicate the times at which a stronger tetanus (100 Hz, 0.5 sec) was applied. (*B*) Summary of experiments in which either weak tetani (n = 8) or single strong shocks (n = 7) were given to the test pathway prior to induction of LTP (arrow indicates time at which 100 Hz, 0.5 sec tetanus was given). (*C*) The total potentiation evoked in control and test pathways from same experiments illustrated in *B*, demonstrating that saturation of LTP mechanisms cannot account for the differences between the two pathways (control, 41 ± 5%; test, 7 ± 4%; $P < .01$; n = 15). The total potentiation was calculated by averaging all EPSP slope values 25–30 min following LTP induction and comparing this value with the average of all EPSP slope values during the initial 10 min baseline before any manipulations were performed. (*D*) Example of an experiment demonstrating that the inhibition of LTP induction by prior synaptic activity is transient. Small arrows indicate time of weak tetanus (30 Hz, 0.15 sec); large arrows indicate time of strong tetanus (100 Hz, 0.5 sec). (Reprinted with permission from Huang et al., 1992)

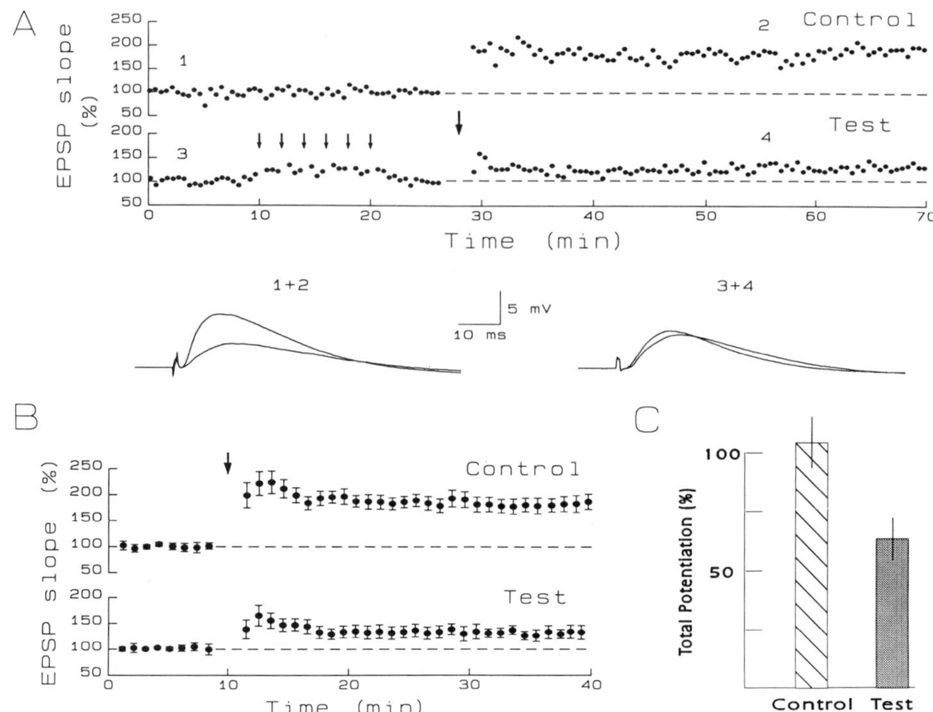

Figure 8.7 Inhibition of LTP induction in a single cell by pairing depolarization with synaptic stimulation prior to LTP induction. (*A*) Plot of a typical experiment in which depolarizing current pulses were paired with afferent stimulation (small arrows) in the test pathway so as to activate NMDA receptors and cause modest STP in this set of synapses. At the large arrow, LTP was induced by depolarizing the cell to approximately −10 mV and applying 25 stimuli to both the control and test paths. When compared to the control synapses, LTP was significantly reduced in the test synapses which had been activated previously. Sample EPSPs (average of four successive sweeps) shown below the graph were taken at the times indicated by the numbers above the graphs. (*B*) Summary of this sort of experiment (n = 13) showing that LTP was reduced, but not completely blocked, by prior synaptic activity. (*C*) The total potentiation was also decreased significantly in the test synapses when compared to control synapses on the same cells. (Reprinted with permission from Huang et al., 1992)

proposal, the inhibition of LTP induced by repetitive weak tetani could be overcome if the LTP-inducing tetanus was given repetitively with increased stimulation strength. Figure 8.7 shows a different experiment demonstrating this point in a single cell. In the test pathway, depolarizing current pulses were paired with afferent stimulation prior to LTP induction. This caused no long-lasting synaptic enhancement but did decrease the magnitude of LTP in that pathway when compared to a control pathway to the same cell (figure 8.7B).

These results provide physiological relevance to the observation that chronic activation of NMDA receptors, either by application of low concentrations of NMDA itself (Izumi et al., 1992) or by synaptic stimulation in Mg-free medium (Coan et al., 1989), can inhibit LTP. More important, they suggest that the threshold for eliciting LTP, even in a single cell, may not be constant from synapse to synapse but can vary significantly according to the history of prior synaptic activity. One prediction is that very active inputs would have a higher LTP threshold than quiescent inputs onto the same cell. The synapse specificity of this activity-dependent adjustment of the LTP threshold distinguishes this experimental phenomenon from the theoretical mechanism (Bienenstock et al., 1982; Bear et al., 1987) by which the threshold for synaptic modification is adjusted simultaneously at all inputs onto a single cell. The consequences of this distinction for the development of neural circuits and the storage of information by neural networks are likely to be quite important and will need to be determined.

Based on the work demonstrating the importance of the degree of NMDA receptor activation and consequent postsynaptic Ca^{2+} rise in controlling the generation of STP versus LTP, an obvious mechanism to account for the adjustment of the threshold for LTP induction is some modification of NMDA receptor function. Manipulations that acutely enhance NMDA currents would decrease the LTP threshold (Aniksztejn et al., 1992; Ben-Ari et al., 1992), while any downregulation of NMDA receptors would increase this threshold. Biochemical processes that could contribute to the modulation of NMDA receptor function include protein kinases (MacDonald et al., 1989; Chen and Huang, 1991; Kelso et al., 1992) and presumably phosphatases, cytoskeletal elements such as actin filaments (Rosenmund and Westbrook, 1992), or the release of nitric oxide (Manzoni et al., 1992), although it seems unlikely that nitric oxide would remain available for sufficient periods of time to account for all of the observed effects of prior synaptic activity. Mechanisms that involve downstream events independent of NMDA receptor modification may also be envisioned. For example, repeatedly activating NMDA receptors in a manner sufficient to generate STP may result in a transient and partial depletion of enzymes required for LTP induction. Higher or longer duration rises in Ca^{2+} may then be required to induce LTP until the proteins are recycled or replenished. A role for nitric oxide, independent of any direct effect on NMDA receptors, must also be considered, since recent experiments suggest that nitric oxide release is responsible for the inhibition of LTP caused by application of low concentrations of NMDA (Izumi et al., 1992).

CONCLUSIONS

Evidence has been presented that activation of NMDA receptors does not necessarily result in LTP but, via different patterns of synaptic activity, can also cause a repertoire of distinct changes in synaptic efficacy. Both the duration and direction of synaptic modification are influenced by the degree, pattern, and history of NMDA receptor activation. Among the important mechanistic questions to emerge from this additional complexity is whether NMDA receptor stimulation sufficient to generate LTP also always results in activation of the processes responsible for the other forms of synaptic modification (e.g., LTD, STP). If this is the case, it would suggest the existence of a serial sequence of biochemical steps, perhaps differentially activated as Ca^{2+} levels rise in the postsynaptic cell. An alternative possibility is the existence of parallel biochemical cascades accessed by distinct patterns of NMDA receptor activity. Whatever the mechanistic relationship between the various forms of NMDA receptor–dependent synaptic modification, one practical implication of these findings is that "simple" experiments, such as applying a compound and asking whether or not it blocks LTP, may actually involve manipulation of several distinct physiological processes, any one of which can influence whether LTP is generated.

Hopefully, the functional implications of these findings are apparent. As several neural network models indicate, there are distinct advantages to using Hebbian synaptic modifications that are transient (von der Malsburg and Schneider, 1986; Sporns et al., 1991) or, as mentioned above, of being able to adjust the threshold for synaptic modification (Bienenstock et al., 1982; Bear et al., 1987). In addition, the presented results indicate that the type of synaptic modification caused by fixed patterns of synaptic activity can change significantly as NMDA receptor function is adjusted. This may occur during development (Hestrin, 1992) or conceivably when other neurotransmitter systems are active (Markram and Segal, 1990; Chen and Huang, 1991; Aniksztejn et al., 1992) during different behavioral states.

ACKNOWLEDGMENTS

Work in the author's laboratory was supported by grants from the National Institute of Mental Health, the Klingenstein Foundation, the McKnight Endowment Fund for Neuroscience, the Sloan Foundation, and the National Alliance for Research on Schizophrenia and Depression.

REFERENCES

Abraham, W. C., and Wickens, J. R. (1991) Heterosynaptic long-term depression is facilitated by blockade of inhibition in area CA1 of the hippocampus. *Brain Res.* 546:336–340.

Adams, S. R., Kao, J. P. Y., and Tsien, R. Y. (1989) Biologically useful chelators that take up Ca^{2+} upon illumination. *J. Am. Chem. Soc.* 111:7957–7968.

Aniksztejn, L., and Ben-Ari, Y. (1991) Novel form of long-term potentiation produced by a K^{+} channel blocker in the hippocampus. *Nature* 349:67–69.

Aniksztejn, L., Otani, S., and Ben-Ari, Y. (1992) Quisqualate metabotropic receptors modulate NMDA currents and facilitate induction of long-term potentiation through protein kinase C. *Eur. J. Neurosci.* 4:500–505.

Anwyl, R., Mulkeen, D., and Rowan, M. J. (1989) The role of *N*-methyl-D-aspartate receptors in the generation of short-term potentiation in the rat hippocampus. *Brain Res.* 503: 148–151.

Arai, A., Larson, J., and Lynch, G. (1990) Anoxia reveals a vulnerable period in the development of long-term potentiation. *Brain Res.* 511:353–357.

Asztely, F., Hanse, E., Wigström, H., and Gustafsson, B. (1991) Synaptic potentiation in the hippocampal CA1 region induced by application of *N*-methyl-D-aspartate. *Brain Res.* 558: 153–156.

Baudry, M., and Davis, J. (eds.) (1991) *Long-term Potentiation: A Debate of Current Issues.* Cambridge, Mass.: MIT Press.

Bear, M. F., Cooper, L. N., and Ebner, F. E. (1987) A physiological basis for a theory of synapse modification. *Science* 237:42–48.

Ben-Ari, Y., Aniksztejn, L., and Bregestovski, P. (1992) Protein kinase C modulation of NMDA currents: An important link for LTP induction. *Trends Neurosci.* 15:333–339.

Bienenstock, E. L., Cooper, L. N., and Munro, P. W. (1982) Theory for the development of neuron selectivity: Orientation specificity and binocular interaction in visual cortex. *J. Neurosci.* 2:32–48.

Brocher, S., Artola, A., and Singer, W. (1992) Intracellular injection of Ca^{2+} chelators blocks induction of long-term depression in rat visual cortex. *Proc. Natl. Acad. Sci. USA* 89:123–127.

Charriaut-Marlangue, C., Otani, S., Creuzet, C., Ben-Ari, Y., and Loeb, J. (1991) Rapid activation of hippocampal casein kinase II during long-term potentiation. *Proc. Natl. Acad. Sci. USA* 88:10232–10236.

Chen, L., and Huang, L. Y. M. (1991) Sustained potentiation of NMDA receptor-mediated glutamate responses through activation of protein kinase C by a μ opioid. *Neuron* 7:319–326.

Chen, L., and Huang, L. Y. M. (1992) Protein kinase C reduces Mg^{2+} block of NMDA-receptor channels as a mechanism of modulation. *Nature* 356:521–523.

Coan, E. J., Irving, A. J., and Collingridge, G. L. (1989) Low-frequency activation of the NMDA receptor system can prevent the induction of LTP. *Neurosci. Lett.* 105:205–210.

Cohen, P., and Klee, C. B. (1988) *Calmodulin.* New York: Elsevier.

Colino, A., Huang, Y.-Y., and Malenka, R. C. (1992) Characterization of the integration time for the stabilization of long-term potentiation in area CA1 of the hippocampus. *J. Neurosci.* 12:180–187.

Collingridge, G. L., Kehl, S. J., and McLennan, H. (1983) Excitatory amino acids in synaptic transmission in the Schaffer collateral-commissural pathway of the rat hippocampus. *J. Physiol, (Lond).* 334:33–46.

Dudek, S. M., and Bear M. F. (1992) Homosynaptic long-term depression in area CA1 of the hippocampus and effects of *N*-methyl-D-aspartate receptor blockade. *Proc. Natl. Acad. Sci. USA* 89:4363–4367.

Dunwiddie, T., and Lynch, G. (1978) Long-term potentiation and depression of synaptic responses in the rat hippocampus: Localization and frequency dependency. *J. Physiol.* 276:353–367.

Gerber, G., Kangrga, I., Ryu, P. D., Larew, J. S. A., and Randic, M. (1989) Multiple effects of phorbol esters in the rat spinal dorsal horn. *J. Neurosci.* 9:3606–3617.

Grover, L. M., and Teyler, T. J. (1990) Two components of long-term potentiation induced by different patterns of afferent activation. *Nature* 347:477–479.

Gustafsson, B., Wigström, H., Abraham, W. C., and Huang, Y.-Y. (1987) Long-term potentiation in the hippocampus using depolarizing current pulses as the conditioning stimulus to single volley synaptic potentials. *J. Neurosci.* 7:774–780.

Gustafsson, B., Asztely, F., Hanse, E., and Wigström, H. (1989) Onset characteristics of long-term potentiation in the guinea-pig hippocampal CA1 region in vitro. *Eur. J. Neurosci.* 1:382–394.

Hanse, E., and Gustafsson, B. (1992) Postsynaptic, but not presynaptic, activity controls the early time course of long-term potentiation in the dentate gyrus. *J. Neurosci.* 12:3226–3240.

Hestrin, S. (1992) Developmental regulation of NMDA receptor-mediated synaptic currents at a central synapse. *Nature* 357:686–689.

Hestrin, S., Nicoll, R. A., Perkel, D. J., and Sah, P. (1990) Analysis of excitatory synaptic action in pyramidal cells using whole-cell recording from rat hippocampal slices. *J. Physiol.* 422:203–225.

Hirsch, J. C., and Crepel, F. (1992) Postsynaptic calcium is necessary for the induction of LTP and LTD of monosynaptic EPSPs in prefrontal neurons: an in vitro study in the rat. *Synapse* 10:173–175.

Huang, Y.-Y., and Malenka, R. C. (1993) Examination of TEA-induced synaptic enhancement in area CA1 of the hippocampus: The role of voltage-dependent Ca^{2+} channels in the induction of LTP. *J. Neurosci.* 13:568–576.

Huang, Y.-Y., Colino, A., Selig, D. K. and Malenka, R. C. (1992) The influence of prior synaptic activity on the induction of long-term potentiation. *Science* 255:730–733.

Ito, M. (1989) Long-term depression. *Annu. Rev. Neurosci.* 12:85–102.

Izumi, Y., Clifford, D. B., and Zorumski, C. F. (1992) Inhibition of long-term potentiation by NMDA-mediated nitric oxide release. *Science* 257:1273–1276.

Kauer, J. A., Malenka, R. C., and Nicoll, R. A. (1988) NMDA application potentiates synaptic transmission in the hippocampus. *Nature* 334:250–252.

Kelso, S. R., Nelson, T. E., and Leonard, J. P. (1992) Protein kinase C-mediated enhancement of NMDA currents by metabotropic glutamate receptors in *Xenopus* oocytes. *J. Physiol.* 449:705–718.

Konnerth, A., Dreessen, J., and Augustine, G. J. (1992) Brief dendritic calcium signals initiate long-lasting synaptic depression in cerebellar Purkinje cells. *Proc. Natl. Acad. Sci. USA* 89:7051–7055.

Korn, H., and Faber, D. F. (1991) Quantal analysis and synaptic efficacy in CNS. *Trends Neurosci.* 14:439–445.

Kullman, D. M., Perkel, D. J., Manabe, T., and Nicoll, R. A. (1992) Ca^{2+} entry via postsynaptic voltage-sensitive Ca^{2+} channels can transiently potentiate excitatory synaptic transmission in the hippocampus. *Neuron* 9:1175–1183.

Larson, J., and Lynch, G. (1988) Role of *N*-methyl-D-aspartate receptors in the induction of synaptic potentiation by burst stimulation patterned after the hippocampal θ-rhythm. *Brain Res.* 441:111–118.

Larson, J., Wong, D., and Lynch, G. (1986) Patterned stimulation at the theta frequency is optimal for the induction of hippocampal long-term potentiation. *Brain Res.* 368:347–350.

Linden, D. J., Dickinson, M. H., Smeyne, M., and Connor, J. A. (1991) A long-term depression of AMPA currents in cultured cerebellar Purkinje neurons. *Neuron* 7:81–89.

Lisman, J. (1989) A mechanism for the Hebb and the anti-Hebb processes underlying learning and memory. *Proc. Natl. Acad. Sci. USA* 86:9574–9578.

Lynch, G., Larson, J., Kelso, S., Barrionuevo, G., and Schottler, F. (1983) Intracellular injections of EGTA block induction of hippocampal long-term potentiation. *Nature* 305: 719–721.

MacDonald, J. F., Mody, I., and Salter, M. W. (1989) Regulation of *N*-methyl-D-aspartate receptors revealed by intracellular dialysis of murine neurons in culture. *J. Physiol.* 414: 17–34.

Madison, V. D., Malenka, R. C., and Nicoll, R. A. (1991) Mechanisms underlying long-term potentiation of synaptic transmission. *Annu. Rev. Neurosci.* 14:379–397.

Malenka, R. C. (1991) Postsynaptic factors control the duration of synaptic enhancement in area CA1 of the hippocampus. *Neuron* 6:53–60.

Malenka, R. C., Kauer, J. A., Zucker, R. J., and Nicoll, R. A. (1988) Postsynaptic calcium is sufficient for potentiation of hippocampal synaptic transmission. *Science* 242:81–84.

Malenka, R. C., Kauer, J. A., Perkel, D. J., Mauk, M. D., Kelly, P. T., Nicoll, R. A., and Waxham, M. N. (1989) An essential role for postsynaptic calmodulin and protein kinase activity in long-term potentiation. *Nature* 340:554–557.

Malenka, R. C., Lancaster, B., and Zucker, R. S. (1992) Temporal limits on the rise in postsynaptic calcium required for the induction of long-term potentiation. *Neuron* 9:121–128.

Malinow, R., Madison, D. V., and Tsien, R. W. (1988) Persistent protein kinase activity underlies long-term potentiation. *Nature* 335:820–824.

Malinow, R., Schulman, H., and Tsien, R. W. (1989) Inhibition of postsynaptic PKC or CaMKII blocks induction but not expression of LTP. *Science* 245:862–866.

Manzoni, O., Prezeau, L., Marin, P., Deshager, S., Bockaert, J., and Fagni, L. (1992) Nitric oxide-induced blockade of NMDA receptors. *Neuron* 8:653–662.

Markram, H., and Segal, M. (1990) Long-lasting facilitation of excitatory postsynaptic potentials in the rat hippocampus by acetylcholine. *J. Physiol.* 427:381–393.

Mayer, M. L., MacDermott, A. B., Westbrook, G. L., Smith, S. J., and Barker, J. L. (1987) Agonist- and voltage-gated calcium entry in cultured mouse spinal cord neurons under voltage clamp measured using arsenazo III. *J. Neurosci.* 7:3230–3244.

McGuinness, N., Anwyl, R., and Rowan, M. (1991) Inhibition of *N*-methyl-D-aspartate induced short-term potentiation in the rat hippocampal slice. *Brain Res.* 562:335–338.

McNaughton, B. L. (1982) Long-term synaptic enhancement and short-term potentiation in rat fascia dentata act through different mechanisms. *J. Physiol.* 324:249–262.

Meyer, T., Hanson, P. I., Stryer, L., and Schulman, H. (1992) Calmodulin trapping by calcium-calmodulin-dependent protein kinase. *Science* 256:1199–1202.

Mulkey, R. M., and Malenka, R. C. (1992) Mechanisms underlying homosynaptic long-term depression in area CA1 of the hippocampus. *Neuron* 9:967–975.

Perkel, D. J., and Nicoll, R. A. (1992) Effects of presynaptic modulation of synaptic transmission and LTP on dual-component excitatory synaptic currents in the hippocampus. *Soc. Neurosci. Abstr.* 18:403.

Pockett, S., Brookes, N. H., and Bindman, L. J. (1990) Long-term depression at synapses in slices of rat hippocampus can be induced by bursts of postsynaptic activity. *Exp. Brain Res.* 80:196–200.

Powell, C. M., Johnston, D., and Sweatt, J. D. (1992) Persistent PKC activation in the maintenance phase of NMDA receptor-independent LTP (LTP_K). *Soc. Neurosci. Abstr.* 18:760.

Raastad, M., Storm, J. F., and Andersen, P. (1992) Putative single quantum and single fibre excitatory postsynaptic currents show similar amplitude range and variability in rat hippocampal slices. *Eur. J. Neurosci.* 4:113–117.

Racine, R. J., and Milgram, N. W. (1983) Short-term potentiation phenomena in the rat limbic forebrain. *Brain Res.* 260:201–216.

Rosenmund, C., and Westbrook, G. L. (1992) Calcium-induced actin depolymerization reduces NMDA channel activity. *Soc. Neurosci. Abstr.* 18:650.

Sastry, B. R., Goh, J. W., and Auyeung, A. (1986) Associative induction of posttetanic and long-term potentiation in CA1 neurons of rat hippocampus. *Science* 232:988–990.

Schulman, H., Hanson, P. I., and Meyer, T. (1992) Decoding calcium signals by multifunctional CaM kinase. *Cell Calcium* 13:401–411.

Silva, A., Stevens, C. F., Tonegawa, S., and Wang, Y. (1992) Deficient hippocampal long-term potentiation in α-calcium-calmodulin kinase II mutant mice. *Science* 257:201–206.

Sporns, O., Tononi, G., and Edelman, G. M. (1991) Modeling perceptual grouping and figure-ground segregation by means of active reentrant connections. *Proc. Natl. Acad. Sci. USA* 88:129–133.

Stanton, P. K., and Sejnowski, T. J. (1989) Associative long-term depression in the hippocampus induced by hebbian covariance. *Nature* 339:215–218.

von der Malsburg, C., and Schneider, W. (1986) A neural cocktail-party processor. *Biol. Cybern.* 54:29–40.

Wickens, J. R., and Abraham, W. C. (1991) The involvement of L-type calcium channels in heterosynaptic long-term depression in the hippocampus. *Neurosci. Lett.* 130:128–132.

Yamazaki, M., Mori, M., Araki, K., Mori, K. J., and Mishina, M. (1992) Cloning, expression and modulation of a mouse NMDA receptor subunit. *FEBS Letters* 300:39–45.

Zucker, R. S. (1989) Short-term synaptic plasticity. *Annu. Rev. Neurosci.* 12:13–31.

III Relationships between Long-Term Potentiation and Long-Term Depression

9 Mechanisms of Synaptic Plasticity in the Cerebellum

Francis Crépel, Hervé Daniel, Nathalie Hémart, and Danielle Jaillard

In keeping with the Marr-Albus theory of motor learning in the cerebellum, Ito and coworkers demonstrated in an elegant series of in vivo experiments (Ito et al., 1982; Ito, 1984, 1989) that, in rabbit cerebellum, conjunctive stimulation of parallel fibers (PFs) and climbing fibers (CFs) leads to a long-term depression (LTD) of synaptic transmission at PF-Purkinje cell (PC) synapses. The fact that only those PF-PC synapses activated in conjunction with CFs are affected (Ito, 1984) indicated that the changes in synaptic strength are restricted to the activated synapses. Moreover, Kano and Kato (1987) demonstrated that pairing CF input with glutamate (Glu) or quisqualate (QA) application on PC dendrites also induces subsequent LTD of synaptic transmission between PFs and PCs, whereas other excitatory amino acids (EAAs) are ineffective in this respect.

Involvement of calcium (Ca^{2+}) in induction of LTD was initially suggested by the observation that stellate cell inhibition prevents LTD from occurring (Ekerot and Kano, 1985), probably by blocking Ca^{2+}-dependent plateau potentials in PC dendrites following their activation by CFs (Ekerot and Oscarsson, 1981). Accordingly, Ito proposed that the efficacy of PF-PC synapses is decreased as result of both the activation of AMPA receptors of PCs and the Ca^{2+} influx which occurs in these cells during their activation by CFs (Ito, 1986, 1989).

Indeed, in more recent experiments in rat cerebellar slices, we showed that LTD of PF-mediated excitatory postsynaptic potentials (EPSPs) is consistently induced by pairing these synaptic responses with Ca^{2+} spikes directly induced in the postsynaptic cell by depolarizing current pulses, whereas no LTD occurs if PFs are not stimulated during the period of Ca^{2+} spike firing. In contrast, when only sodium spikes are induced in PCs during the pairing protocol, LTD is no longer observed and is replaced by LTP of PF-mediated EPSPs (Crépel and Jaillard, 1991).

From these results, it seems rather safe to conclude that LTD of synaptic transmission at PF-PC synapses is indeed triggered by an entry of Ca^{2+} in the postsynaptic cell through voltage-gated Ca^{2+} channels opened by cell depolarization, a result recently confirmed by combining the whole-cell recording technique with fluorometric measurements of the intracellular Ca^{2+} concentration (Konnerth et al., 1992).

Moreover, as early as 1988, we proposed that Ca^{2+} acts through the activation of protein kinase C (PKC) to desensitize Glu receptors of PCs (Crépel and Krupa, 1988), and this, for two reasons. First, one knows that PKC I is very abundant in PCs (Nishizuka, 1986; Hidaka et al., 1988). Second, metabotropic Glu receptors are located at PF-PC synapses and are known to be coupled with phospholipase C (PLC) (Sladeczek et al., 1985; Nicoletti et al., 1986; Recasens et al., 1987; Sugiyama et al., 1987). It was therefore tempting to postulate that the cascade of events leading to the desensitization of AMPA receptors during LTD involves a coactivation of PKC of PCs by Ca^{2+} entry through voltage-gated ionic channels, and by diacylglycerol (DAG) produced by the activation of metabotropic Glu receptors by Glu released by PFs.

Indeed, a selective LTD of the responsiveness of PCs to Glu (i.e., aspartate-induced responses were unaffected) was obtained in about 25% of the tested cells when iontophoresis of this agonist was paired with a strong depolarization of PCs giving rise to Ca^{2+} spike firing, whereas sodium spike firing of PCs did not have such a depressant effect (Crépel and Krupa, 1988). Moreover, in about 40% of PCs, there was also a selective decrease of their responsiveness to Glu and QA in the presence of phorbol esters known to activate PKC, whereas inactive analogs were without any effect (Crépel and Krupa, 1988,1990). Finally, in a more recent series of experiments (Crépel and Jaillard, 1990), LTD of PF-mediated EPSPs following their pairing with Ca^{2+} spikes was nearly totally prevented by polymixin B, a potent blocker of PKC and to a lesser extent of calmodulin (CAM)-dependent kinase (Mazzei et al., 1982), thus confirming the involvement of PKC in LTD induction. These data in favor of a role of PKC in LTD induction have been recently confirmed and extended in cultured PCs (Linden and Connor, 1991).

On the other hand, there is now experimental evidence that LTD also depends on another cascade of events beyond Ca^{2+} entry into PCs. It is known that Ca^{2+} can induce the formation of nitric oxide (NO) from arginine by activating a CAM-dependent NO synthase (Garthwaite et al., 1988, 1989; Ross et al., 1990). Nitric oxide is highly diffusible and thus activates soluble guanylate cyclase (Tremblay et al., 1988) in cells where it is produced, as well as in surrounding cellular elements (Garthwaite et al., 1989; Ross et al., 1990). This cascade of events can therefore activate cGMP-dependent protein kinases in PCs where this enzyme is particularly abundant (Lohmann et al., 1981) as well as in PFs and glial cells.

In intracellularly recorded PCs, N_G-methyl-L-arginine (NMLA), a potent inhibitor of NO synthesis (Knowles et al., 1989), significantly reduced LTD of PF-mediated EPSPs following their pairing with Ca^{2+} spikes, and that the same effect was obtained by methylene blue, which acts by trapping NO (Crépel and Jaillard, 1990). The same effect of NMLA on LTD of PF-mediated EPSPs was observed more recently in patch-clamped PCs in thin slices in vitro, and was reversed by an excess of arginine. Furthermore, bath application of the NO donor sodium nitroprusside, as well as bath application of 8-bromoguanosine 3′:5′ cyclic monophosphate (8-bromo-cGMP) were able to reproduce a LTD-like phenomenon that outlasted the washout of these compounds. Finally, LTD of PF-mediated EPSPs also occurred when SIN-1 or guanosine 3′:5′ cyclic monophosphate (cGMP) were directly dialyzed into PCs by the patch pipette, and this effect partially occluded that induced by pairing PF-mediated EPSPs with Ca^{2+} spikes. These results (Daniel et al., 1993) as well as indirect evidences at a multicellular level previously obtained by Ito and Karachot (1990) show that NO indeed plays a role in LTD induction, and demonstrate for the first time that its site of action is probably the soluble guanylate cyclase of PCs. This is puzzling since NO synthase is not detected in PCs (Bredt et al., 1990; Southam et al., 1992), but it should be remembered that this enzyme is present in basket cells, which are also activated by PFs during coactivation protocols. Furthermore, following a large entry of Ca^{2+} in PCs during pairing experiments, it is plausible that the resulting potassium efflux from PCs through Ca^{2+}-dependent potassium conductances contributes to depolarization of neighboring cellular elements to such an extent that their NO synthase is largely activated and produces enough NO to reach nearby PCs, where the next step is activation of the soluble guanylate cyclase (figure 9.1).

The fact that LTD involves two different routes, via NO and PKC, respectively (figure 9.1), might also explain why no effect of NO blockers was seen in experiments on LTD performed in dissociated cell cultures (Linden and Connor, 1992) since, in particular, the putative NO donors (PFs and basket cells) were possibly only scarcely represented. It is therefore conceivable that in such conditions, cultured PCs develop only the PKC route and are thus also insensitive to bath application of NO donors.

Finally, the effects on PF-mediated EPSPs of coactivation of metabotropic glutamate receptors and of voltage-gated Ca^{2+} channels of PCs by bath application of 50 μM *trans*-1-amino-cyclopentyl-1,3-dicarboxylate (*trans*-ACPD) and by direct depolarization of the cells, respectively, was also studied (Daniel et al., 1992). Surprisingly, in the presence of 50 μM *trans*-ACPD in the bath, a large LTD of PF-mediated EPSPs was observed after washout of this drug, whether or not PFs were stimulated in conjunc-

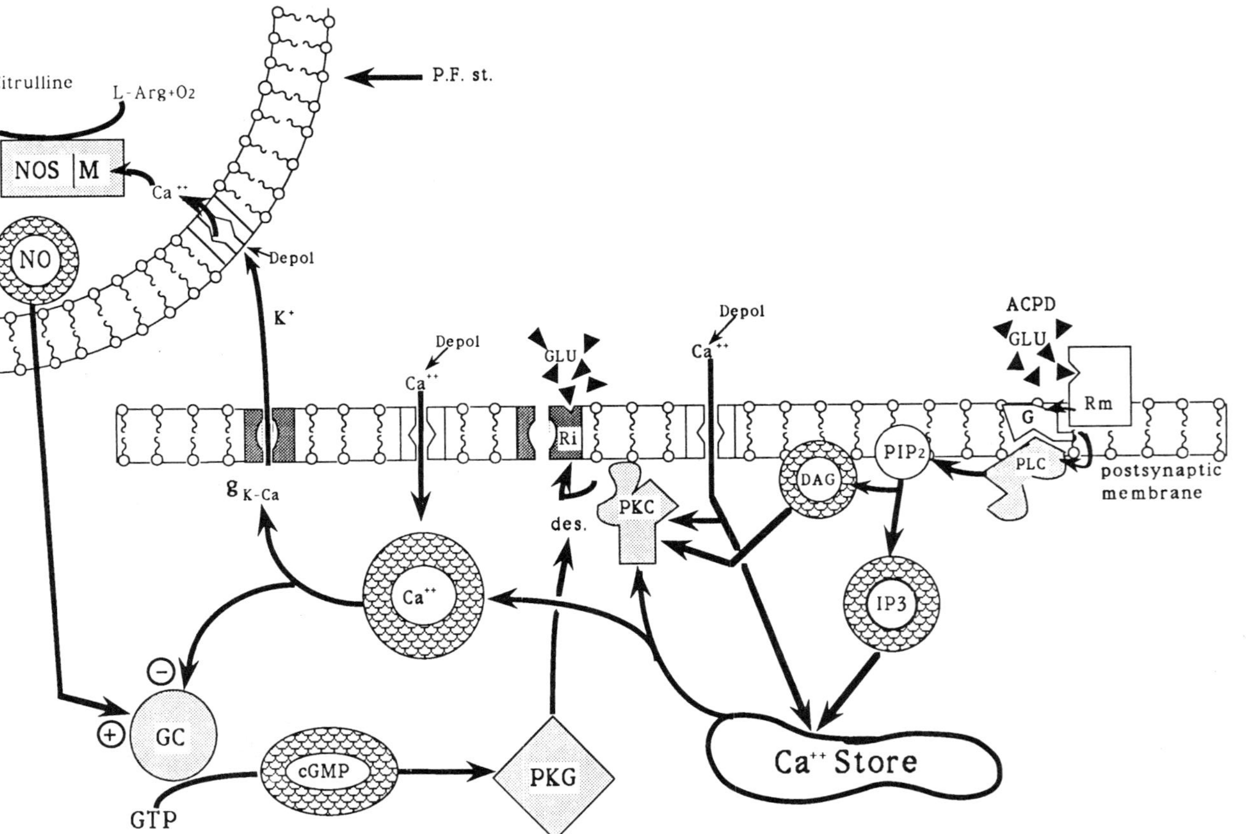

Figure 9.1 Schematic diagram of the signal transduction processes involving the PKC and the NO pathway that are presumed to underlie cerebellar long-term depression. ACPD, *trans*-1-aminocyclopentane-1,3-dicarboxylate; cGMP, cyclic guanosine monophosphate; DAG, diacylglycerol; depol, depolarization; des, desensitization; G, G protein; GC, guanylate cyclase; gK-Ca, calcium-dependent potassium conductance; GLU, glutamate; GTP, guanosine triphosphate; IP3, inositol-1,4,5-triphosphate; L-Arg, L-arginine; M, calmodulin; NO, nitric oxide; NOS, nitric oxide synthase; PF, parallel fiber; PIP_2, phosphatidyl inositol 4,5-biphosphate; PKC, protein kinase C; PKG, protein kinase G; PLC, phospholipase C; Ri, ionotropic receptor; Rm, metabotropic receptor.

tion with Ca^{2+} spike firing, as in previously described experiments. As *trans*-ACPD alone induced fully reversible depressions of EPSPs, coactivation of metabotropic Glu receptors and of voltage-gated Ca^{2+} channels is therefore likely to be sufficient to induce LTD of PF-mediated EPSPs. Thus the present results indicate that at least two forms of LTD induction might coexist in PCs, one depending primarily on the activation of ionotropic AMPA receptors and the other independent of this class of receptors, but linked to a strong activation of metabotropic Glu receptors.

REFERENCES

Bredt, D. S., Hwang, P. M., and Snyder, S. H. (1990) Localisation of nitric oxide synthase indicating a neural role for nitric oxide. *Nature* 347:768–770.

Crépel, F., and Jaillard, D. (1990) Protein kinases, nitric oxide and long-term depression of synapses in the cerebellum. *NeuroReport* 1:133–136.

Crépel, F., and Jaillard, D. (1991) Pairing of pre and postsynaptic activities in cerebellar Purkinje cells induces long-term changes in synaptic efficacy. An in vitro study. *J. Physiol. (Lond.)* 432:123–141.

Crépel, F., and Krupa, M. (1988) Activation of protein kinase C induces a long-term depression of glutamate sensitivity of cerebellar Purkinje cells. An in vitro study. *Brain Res.* 458:397–401.

Crépel, F., and Krupa, M. (1990) Modulation of the responsiveness of cerebellar Purkinje cells to excitatory amino acids. In *Excitatory Amino Acids and Neuronal Plasticity*, Y. Ben-Ari (ed.). New York: Plenum Press, pp. 323–329.

Daniel, H., Hémart, N., Jaillard, D., and Crépel, F. (1992) Coactivation of metabotropic glutamate receptors and voltage-gated calcium channels induces long-term depression in cerebellar Purkinje cells in vitro. *Exp. Brain Res.* 90:327–331.

Daniel, H., Hémart, N., Jaillard, D., and Crépel, F. (1993) Long-term depression requires nitric oxide and guanosine 3′:5′ cyclic monophosphate production in rat cerebellar Purkinje cells. *Eur. J. Neurosci.* 5:1079–1082.

Ekerot, C. F., and Kano, M. (1985) Long-term depression of parallel fibre synapses following stimulation of climbing fibres. *Brain Res.* 342:357–360.

Ekerot, C. F., and Oscarsson, O. (1981) Prolonged depolarization elicited in Purkinje cell dendrites by climbing fibre impulses in the cat. *J. Physiol. (Lond.)* 318:207–221.

Garthwaite, J., Charles, S. L., and Chess-Williams, R. (1988) Endothelium-derived relaxing factor release on activation of NMDA receptors suggests role as intercellular messenger in the brain. *Nature* 336:385–388.

Garthwaite, J., Southam, E., and Anderton, M. (1989) A kainate receptor linked to nitric oxide synthesis from arginine. *J. Neurochem.* 53:1952–1954.

Hidaka, H., Tanaka, T., Onoda, K., Hagiwara, M., Watanabe, M., Ohta, H., Ito, Y., Tsurudome, M., and Yoshida, T. (1988) Cell-specific expression of protein kinase C isozymes in the rabbit cerebellum. *J. Biol. Chem.* 263:4523–4526.

Ito, M. (1984) *The Cerebellum and Neural Control*. New York: Raven Press.

Ito, M. (1986) Long-term depression as a memory process in the cerebellum. *Neurosci. Res.* 3:531–539.

Ito, M. (1989) Long-term depression. *Annu. Rev. Neurosci.* 12:85–102.

Ito, M., and Karachot, L. (1990) Messengers mediating long-term desensitization in cerebellar Purkinje cells. *NeuroReport* 1:129–132.

Ito, M., Sakurai, M., and Tongroach, P. (1982) Climbing fibre induced depression of both mossy fibre responsiveness and glutamate sensitivity of cerebellar Purkinje cells. *J. Physiol. (Lond.)* 324:113–134.

Kano, M., and Kato, M. (1987) Quisqualate receptors are specifically involved in cerebellar synaptic plasticity. *Nature* 325:276–279.

Knowles, R. G., Palacios, M., Palmer, R. M. J., and Moncada, S. (1989) Formation of nitric oxide from L-arginine in the central nervous system: A transduction mechanism for stimulation of the soluble guanylate cyclase. *Proc. Natl. Acad. Sci. USA* 89:5159–5162.

Konnerth, A., Dreessen, J., and Augustine, G. J. (1992) Brief dendritic calcium signals initiate long-lasting synaptic depression in cerebellar Purkinje cells. *Proc. Natl. Acad. Sci. USA* 89:7051–7055.

Linden, D. J., and Connor, J. A. (1991) Participation of postsynaptic PKC in cerebellar long-term depression in culture. *Science* 254:1656–1659.

Linden, D. J., and Connor, J. A. (1992) Long-term depression of glutamate currents in cultured cerebellar Purkinje neurons does not require nitric oxide signalling. *Eur. J. Neurosci.* 4:10–15.

Lohmann, S. M., Walter, V., Miller, P. E., Greengard, P., and Camilli, P. D. (1981) Immunohistochemical localization of cyclic GMP-dependent protein kinase in mammalian brain. *Proc. Natl. Acad. Sci. USA* 78:653–657.

Mazzei, G. J., Katoh, N., and Kuo, J. F. (1982) Polymyxin B is a more selective inhibitor for phospholipid-sensitive Ca^{2+}-dependent protein kinase than for calmodulin-sensitive Ca^{2+}-dependent protein kinase. *Biochem. Biophys. Res. Commun.* 109:1129–1133.

Nicoletti, F., Meek, J. M., Iadarola, M. J., Chuang, D. M., Roth, B. L., and Costa, E. (1986) Coupling of inositol phospholipid metabolism with excitatory amino acid recognition sites in rat hippocampus. *J. Neurochem.* 46:40–46.

Nishizuka, Y. (1986) Studies and perspectives of protein kinase C. *Science* 233:305–311.

Recasens, M., Sassetti, I., Nourigat, A., Sladeczek, F., and Bockaert, J. (1987) Characterization of subtypes of excitatory amino acid receptors involved in the stimulation of inositol phosphate synthesis in rat-brain synaptoneurosomes. *Eur. J. Pharmacol.* 141:87–93.

Ross, C. A., Bredt, D., and Snyder, S. H. (1990) Messenger molecules in the cerebellum. *Trends Neurosci.* 13:216–222.

Sladeczek, F., Pin, J. P., Recasens, M., Bockaert, J., and Weiss, S. (1985) Glutamate stimulates inositol phosphate formation in striatal neurons. *Nature* 317:717–719.

Southam, E., Morris, R., and Garthwaite, J. (1992) Sources and targets of nitric oxide in rat cerebellum. *Neurosci. Lett.* 137:241–244.

Sugiyama, H., Ito, I., and Hirono, C. (1987) A new type of glutamate receptor linked to inositol phospholipid metabolism. *Nature* 325:531–533.

Tremblay, J., Gerzer, R., and Hamet, P. (1988) Cyclic GMP in cell function. In *Advances in Second Messengers and Phosphoprotein Research*, Vol. 22. New York: Raven Press, pp. 319–368.

10 Long-Term Enhancement of Inhibitory Synaptic Transmission in the Central Nervous System

Stéphane Charpier, Yoichi Oda, and Henri Korn

Despite a considerable amount of investigation of long-term potentiation (LTP), the question of whether this process occurs at inhibitory synapses remains controversial. Plasticity of inhibition was predicted by Brindley (1967), and a post-tetanic potentiation (PTP) of inhibitory potentials (IPSPs) has been described (Waziri et al., 1969). Yet confusing, if not conflicting, data followed these early indications. For example, Bliss and Lømo (1973) suggested a shift in tonic inhibition as a possible mechanism for enhancement of synaptic excitation (see also Steward et al., 1990). In hippocampal slices LTP seems to be associated with a decrease in IPSP amplitude (Yamamoto and Chujo, 1978; Misgeld et al., 1979), but this could result from the masking effect of a simultaneously increased depolarization (Teyler and Di Scenna, 1987). Local inhibition can bias the degree of activity required to induce excitatory LTP (Alger and Nicoll, 1982; Douglas et al., 1982; Wigström and Gustafsson, 1985), the initiation of which is facilitated in CA1 by the presence of $GABA_A$ blockers (Wigström and Gustafsson, 1983).

Results concerning specifically inhibitory junctions are scarce, presumably because it is difficult to evoke inhibitory responses in the CNS without introducing parallel excitation. When observed, enhanced inhibition during LTP has been attributed to increased excitation and firing of the inhibitory interneurons (Misgeld et al., 1979; Buszaki and Eidelberg, 1982; Abraham et al., 1987; Chavez-Noriega et al., 1989). Also in the hippocampus, a decrease in GABAergic transmission (Yamamoto and Chujo, 1978; Haas and Rose, 1982) has been postulated (Misgeld et al., 1979; Scharfman and Sarvey, 1985; Stelzer et al., 1987), but no modifications of synaptic chloride currents have been detected (Griffith et al., 1986).

The Mauthner (M) cell of teleosts allows this problem to be addressed because inhibition can be evoked selectively in this system. The posterior branch of the eighth nerve produces a disynaptic inhibition of the contralateral M-cell, relayed through crossed second-order vestibular interneurons (Korn et al., 1990). Their terminals are glycinergic (Faber and

Korn, 1988), they are grouped on the soma (Triller and Korn, 1986), and indications have been obtained that their synapses are modifiable (Wolszon and Faber, 1989; Mintz et al., 1989; Mintz and Korn, 1991).

Although the M-cell is a reticulospinal neuron that processes sensory information, mixed excitatory (i.e., electrotonic and chemical) synapses onto its lateral dendrite exhibit a typical NMDA-dependent LTP (Yang et al., 1990). In addition, we have found that the glycinergic inhibition in the teleost M-cell shows LTP following "classic" tetanization of that pathway. This enhancement occurs at both stages of the inhibitory circuit, first at the synapses between primary afferents onto the commissural interneurons, and second, at their connections with the M-cell. Some of the evidence that establishes the reality of this new form of LTP has been presented before (Oda and Korn, 1991; Korn et al., 1992). As shown here, this conclusion is supported by results obtained with paired recordings at identified inhibitory connections.

MATERIAL AND METHODS

Recordings of inhibitory postsynaptic potentials were obtained in vivo from the M-cell soma of goldfish (*Carassius auratus*) anesthetized with MS 222 and immobilized with D-tubocurarine, with low-resistance KCl (2.7 M) or KOAc (4 M)-filled microelectrodes. Single-electrode voltage-clamp techniques (Faber and Korn, 1987) were used to collect inhibitory synaptic currents.

Our protocols were based on the neuronal circuit illustrated in figure 10.1. Test and conditioning stimuli were applied to the posterior branch of the contralateral eighth (VIIIth) nerve through bipolar steel electrodes with tip diameters and separations of about 10 and 50 μm, respectively. Test stimuli, applied every 2 sec, were typically below the threshold (T) for firing the other M-cell. Tetani were short trains (12–20 pulses) at 300–500 Hz, repeated once every 2–4 sec for 1–3 min, using intensities that were not always equal to that of the test stimulus.

Inhibitory responses were quantified on averaged sweeps (N = 4–20), using two methods. The first involved comparisons of peak IPSP or IPSC magnitudes. The second was based on measurements of the reduction in the antidromic spike height due to the inhibitory shunt. Since the antidromic action potential propagates passively into the soma, any conductance change can be calculated as $r' = (V/V') - 1$, where V and V' are spike amplitudes in the absence and presence of inhibition, respectively. This expression represents the ratio (or fractional conductance) $r' = G_{IPSP}/G_m$, the two terms being the inhibitory and resting conductances (Faber and Korn, 1982).

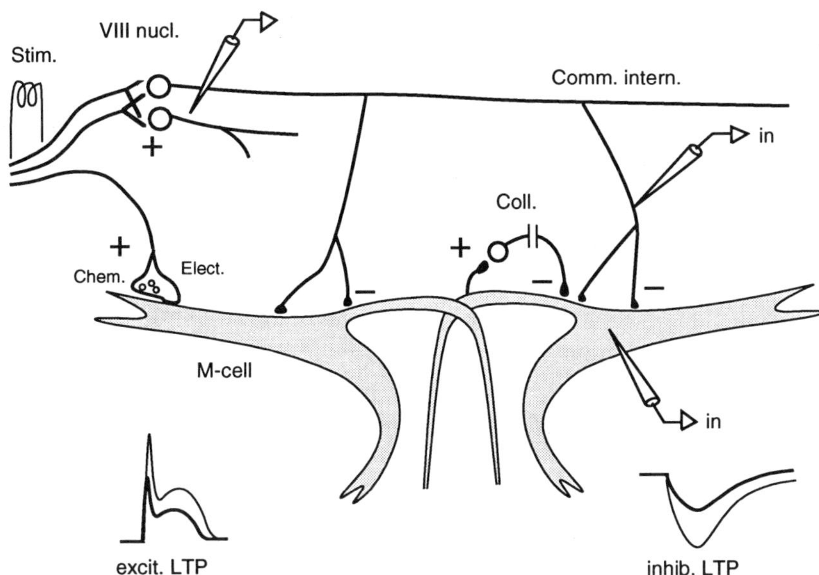

Figure 10.1 Scheme of the experimental set-up and of the Mauthner cell–related networks which manifest long term changes in synaptic strength. Recording microelectrodes were inserted into the vestibular nuclei (VIII nucl), in second order commissural inhibitory axons (Comm. intern.) and in the M-cell that was contralateral to the stimulated (Stim.) eighth nerve primary afferents. The diagram indicates that both M-cells are inhibited disynaptically after stimulations of the posterior branch of the eighth nerve, via the interneurons whose cell bodies are in the vestibular nuclei. Eighth nerve fibers excite the ipsilateral M-cell dendrite by way of mixed electrotonic (Elect.) and chemical (Chem.) synapses, both of which exhibit LTP after classic tetanic stimulations. As also indicated below the diagram, the same protocol potentiates inhibition of the contralateral M-cell, whereas the nontetanized recurrent collateral inhibition (Coll.) is unaffected. Symbols (+) and (−) refer to the corresponding sign of synaptic function. (From Korn et al., 1992)

The specificity of the tetanization effects and the stability of chloride loading were assessed by monitoring the full-sized recurrent collateral inhibition, which involves a second set of interneurons and is produced by antidromic activation of the M-axon in the spinal cord. The stationarity of G_m was also verified by measuring the antidromic spike, or, in voltage clamp, responses to command pulses. Results reported are from cells in which these parameters and the resting potential (−70 to −90 mV) varied by no more than 10%.

In some experiments, (1) the calcium chelator BAPTA in KOAc electrodes (3–10 mM) was continuously injected iontophoretically or, (2) the excitatory amino acid receptor antagonists CNQX (50 μM) and APV (100 μM) were applied by superfusion in saline.

Simultaneous intracellular recordings from the M-cell and single presynaptic interneurons were also obtained, using KCl (3 M)-containing microelectrodes coated with a silver paint that was grounded to minimize coupling artefacts. The second-order vestibular interneurons were identified on the basis of two criteria. One was the presence of a passive hyperpolarizing potential (PHP) due to a field effect (Korn and Faber, 1975), indicating unambiguously that these cells are inhibitory to the M-cell (Korn and Faber, 1976). The second was a short latency (0.3–0.8 msec) EPSP upon stimulation of the contralateral eighth nerve (Zottoli and Faber, 1980; Triller and Korn, 1981). Cl^- was iontophoretically injected into the postsynaptic cell through the recording electrode until large and stable depolarizing collateral IPSPs (V_{coll}) were recorded. As in previous studies (Korn and Faber, 1976; Faber and Korn, 1982), unitary IPSPs (V_{IPSP}) were evoked in the M-cell by direct presynaptic spikes, at the rate of 2 per second, and they appeared as depolarizing potentials. Assessments of unitary inhibitory conductances (G_{IPSP}) were based on the relationship

$$G_{IPSP} = [V_{IPSP}/(2V_{coll} - V_{IPSP})]G_m,$$

where G_m is the resting input conductance and is about 6.08×10^{-6} S (Faber and Korn, 1982).

RESULTS

In this study, we have observed modifications of synaptic efficacy at the level of (1) the primary afferents synapsing on the commissural cells and, (2) the synapses of these interneurons with their target.

Overall, in Cl^--loaded M-cells, the VIIIth nerve evoked IPSPs monitored in the current clamp mode appear as depolarizing, and they exhibit two components; the first is somatic in origin and the second is generated in the lateral dendrite (Faber and Korn, 1978). As shown in figure 10.2, both components were significantly increased after VIIIth nerve tetanization. This enhancement was not secondary to modifications of G_m or of the intracellular Cl^- concentration, as indicated by the stability of the antidromic spike and of the collateral IPSP. The maximum values of the potentiated IPSPs were reached at about 10 min after the onset of the tetanus, and they averaged $+83 \pm 20\%$ and $+104 \pm 27\%$ ($n = 13$, $m \pm$ SEM), for the first and second component, respectively. Although the IPSPs had a tendency to fluctuate, the potentiation persisted throughout the recording session (the longest of which lasted 180 min). Twelve of 16 cells showed such a facilitation which, in two instances was preceded by a brief period (3–5 min) of post-tetanic depression. In all other cases, the initial phase of the LTP was signaled by an obvious increase of the spontaneous inhibitory synaptic noise (not shown).

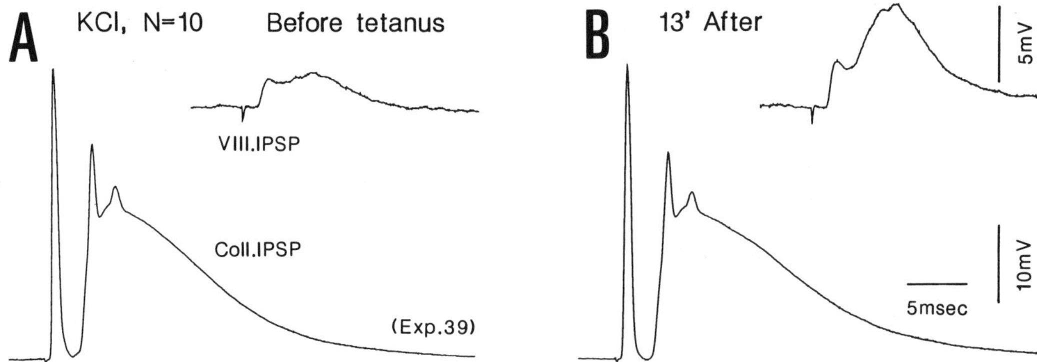

Figure 10.2 Prolonged enhancement of inhibitory synaptic efficacy. (*A*) Control antidromic spike followed by a depolarizing collateral IPSP, recorded with a KCl-filled microelectrode. *Inset*: Typical eighth nerve evoked IPSP. (*B*) Same responses as in (*A*), obtained 13 min after the onset of a tetanus applied to the contralateral eighth nerve (trains of 30 pulses, at 500 Hz, every 2 sec during 90 sec, intensity (i) = 10 × test strength). Note that both components of the Cl^--dependent IPSP were still increased, whereas the collateral IPSP was unchanged (all sweeps are averages of 10 consecutive traces).

VOLTAGE-CLAMP ANALYSIS

To minimize variations in IPSC amplitudes due to possible shifts in Cl^- loading of the M-cell, inhibitory conductances were measured using the single-electrode voltage-clamp technique (Faber and Korn, 1987, 1988). A first series of experiments confirmed that the VIIIth nerve evoked and the Cl^--dependent recurrent collateral inhibitory postsynaptic currents (IPSCs) have similar voltage dependence and reversal potential (figure 10.3). In three of five Cl^- loaded cells studied, one of which was held for more than 2 hr, the synaptic conductance associated with each component of the potentiated IPSCs, corresponding to that of the IPSPs, increased after the tetanus by about 75% ± 25% (SEM). No changes in the IPSC equilibrium potential were observed, strongly suggesting that this enhancement was not associated with the development of an additional current. The resting conductance G_m was derived from current voltage relationships (Faber and Korn, 1988) and remained also unaffected.

Inhibitory LTP in voltage clamp is shown by the sample recordings of figure 10.4: following a control period of about 5 min, the VIIIth nerve was tetanized and the test IPSC rose to 25%. This increase could be larger (up to 100%) regardless of whether its size was expressed as a fraction of the collateral current (for normalization purposes; see Faber and Korn, 1982), or in absolute values (figure 10.5, left and right ordinates, respectively). They remained larger than before conditioning until the penetration was lost.

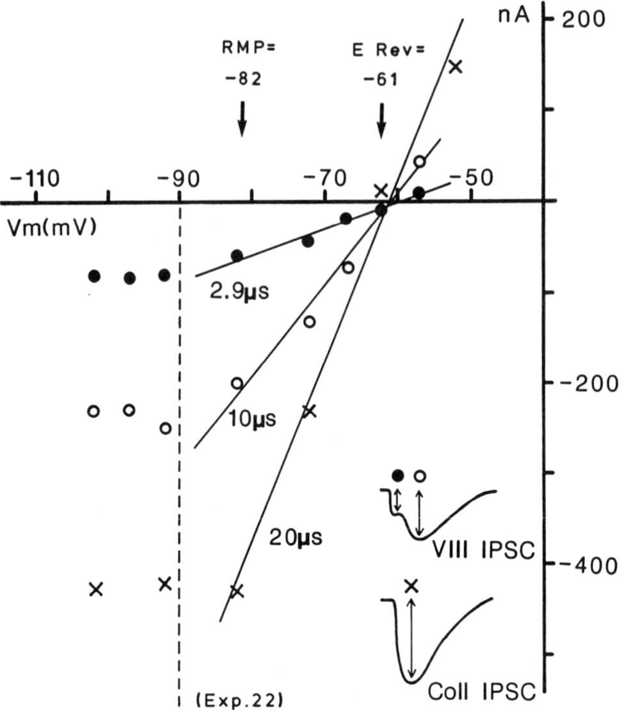

Figure 10.3 Similar properties of the vestibular-evoked and of the recurrent collateral synaptic currents. Current-voltage relationships of the first (filled circles, somatic) and the second (open circles, dendritic) components of the VIIIth nerve induced, and of the collateral (X) IPSCs. Potentials were recorded during a single voltage clamp experiment (chopping frequency 9.3 kHz) from a chloride-loaded M-cell. The regression lines, which were fit with least square statistics on the linear parts of the curves (data points to right of vertical dashed line), yield the indicated inhibitory conductances. Note that the currents have the same reversal potential, and that at more hyperpolarized potentials the slope of the plots decrease, indicating a nonlinear function of membrane potential.

MODIFICATIONS OF SYNAPTIC CONDUCTANCES DURING LTP

Measurements of inhibitory shunts, which are independent of Cl^- loading (Faber and Korn, 1982), allowed us to use KOAc microelectrodes to follow the LTP more reliably. For this purpose, the size of the antidromic spike, which propagates passively along the somatodendritic membrane (Furshpan and Furukawa, 1962), was measured in the control and during the VIIIth nerve–evoked IPSP (see Methods). In these experiments, the peak of the test antidromic spike was timed to occur about 3 msec after that of the presynaptic inhibitory volley, which corresponds to the early part of the IPSP falling phase (Korn and Faber, 1976). This test was used because without Cl^- loading, the IPSP does not appear as a potential change in the M-cell, since this ion's equilibrium potential lies close to the resting potential (Furukawa and Furshpan, 1963).

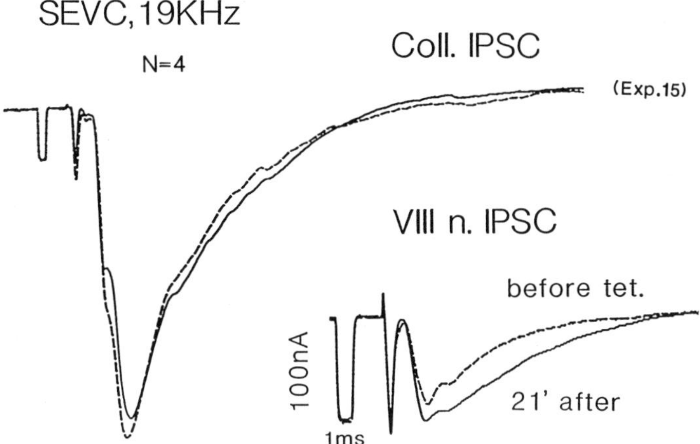

Figure 10.4 Selectivity of potentiation. Superimposed inhibitory currents obtained before, and 21 min after tetanization of the contralateral vestibular nerve (same experiment as that of figure 10.5). While vestibular evoked IPSCs (right) were enhanced (dashed lines = control recordings), those evoked by activation of the recurrent collateral pathways were almost unchanged (left).

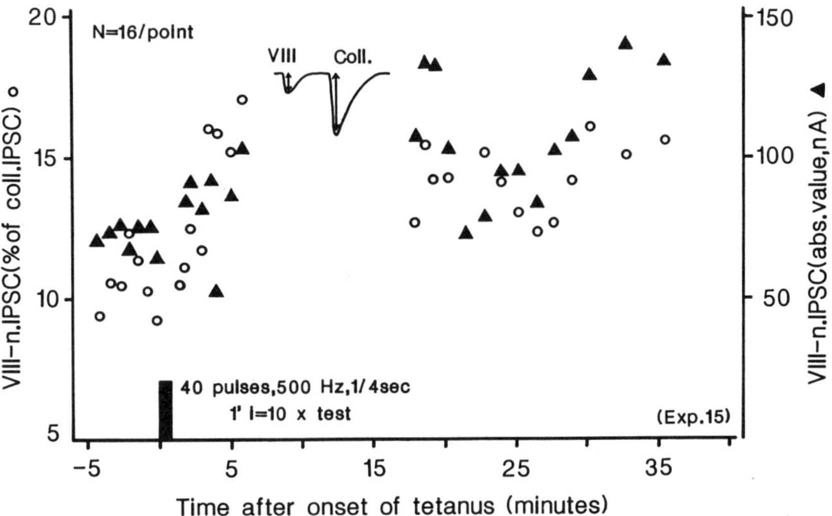

Figure 10.5 LTP of VIIIth nerve–evoked IPSCs. Amplitudes of inhibitory currents, expressed as percentage of the recurrent contralateral IPSC (left ordinate, circles) and in absolute value (right side, triangles) are plotted against time; each data point is the average of 16 successive responses (the recording was switched to the current clamp mode between +8 to +17 min).

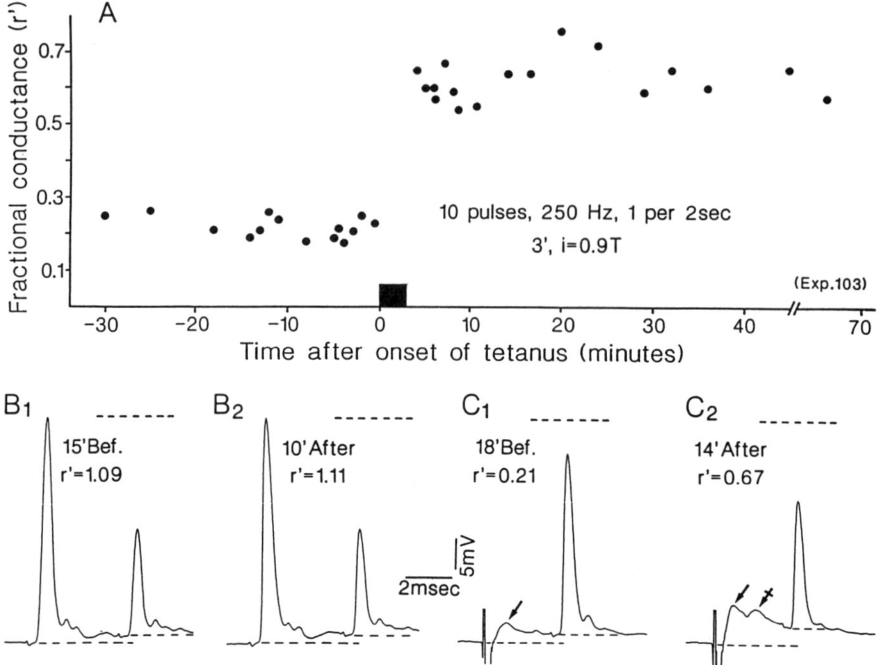

Figure 10.6 Time course of postsynaptic conductance changes during inhibitory LTP. (*A*) Plot of the fractional conductance (r') following contralateral eighth nerve stimulation (with the indicated parameters), versus time. r' was calculated using the reduction of a test antidromic spike height as indicated in text. (B_1–C_2) Synapse specificity and presynaptic component of LTP. (B_1–B_2) Lack of an effect of eighth nerve tetanization on collateral inhibition. The uppermost horizontal dashed lines indicate control action potential amplitudes, and lower ones are continuations of baselines used for measuring the related potentials. (C_1–C_2) Increased eighth nerve–evoked inhibition after tetanization. The responses consist of the intracellularly recorded presynaptic volley (arrows) followed by a small depolarizing IPSP (crossed arrow in C_2), the latter producing the shunt. In this experiment LTP, which appeared as an increased IPSP and fractional conductance was associated with an enhanced volley. (From Korn et al., 1992)

In figure 10.6, the ratio, r', of inhibitory to resting conductance was constant for 30 min before induction of LTP. This parameter increased immediately after the tetanus by about 200%, and it remained elevated until the end of the experiment (figure 10.6A). It can also be noted that the enhancement was specific to the tetanized pathway, as r' collateral inhibition remained stationary (figure 10.6B_1–B_2), varying by no more than 10% about a mean of 1.16.

The average increase of the fractional conductance was 119.6% ($\pm$11.7 SEM, n = 17, range 63–233%). This potentiation, which was observed in every experiment was somewhat larger than that obtained in voltage clamp with a more restricted sample. Initial values for r', which is proportional to the number of activated interneurons (see Discussion) ranged from 0.04 to 0.45, and they were not correlated with the magnitude of the LTP. Similarly, there was no relation between the degree of potentiation and the intensity of the conditioning stimulus, which varied from 1 to 10 times the test strength.

DIFFERENT LOCI INVOLVED IN INHIBITORY LTP

Excitatory Synapses onto Inhibitory Interneurons

A major advantage of the M-cell system is that the presynaptic inhibitory volley can be recorded extracellularly, close to the terminals, as the so-called extrinsic hyperpolarizing potential (or EHP; Furukawa and Furshpan, 1963). Furthermore, it is also observed with intracellular recordings as an early positivity whose amplitude and time course parallel those of the EHP. Thus, measurements of the volley allowed us to distinguish between an increased excitation of inhibitory cells and changes at their terminal synapses.

A frequent finding was that the presynaptic volley was modified during LTP: in 10 of 13 experiments this augmentation averaged 83 $\pm$ 15.6% (mean and SEM), while the fractional conductance r' was increased by 155 $\pm$ 31%. This is illustrated in figure 10.6C_1–C_2, where both the volley and the depolarizing IPSP increased after the tetanus.

This observation suggested a facilitated transmission at excitatory connections between VIIIth nerve primary afferents and the commissural interneurons. At this level synapses are mixed, electrotonic and chemical (Korn et al., 1977; see also Zottoli and Faber, 1980). Hence the question was raised whether these connections have the same plastic properties as the mixed synapses established by VIIIth nerve fibers on another target, the M-cell dendrite, where they can express LTP (Yang et al., 1990). This notion was confirmed by extracellular recordings of the typical eighth nerve–evoked field potential in the ipsilateral vestibular nucleus (Korn et al., 1977). The early (electrical) and late (chemical) postsynaptic components of this field were potentiated after a tetanus for the whole remaining duration of the experiments, which could exceed 3 hr (not shown).

LTP at Inhibitory Synapses

Indirect Evidence In three experiments however, the presynaptic volley did not change after the tetanus and yet there was an elevated inhibitory shunt in the M-cell, with increments of 52%, 80%, and 150% (figure 10.7). Therefore, in another series of experiments, we compared the input-output relations of the inhibitory pathway by varying the stimulus inten-

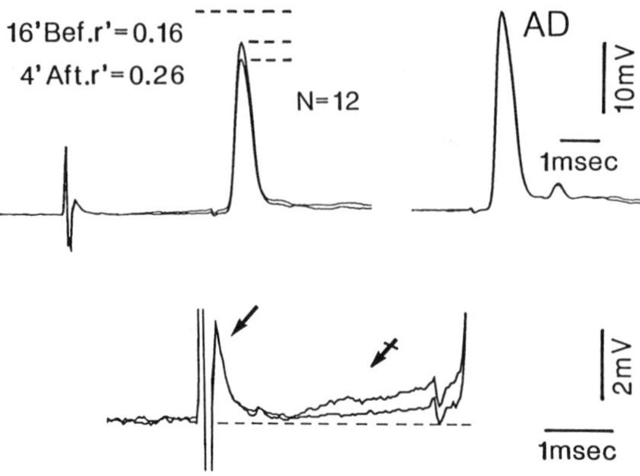

Figure 10.7 Evidence for an increased synaptic gain. (*Above*, from left to right) Shunt of the control spike before ($r' = 0.16$) and after ($r' = 0.26$) conditioning, at the indicated times, and superimposed control action potentials (AD). (*Below*) Same as above, at higher magnification. The presynaptic volley (arrow) remained constant in amplitude but was nevertheless followed by an enhanced, slowly rising depolarizing IPSP (crossed arrow). Recordings obtained with KOAc microelectrode (RP, −91 mV; trains of 20 pulses at 500 Hz, 1 per 2 sec, during 90 sec, i = 1.6 × test).

sity before and after tetanization. The results showed that the slope of the r'-volley plots became significantly steeper and that when the size of the presynaptic volley matched the magnitude of that in the control one, r' was still enhanced (not illustrated). Such shifts persisted during periods when LTP was stable. The corresponding maximum increase in gain averaged 75.25% ± 34% (m ± SD, n = 4).

Advantage was also taken of the fact that transmission at the first-order synapses is mixed, the electrotonic PSP alone being sufficient to fire the inhibitory interneurons (Faber and Korn, 1978). Thus, CNQX and APV were both applied throughout in five experiments, in order to (1) block potentiation at this level, and (2) "bypass" the first relay by reducing it to a purely electrical transmission line. The effectiveness of this procedure was demonstrated by the absence of changes in the volleys after tetanization. Yet, as shown in figure 10.8, there still was an inhibitory LTP, which in this case was stable from 20 min on, the enhancement averaging then 42% (n = 13) and 37% (n = 8) for low and high test intensities, respectively.

Direct Evidence The most compelling indication that synaptic efficacy of glycinergic synapses is increased persistently by presynaptic trains was obtained with paired recordings. The amplitudes and underlying conductances of unitary IPSPs evoked by presynaptic interneurons were both affected, as exemplified by the sample recordings and the plot of figure 10.9. In this experiment, the mean synaptic conductance produced by

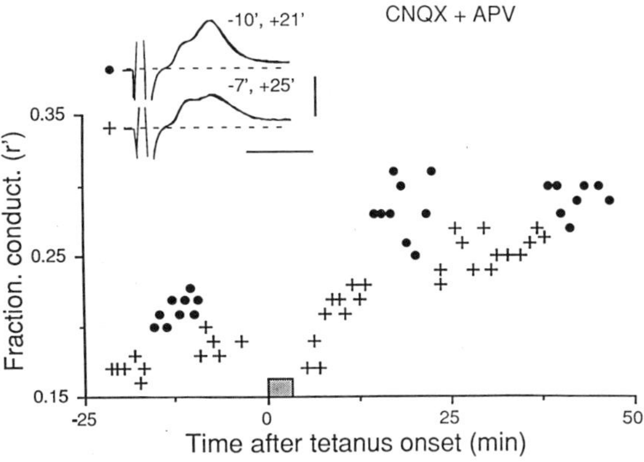

Figure 10.8 LTP at the level of the inhibitory synapse. Plot of fractional conductance (r') expressed against time, from a CNQX-APV treated preparation. Two test strengths were used for stimulations of the contralateral eighth nerve (crosses, circles), and their effects on the postsynaptic cell were both potentiated. Each data point is the average of at least 10 responses (trains of 15 pulses at 500 Hz every 4.5 sec for 150 sec, intensity = higher test strength). *Inset*: Stability of the biphasic presynaptic volleys recorded intracellularly with a KOAc microelectrode. Superimposed averaged responses obtained before (N = 10) and during (N = 10) LTP, at the indicated times. Volleys remained unaffected for both the stronger (above) and lower (below) intensities. Calibrations: 5 mV, 1 msec. (From Korn et al., 1992)

direct spikes in a commissural interneuron rose from 60 to 121 nS. A similar potentiation was observed at 7 of 18 pairs, whereas in one case the IPSPs decreased, and they remained stable in the others. As illustrated here, the increase in synaptic strength was already apparent at the break of the conditioning VIIIth nerve tetanization, and reached its maximum 8–11 minutes after the first train (m = 64.2% ± 31.7% SD). Despite this limited sample, there was a wide range of potentiation (19–102%). The apparent initial quantal content in the control IPSPs also varied greatly, from 1.8 to 12% of V_{coll} (a single exocytotic event at the concerned endings typically produces in the M-cell a voltage change of about 1% that of the full-sized collateral IPSP; Korn et al., 1982, 1986).

Figure 10.10 is from another Cl^--loaded M-cell, which had a stable collateral IPSP during the 18 min of paired recording (figure 10.10A). Here, the unitary conductance was increased by 92% after VIIIth nerve conditioning. Inspection of the superimposed sweeps (figure 10.10B) indicates that the kinetics of the evoked IPSPs (time to peak and decay) were unaffected, and their values were comparable to those previously published (Korn and Faber, 1975, 1976).

In the course of this study, we encountered presynaptic "silent" cells. When excited, these PHP-cells do not produce measurable responses in the M-cell. Evidence suggests that at these synaptic connections transmitter is

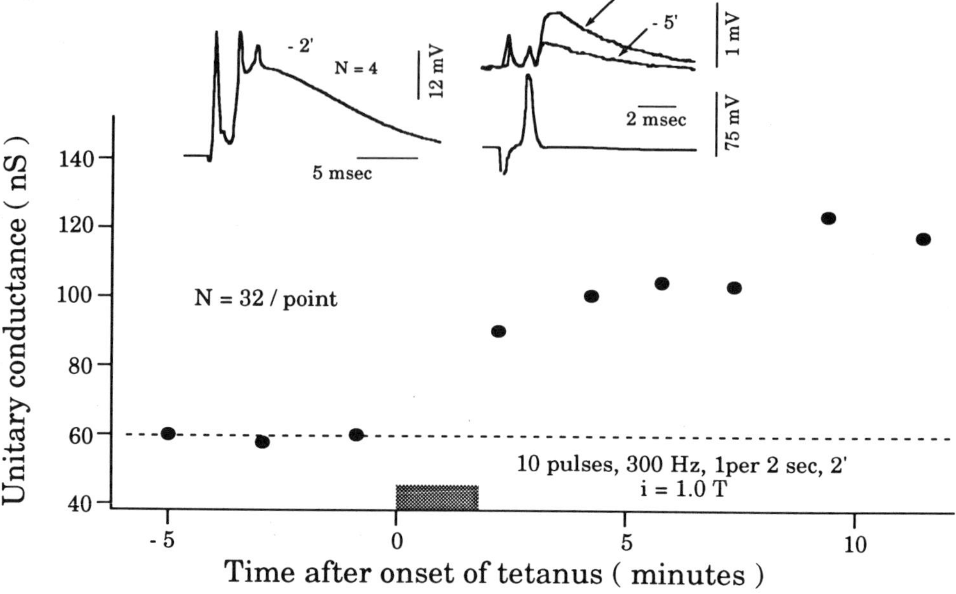

Figure 10.9 Inhibitory LTP at an identified connection. Plot of the unitary conductance (ordinate), calculated as indicated in text, versus time. During the recording session, the unitary conductance was increased to a maximum of 202% relative to the control (horizontal dashed line) 9 min after tetanization. *Inset, left*: Response of the M-cell to spinal stimulation showing the antidromically evoked collateral IPSP. *Right*: Unitary IPSPs (*upper*) evoked by single presynaptic spikes (*lower*) recorded 5 min before and 2 min after onset of the tetanus, respectively (32 superimposed sweeps for each of the illustrated periods).

released, but that the associated postsynaptic receptors are unresponsive (Korn and Faber, 1990; Faber et al., 1991). A striking finding was that tetanization of the VIIIth nerve converted them into fully functional inhibitory interneurons (S. Charpier, J. C. Behrends and H. Korn, in preparation).

DISCUSSION

During inhibitory LTP of the M-cell, an increase of the presynaptic volley reflects a potentiation of primary excitatory afferent synapses, with the consequence that more interneurons are fired by a fixed stimulus. However, the novel finding concerns the persistent facilitation of the inhibitory synapses. Its demonstration required, first, circumventing the chemical component of the first step in the disynaptic pathway using glutamate antagonists, so that electrotonic junctions alone secured transmission and, second, directly activating the inhibitory cell's axon while recording simultaneously from the M-cell. The mechanism of this LTP is still unclear and it may involve activation of parallel networks such as, for example, serotonergic ones (Mintz et al., 1989; Mintz and Korn, 1991) which modulate

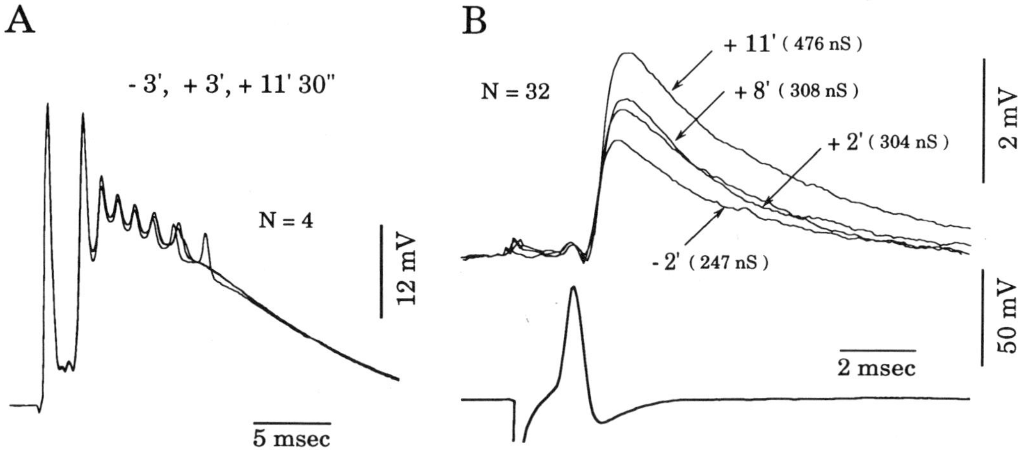

Figure 10.10 Potentiation of unitary IPSPs and associated conductance changes. (*A*) Superimposed averaged collateral depolarizing IPSPs which were large enough to generate spikes. Their stability indicates that the chloride loading of the M-cell was constant during the experiment. (*B*) Averaged unitary IPSPs (*upper*) produced in the M-cell by directly evoked presynaptic spikes (*lower*) in a simultaneously impaled presynaptic interneuron. The indicated conductance changes were derived as explained in text (conditioning stimulation of the eighth nerve: trains of 10 pulses at 300 Hz, 1 per 3 sec during 90 sec, i = 0.75T).

the strength of glycinergic synapses. Also, while the studied inhibition is glycinergic, involvement of GABA cannot be excluded, since this transmitter has been implicated in the dendritic inhibition of the M-cell (Korn et al., 1990; Petrov et al., 1991).

Quantal analysis has not been attempted yet to assess if inhibitory LTP has a presynaptic or postsynaptic origin, but our data suggest that, as in other instances, induction is postsynaptic and requires Ca^{2+}. In a separate series, the calcium chelator BAPTA was injected into the M-cell, with the result that input-output curves were not displaced. That is, even when the volley increased, LTP, if present, was restricted to the first (vestibular) relay, but the inhibitory synaptic gain was not affected by the conditioning protocol (Korn et al., 1992).

Involvement of Inhibitory Synapses

For a given size of the volley, LTP at the terminal synapses was greater in the absence than in the presence of excitatory amino-acid blockers, suggesting that in normal conditions there may be two components to this potentiation. One may implicate interactions of glycine with glutamate receptors, including those for NMDA, since the latter are present on the M-cell (Wolszon and Faber, 1988). This hypothesis is supported by the observation that conditioning trains are followed by a prolonged (200–300 msec) decaying depolarizing tail, a fraction of which is reduced by the blockers (Korn et al., 1992). In any case, this component would be critical for induction of LTP which persists in the presence of APV-CNQX.

A postsynaptic elevation of Ca^{2+} during the induction phase could result from this ion's influx or internal release; both may be consequences of transmitter interaction with the M-cell membrane similar to that of GABA with cerebellar granule cells (Connor et al., 1987), or they could be produced by depolarization, as in Purkinje cells (Llano et al., 1991; Kano et al., 1992). Finally, there could be a contribution by an unidentified voltage-dependent Ca^{2+} channel; along this line, it may be significant that a fraction only of the above-mentioned depolarizing tail is Cl^- dependent. Regardless, enhancement of M-cell glycinergic responses mediated by cAMP (Wolszon and Faber, 1989), with an increased probability of Cl^- channel opening (as in trigeminal neurons; Song and Huang, 1990) is one of the possible intracellular pathways.

LTP of inhibition is synapse specific, and it can be expressed when a few cells are active. This conclusion was reached by expressing the effectiveness of the VIIIth nerve stimulation in terms of the number of interneurons fired by the test stimulus, which is possible since the typical cell produces an r' of about 0.02 (Faber and Korn, 1982; Korn et al., 1982). Maximal LTPs of inhibition were obtained with stimulus strengths close to those of the test. In several such experiments, LTP resulted from the coactivation of a small pool of cells ranging from 2 to 12 and it was established that a component of the facilitation was synaptic.

Functional Consequences of Inhibitory LTP

In a study of ipsilateral M-cell excitatory LTP it was found that, to be effective, the tetanizing strength had to be above threshold (T) for orthodromic activation of that neuron (Yang et al., 1990). In contrast, the contralateral inhibition is potentiated by weaker stimuli. This selective enhancement of inhibition, which has been predicted for the hippocampus as well (Buszaki and Eidelberg, 1982), is in part the consequence of the network design: it is expected that a weak input will be more potent in activating high-resistance interneurons (Faber and Korn, 1988).

This may have significant functional consequences for the operation of the M-cell circuits. Specifically, weak repetitive sensory inputs evoked by auditory stimuli may preferentially depress the responsiveness of the contralateral cell; stronger ones will further shift the balance in favor of the ipsilateral neuron by increasing its excitation. Such a "push-pull" process may well preset the directionality of the startle response triggered by a single action potential in one M-cell (Foreman and Eaton, 1991). Finally, feedforward inhibition is a common feature of CNS networks; its modification, in a manner analogous to that described here, can be expected to sharpen and/or restrict the focus of excitation in these circuits.

As summarized in figure 10.1 (bottom) this study demonstrates that in vivo learning can alter the balance between excitation and inhibition within a network by modifying one or both of them. Also, if LTP is an appropriate paradigm of learning (Berger, 1984; Rose and Dunwiddie, 1986), then the role of inhibition can no longer be taken as one that only regulates the plasticity of excitation, for instance by setting the threshold for its modification (Yang and Faber, 1991). Indeed, it plays an active role in the maintenance of associative learning. More generally, it can be suggested that prolonged reinforcement of inhibition at either stage may underlie habituation of some behavioral responses, such as that of the M-cell–mediated escape response. Similar long-term potention of GABAergic transmission has recently also been reported in vitro in rat visual cortex (Komatsu and Iwakini, 1993).

ACKNOWLEDGMENTS

We thank J. Behrends for his help during this work, M. Titmus for carrying with us experiments with the glutamate antagonists, and D. S. Faber for critical evaluation of the latest data. This research was supported in part by grant no. 92-058 from the Direction des Recherches, Études et Techniques.

REFERENCES

Abraham, W. C., Gustafsson, B., and Wigström, H. (1987) Long-term potentiation involves enhanced synaptic excitation relative to synaptic inhibition in guinea-pig hippocampus. *J. Physiol.* 394:367–380.

Alger, B. E., and Nicoll, R. A. (1982) Pharmacological evidence for two kinds of GABA receptor on rat hippocampal pyramidal cells studied in vitro. *J. Physiol.* 328:125–141.

Berger, T. W. (1984) Long-term potentiation of hippocampal synaptic transmission affects rate of behavioral learning. *Science* 224:627–630.

Bliss, T. V. P., and Lømo, T. (1973) Long-lasting potentiation of synaptic transmission in the dentate area of the anaesthetized rabbit following stimulation of the perforant path. *J. Physiol.* 232:331–356.

Brindley, F. R. S. (1967) The classification of modifiable synapses and their use in models for conditioning. *Proc. R. Soc. B* 168:361–376.

Buzsaki, G., and Eidelberg, E. (1982) Direct afferent excitation and long-term potentiation of hippocampal interneurons. *J. Neurophysiol.* 48:597–607.

Chavez-Noriega, L. E., Bliss, T. V. P., and Halliwell, J. V. (1989) The EPSP-spike (E-S) component of long-term potentiation in the rat hippocampal slice is modulated by GABAergic but not cholinergic mechanisms. *Neurosci. Lett.* 104:58–64.

Connor, J. A., Tseng, H.-Y., and Hockberger, P. E. (1987) Depolarization- and transmitter-induced changes in intracellular Ca^{2+} of rat cerebellar granule cells in explant cultures. *J. Neurosci.* 7:1384–1400.

Douglas, R. M., Goddard, G. V., and Riives, M. (1982) Inhibitory modulation of long-term potentiation: Evidence for a postsynaptic locus of control. *Brain Res.* 240:259–272.

Faber, D. S., and Korn, H. (1978) Electrophysiology of the Mauthner cell: Basic properties, synaptic mechanisms, and associated networks. In *Neurobiology of the Mauthner cell,* D. S. Faber and H. Korn (eds.). New York: Raven Press, pp. 47–131.

Faber, D. S., and Korn, H. (1982) Transmission at a central inhibitory synapse. I. Magnitude of unitary postsynaptic conductance change and kinetics of channel activation. *J. Neurophysiol.* 48:654–678.

Faber, D. S., and Korn, H. (1987) Voltage-dependence of glycine activated Cl^- channels: A potentiometer for inhibition? *J. Neurosci.* 7:807–811.

Faber, D. S., and Korn, H. (1988) Unitary conductance changes at teleost Mauthner cell glycinergic synapses: A voltage-clamp and pharmacologic analysis. *J. Neurophysiol.* 60: 1982–1999.

Faber, D. S., Lin, J. W., and Korn, H. (1991) Silent synaptic connections and their modifiability. *Ann. N. Y. Acad. Sci.* 627:151–164.

Foreman, M. B., and Eaton, R. C. (1991) Bilateral Mauthner lesions cause increases in EMG measures of escape responses. *Soc. Neurosci. Abstr.* 17:84.8.

Furshpan, E. J., and Furukawa, T. (1962) Intracellular and extracellar responses of the several regions of the Mauthner cell of the goldfish. *J. Neurophysiol.* 25:732–771.

Furukawa, T., and Furshpan, E. J. (1963) Two inhibitory mechanisms in the Mauthner neurons of goldfish. *J. Neurophysiol.* 26:140–176.

Griffith, W. H., Brown, T. H., and Johnston, D. (1986) Voltage-clamp analysis of synaptic inhibition during long-term potentiation in hippocampus. *J. Neurophysiol.* 55:767–775.

Haas, H. L., and Rose, G. (1982) Long-term potentiation of excitatory synaptic transmission in the rat hippocampus: The role of inhibitory processes. *J. Physiol.* 329:541–552.

Kano, M., Rexhausen, U., Dreessen, J., and Konnerth, A. (1992) Synaptic excitation produces a long-lasting rebound potentiation of inhibitory synaptic signals in cerebellar Purkinje cells. *Nature* 356:601–604.

Komatsu, Y. and Iwakini, M. (1993) Long-term modification of inhibitory synaptic transmission in developing visual cortex. *NeuroReport* 4:907–910.

Korn, H., and Faber, D. S. (1975) An electrically mediated inhibition in goldfish medulla. *J. Neurophysiol.* 38:452–471.

Korn, H., and Faber, D. S. (1976) Vertebrate central nervous system: Same neurons mediate both electrical and chemical inhibitions. *Science* 195:1166–1169.

Korn, H., and Faber, D. S. (1990) Quantitative electrophysiological studies and modelling of central glycinergic synapses. In *Glycine Neurotransmission,* O. P. Ottersen and J. Storm-Mathisen (eds.). New York: Wiley, pp. 139–170.

Korn, H., Sotelo, C., and Bennett, M. V. L. (1977) The lateral vestibular nucleus of the toadfish *Opsanus tau*: Ultrastructural and electrophysiological observations with special reference to electrotonic transmission. *Neuroscience* 2:851–884.

Korn, H., Faber, D. S., and Triller, A. (1986) Probabilistic determination of synaptic strength. *J. Neurophysiol.* 55:402–421.

Korn, H., Mallet, A., Triller, A., and Faber, D. S. (1982) Transmission at a central inhibitory synapse. II. Quantal description of release, with a physical correlate for binomial n. *J. Neurophysiol.* 48:679–707.

Korn, H., Faber, D. S., and Triller, A. (1990) Convergence of morphological physiological, and immunocytochemical techniques for the study of single Mauthner cells. In *Handbook of Chemical Neuroanatomy, Vol. 8: Analysis of Neuronal Microcircuits and Synaptic Integration,* A. Björklund and T. Hökfelt (eds.). Amsterdam: Elsevier, pp. 403–480.

Korn, H., Oda, Y., and Faber, D. S. (1992) Long-term potentiation of inhibitory circuits and synapses in the central nervous system. *Proc. Natl. Acad. Sci. USA* 89:440–443.

Llano, I., Lereshe, N., and Marty, A. (1991) Calcium entry increases the sensitivity of cerebellar Purkinje cells to applied GABA and decreases inhibitory synaptic currents. *Neuron* 6:565–574.

Mintz, I., and Korn, H. (1991) Serotonergic facilitation of quantal release at central inhibitory synapses. *J. Neurosci.* 11:3359–3370.

Mintz, I., Gotow, T., Triller, A., and Korn, H. (1989) Effects of serotonergic afferents on quantal release at central inhibitory synapses. *Science* 245:190–192.

Misgeld, U., Sarvey, J. M., and Klee, M. R. (1979) Heterosynaptic postactivation potentiation in hippocampal CA3 neurons: Long-term changes of the postsynaptic potentials. *Exp. Brain Res.* 37:217–229.

Oda, Y., and Korn, H. (1991) Long-term potentiation of an inhibitory pathway in the CNS. *Soc. Neurosci. Abstr.* 17:1.

Petrov, T., Seitanidou, T., and Triller, A. (1991) Differential distribution of GABA- and serotonin-containing afferents on an identified central neuron. *Brain Res.* 559:75–81.

Rose, G. M., and Dunwiddie, T. V. (1986) Induction of hippocampal long-term potentiation using physiologically patterned stimulation. *Neurosci. Lett.* 69:244–248.

Scharfman, H. E., and Sarvey, J. M. (1985) Gamma-aminobutyrate sensitivity does not change during long-term potentiation in rat hippocampal slices. *Neuroscience* 15:695–702.

Song, Y., and Huang, L. M. (1990) Modulation of glycine receptor chloride channels by cAMP-dependent protein kinase in spinal trigeminal neurons. *Nature* 348:242–245.

Stelzer, A., Slater, N. T., and ten Bruggencate, G. (1987) Activation of NMDA receptors blocks GABAergic inhibition in an in vitro model of epilepsy. *Nature* 326:698–701.

Steward, O., Tomasula, R., and Levy, W. B. (1990) Blockade of inhibition in a pathway with dual excitatory and inhibitory action unmasks a capability for LTP that is otherwise not expressed. *Brain Res.* 516:292–330.

Teyler, T. J., and Di Scenna, P. (1987) Long-term potentiation. *Annu. Rev. Neurosci.* 10: 131–161.

Triller, A., and Korn, H. (1981) Morphologically distinct classes of inhibitory synapses arise from the same neurons: Ultrastructural identification from crossed vestibular interneurons intracellularly stained with HRP. *J. Comp. Neurol.* 203:131–155.

Triller, A., and Korn, H. (1986) Variability of axonal arborizations hides simple rules of construction: A topological study from HRP intracellular injections. *J. Comp. Neurol.* 253: 500–513.

Waziri, R., Kandel, R., and Fraziel, W. T. (1969) Organization of inhibition in abdominal ganglion of Aplysia. II. Posttetanic potentiation, heterosynaptic depression, and increment in frequency of inhibitory postsynaptic potentials. *J. Neurophysiol.* 32:509–519.

Wigström, H., and Gustafsson, B. (1983) Facilitated induction of hippocampal long-lasting potentiation during blockade of inhibition. *Nature* 301:603–604.

Wolszon, L. R., and Faber, D. S. (1988) Fast EPSP's evoked in the goldfish Mauthner cell by sensory afferents are due to NMDA receptor activation. *Soc. Neurosci. Abstr.* 14:939.

Wolszon, L. R., and Faber, D. S. (1989) The effects of postsynaptic levels of cyclic AMP on excitatory and inhibitory responses of an identified central neuron. *J. Neurosci.* 9:784–797.

Yamamoto, C., and Chujo, T. (1978) Long-term potentiation in thin hippocampal sections studied by intracellular and extracellular recordings. *Exp. Neurol.* 58:242–250.

Yang, X.-D., and Faber, D. S. (1991) Initial synaptic efficacy influences induction and expression of long-term changes in transmission. *Proc. Natl. Acad. Sci. USA* 88:4299–4303.

Yang, X.-D., Korn, H., and Faber, D. S. (1990) Long-term potentiation of electrotonic coupling at mixed synapses. *Nature* 348:542–545.

Zottoli, S. J., and Faber, D. S. (1980) An identifiable class of statoacoustic interneurons with bilateral projections in the goldfish medulla. *Neuroscience* 5:1287–1302.

11 Long-Term Depression of Synaptic Transmission: Cellular Mechanisms and Regulation by Previous Synaptic History

Patric K. Stanton, Eric M. Wexler, Libor Velíšek, and Thomas G. Hedberg

For over a century, it has been hypothesized that memories are persistent changes in the strength of connections between specific subsets of neurons (James, 1890; Tanzi, 1893). With more detailed description of synaptic transmission, the focus has narrowed to long-term experience-dependent changes in synaptic efficacy (Hebb, 1949). The discovery of long-term potentiation (LTP; Bliss and Lømo, 1973), a synaptically activated long-lasting increase in connection strength, has generated a rush of excitement and study aimed at elucidating both the cellular mechanisms of LTP induction and expression, and the role LTP plays in memory.

Imagine taking notes on the same sheet of paper day after day, until its surface is uniformly black. After some time, it would hold no more useful information than when it was blank. An information storage and retrieval system such as the brain could be confronted with a similar dilemma if LTP were its only means of modulating synaptic activity. Any rule that serves only to increment synaptic strength will eventually lead to saturation, thus rendering the system useless. Therefore, it seems unlikely that LTP is the only means of modulating synaptic efficacy. The most straightforward way of addressing this problem is to invoke some mechanism to produce long-term depression (LTD) of synaptic strength. The simplest such mechanism would be a nonspecific decay of all synaptic strength. Although this would prevent saturation, it would also result in the loss of information stored by LTP. Theoretical work suggests a more efficient alternative would be a synapse-specific activity-dependent depression (Sejnowski, 1977; Bienenstock et al., 1982). There are compelling reasons to think that information in networks of neurons is encoded, in part, in the temporal firing patterns of afferent inputs. Since the relative timing of these firing patterns contains information, theoreticians have invoked a number of rules that take into account the phase relation of convergent inputs and utilize a combination of LTP and LTD to store the computational result. Not surprisingly, detailed experimental verification of many of these ideas has lagged behind, and is only now beginning to accumulate.

The earliest reports described heterosynaptic LTD of inactive synapses that converged on neurons receiving LTP-inducing stimuli (Lynch et al., 1977; Levy and Steward, 1979). In these studies, the induction of LTP at one set of synapses also elicited LTD of other synaptic contacts on the same neuron, if those synapses were not stimulated. This has the effect of renormalizing the total synaptic strength of the terminal field to a constant value (von der Malsburg, 1973). More recent studies have described a homosynaptic form of LTD that appears to require the pairing of presynaptic firing with postsynaptic inactivity, usually achieved by stimulating presynaptically while hyperpolarizing the postsynaptic neuron (Stanton and Sejnowski, 1989; Chattarji et al., 1989; Artola et al., 1990). An important consequence of such a pairing requirement is that, in conjunction with the "Hebbian" property of LTP that requires pairing of presynaptic activity with simultaneous postsynaptic depolarization (Gustaffson et al., 1987), it confers on the synapse the ability to compute a covariance function between synaptic contacts on the same neuron (Frégnac et al., 1988; Stanton and Sejnowski, 1989; Chattarji et al., 1989). However, this is a long way from decoding either the information contained in patterned trains of action potentials or the behavioral meaning of the covariance being computed and stored.

The question of the reversibility of LTP has also been debated for some time. Two studies have found that previously potentiated, but not naive, synapses can be depressed by prolonged low-frequency synaptic stimulation in vivo (Staubli and Lynch, 1990) and in vitro (Fujii et al., 1991), while others have described LTD induced by prolonged low-frequency stimulation in hippocampal slices in vitro (Dudek and Bear, 1992; Mulkey and Malenka, 1992; Wexler and Stanton, 1992, 1993; Velíšek et al., 1993). While we cannot say whether this effect is a true reversal of the cellular changes mediating LTP expression or an additional biochemical change, it does suggest that the threshold for induction of LTD is not fixed. Recent data from our laboratory, described below, address this possibility. We have found evidence for an activity-dependent "sliding threshold" for the induction of homosynaptic LTD that supports the theoretical predictions of Bienenstock and coworkers (1982). In addition, we discuss the available data implicating two subtypes of glutamate receptor in the induction of LTD, and the possible common role of intracellular calcium as an LTD trigger. While debate will undoubtedly continue, there is mounting evidence that some form of activity-dependent LTD can occur at physiological synapses. It is, however, still speculative to suggest specific roles for LTD in cortical processing and information storage.

LTD IN THE HIPPOCAMPUS AND THE HEBBIAN COVARIANCE FUNCTION

What is Hebbian Covariance?

Statistically, covariance is a measure of the variance between two parameters, each with its own probability distributions. Consequently, two parameters may be either completely independent, or positively or negatively correlated. In the context of synaptic activity, Hebbian covariance is the neuronal computation of the statistical covariance, with respect to time, between synaptic contacts on the same neuron (Frégnac et al., 1988; Stanton and Sejnowski, 1989; Chattarji et al., 1989). Theoretically, maximal extraction of information encoded in two time-varying signals is achieved with a plasticity rule sensitive to covariance (Sejnowski, 1977; Wilshaw and Dayan, 1990). This has led to experimental attempts to determine whether properties of cortical synaptic plasticity might, in fact, lead to induction of LTP when two signals show positively correlated firing patterns, and LTD with negatively correlated activity. Figure 11.1 illustrates the stimulus paradigms employed in these studies (Stanton and Sejnowski, 1989; Chattarji et al., 1989; Stanton et al., 1991; Christie and Abraham, 1992). In these studies, two separate Schaffer collaterals inputs were coactivated with phasic stimulation (figure 11.1a). One of the two inputs was designated as the conditioning input (figure 11.1b), which was stimulated with high-frequency "theta bursts" used by Larson and coworkers (1986) to elicit homosynaptic LTP. A second test input was stimulated with a low-frequency train (5 Hz) that was placed either in phase (positively correlated) or out of phase (negatively correlated) with the bursts. Consistent with the Hebbian mechanism of LTP induction, which requires pairing of postsynaptic depolarization with glutamate release (Gustaffson et al., 1987), the in-phase stimulation yielded LTP of the test input. Conversely, out-of-phase stimulation yielded homosynaptic LTD. Intracellular studies in hippocampus (Stanton and Sejnowski, 1989) and visual cortex (Artola et al., 1990) have found that pairing of postsynaptic hyperpolarization with presynaptic glutamate release is sufficient to induce homosynaptic LTD.

In the associative experimental paradigm described above, the computation performed is, in effect, the covariance between *two* inputs. An interesting question is how a postsynaptic neuron with a massive convergence of inputs could compute the covariance between each synapse pair. It obviously cannot compute the exact covariance, since there is no way for every pair of synapses to interact independently of all other synapses. As a result, the net interaction is a function of the activity at all other syn-

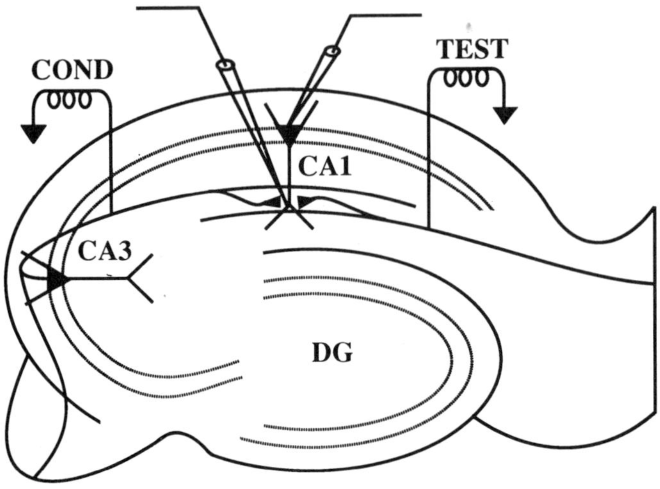

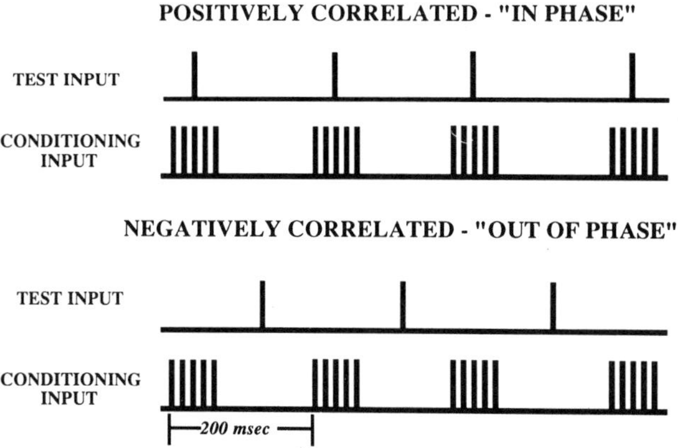

Figure 11.1 Covarying associative stimulus paradigms. (*a*) Schematic diagram of the recording arrangement in the CA1 pyramidal cell somatic and dendritic layers and stimulus sites activating two independent groups of Schaffer collateral/commissural afferents (COND and TEST). (*b*) Schematic diagram of the stimulus paradigm used for Hebbian covariance experiments. Conditioning input stimuli were four trains of bursts, given at 10 sec intervals, each train containing 10 bursts. Each burst consisted of 5 stimuli at 100 Hz, with an interburst interval of 200 msec. Thus, each train lasted 2 sec, with 50 pulses per train. When these inputs were IN PHASE, the test single shocks were superimposed on the middle of each burst of the conditioning input. During OUT OF PHASE stimulation, single shocks were timed symmetrically between the bursts. (Reprinted with permission from Stanton and Sejnowski, 1989)

apses. For a large number of active contacts, the integrated net activity level over time, reflected in membrane potential, is probably used in the covariance computation in lieu of a single synaptic input. Thus, the strength of the synaptic connection being varied is based on the correlation between the synapse in question and the summed effect on membrane potential of all other contacts. This was the rationale for the experimental pairing of presynaptic activity with either postsynaptic depolarization (Gustaffson et al., 1987) or hyperpolarization (Stanton and Sejnowski, 1989; Stanton et al., 1991).

Enhanced Associative LTD following Theta-Frequency Stimulation

A recent in vivo study (Christie and Abraham, 1992) has examined the induction of both heterosynaptic and associative LTD at synapses of the lateral perforant path (LPP) and medial perforant path (MPP) in the dentate gyrus. These studies reported that, at previously unstimulated synapses, only heterosynaptic LTD of LPP synapses was observed when the MPP was tetanized. That is, the application of out-of-phase LPP stimuli as in figure 11.1b did not elicit any further LTD compared to no LPP stimulation. However, an additional associative LTD of equal amplitude with heterosynaptic LTD could be produced after LPP synapses had been stimulated, or "primed," with a brief theta-frequency (5 Hz) stimulus train. It must be emphasized that the heterosynaptic and associative forms of LTD were additive in this study, suggesting that they may represent different cellular mechanisms by which LTD can be induced. Pharmacological experiments further supported this hypothesis. Application of CPP (3-2-carboxypiperazin-4-yl propyl-1-phosphonic acid), an NMDA receptor blocker, during theta stimulation eliminated the priming of associative LTD. Consistent with data from field CA1 (Stanton and Sejnowski, 1989), they found that similar application of CPP during phasic stimulation did not block associative LTD. However, CPP did block the induction of heterosynaptic LTD, indicating that LTD at the same synapses can be induced by separate NMDA and non-NMDA mechanisms.

LOW-FREQUENCY STIMULUS-INDUCED HOMOSYNAPTIC LTD AT SCHAFFER COLLATERAL–CA1 SYNAPSES

There have been a number of reports that prolonged low-frequency stimulation (LFS; 1–5 Hz) can elicit homosynapic LTD of synaptic transmission. However, while some studies indicated that LFS LTD could only be evoked at synapses where LTP had first been induced (Staubli and Lynch, 1990; Fujii et al., 1991), others have described LFS LTD of basal synaptic transmission in naive slices (Dudek and Bear, 1992; Mulkey and Malenka, 1992). Therefore, we systematically compared the abilities of low-frequency synaptic stimulation in field CA1 (LFS, 1 Hz/15 min, 900 pulses) to elicit LTD of synaptic transmission in naive versus previously potentiated slices from rats of different ages. One of two Schaffer collateral inputs to CA1 pyramidal neurons received LFS, while the other was only stimulated once every 30 sec and served as the control. In initial experiments on naive slices from adults, there was a small, but statistically significant, LTD of the population EPSP slope measured 30 min after the end of LFS, when compared to either the pre-LFS baselines, or to control unstimulated pathway EPSPs ($-14.6 \pm 3.7\%$, $n = 7$, $P < .05$; paired t-test). Homosynaptic LTD can, apparently, be induced without any prior stimulus-induced LTP. This does not, of course, rule out the possibility that previous LTP endogenous to the animal's experience-dependent activitation might be necessary for LFS LTD.

A "SLIDING THRESHOLD" FOR INDUCTION OF LTD

Recent work (Huang et al., 1992) has suggested that there is activity-dependent regulation of threshold for the induction of LTP at Schaffer collateral–CA1 synapses. This study found that brief bursts of stimuli that did not produce LTP, but did cause short-term potentiation (STP) of synaptic transmission, prevented a subsequent high-frequency stimulation from inducing LTP. That is, the threshold for inducing LTP was raised at synapses that have been previously active, even if that activation itself did not yield LTP. Theoretical studies have found it useful, or even essential, to employ a variable threshold for the modulation of synaptic strength. Bienenstock and coworkers (1982) employed LTD, in conjunction with LTP, to form a biphasic plasicity rule with some interesting features (figure 11.2). In this model, if the product of presynaptic and postsynaptic activity was over some threshold value, LTP was produced, whereas a product below this threshold elicited LTD. Moreover, the absolute value of this threshold was dynamically changed as a function of overall activity levels in the network. This "sliding threshold" feature had the effect of preventing synapses from either saturating at maximal LTP or depressing to zero synaptic strength, and this was an essential factor in enabling the network to successfully mimic both the effects of visual deprivation studies and the long-term stability of ocular dominance columns.

In light of experiments indicating that the threshold for LTP induction can vary, we hypothesized that previous synaptic activity should make it more likely that LTD would be induced, as well as less likely that LTP be evoked. Since it has been reported that previously inducing LTP at a set of

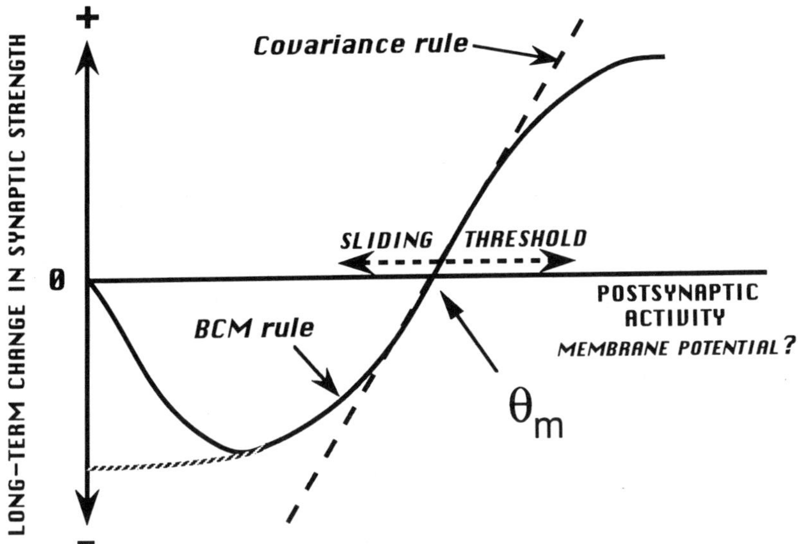

Figure 11.2 The theoretical synaptic strength modification function supported by the data. The learning rule of Bienenstock et al. (1982) is plotted as a solid line (BCM rule), compared to a simple covariance function (dashed line) and a modification rule that does not return to the origin at low levels of postsynaptic activity (hatched line). θ_m is the crossover threshold from LTD to LTP induction, and it is this crossover point that "slides" as a function of previous synaptic activity. See discussion for further details.

synapses increased the likelihood of eliciting LTD with low-frequency stimulation (Staubli and Lynch, 1990; Fujii et al., 1991), we first tested whether LFS would elicit greater average LTD after the induction of LTP. Figure 11.3 illustrates a typical experiment, where LTD was not induced by the first LFS stimulation to naive Schaffer collateral/commissural synapses in field CA1 (solid bar, 1 Hz, test input), but a second, identical LFS applied 30 min after the induction of homosynaptic LTP did elicit robust homosynaptic LTD of test input evoked EPSPs (closed circles). In all experiments, the induction of LTP at one of two synaptic inputs on CA1 pyramidal neurons markedly increased the average LTD produced by a LFS applied 30 min later ($-38.9 \pm 7.2\%$, $n = 9$, Student's t-test, $P < .05$ compared to untetanized controls). These findings confirm that previous induction of LTP does, in fact, lower the threshold for induction of homosynaptic LTD.

It was not clear, however, whether LTP had to be previously induced in order to lower the threshold for LTD induction. Given that Huang and coworkers (1992) showed that inducing STP was sufficient to prevent subsequent LTP, we employed an experimental protocol similar to theirs to determine whether inducing STP could also lower the threshold for LTD. In figure 11.4, LFS (1 Hz) was applied to a control Schaffer collateral

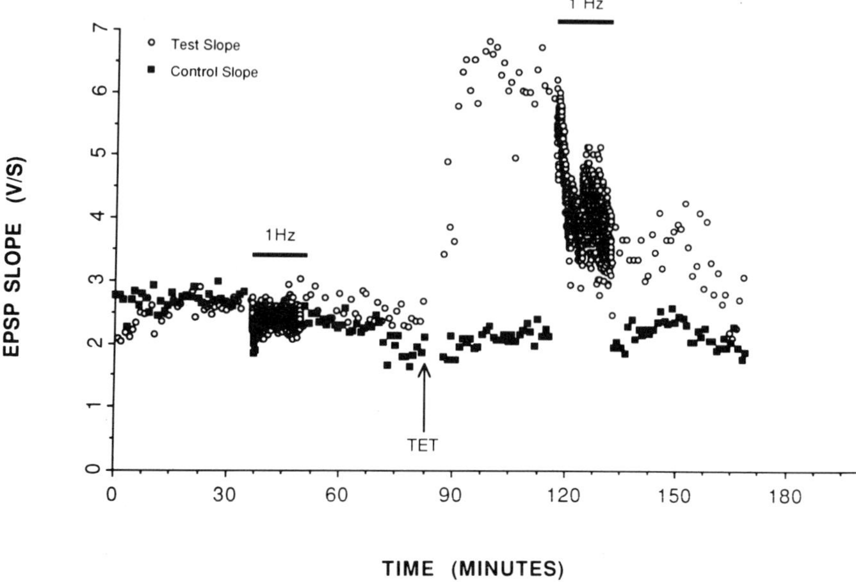

Figure 11.3 Inducing homosynaptic LTP lowers the threshold for long-term depression (LTD) induction at Schaffer collateral–CA1 synapses. EPSP slope (V/sec) is plotted for a test input (open circles) which received low-frequency stimulation (LFS, 1 Hz/15 min) before and after the induction of LTP, and for a control input (closed squares) stimulated only once every 30 sec. When LFS was applied initially (1 Hz, first solid bar), no LTD was elicited. The test input was then tetanized (TET; 4 trains × 100 Hz/500 msec) to induce LTP. 30 min later, the same LFS was applied a second time (1 Hz, second solid bar), and now produced a large homosynaptic LTD of evoked EPSPs.

input and elicited only slight LTD. Thereafter, repeated STPs were produced by brief bursts of Schaffer collateral stimulation (arrows; 30 Hz/.33 sec), and a second LFS given to a second, test input. This LFS now elicited a robust LTD of synaptic transmission, indicating that previous synaptic activity that does not evoke LTP can still make it more likely to induce LTD. Together with the previous LTP study, these experiments supply physiological evidence for a "sliding threshold" for synaptic plasticity, and support the use of such threshold regulation in theoretical neural networks.

We also tested whether NMDA receptor activation alone was sufficient to prime synapses to exhibit enhanced LTD. Huang and colleagues (1992) found that iontophoresis of NMDA alone onto CA1 pyramidal cell apical dendrites was able to block subsequent induction of LTP. We performed analogous experiments with LTD threshold (Wexler and Stanton, 1992, 1993). NMDA was applied iontophoretically to the apical dendrites of CA1 pyramidal neurons prior to LFS (1 Hz/15 min). NMDA application alone did not enhance subsequent induction of LTD, even though the induction of LTP did shift LTD threshold at these same synapses. Thus, we

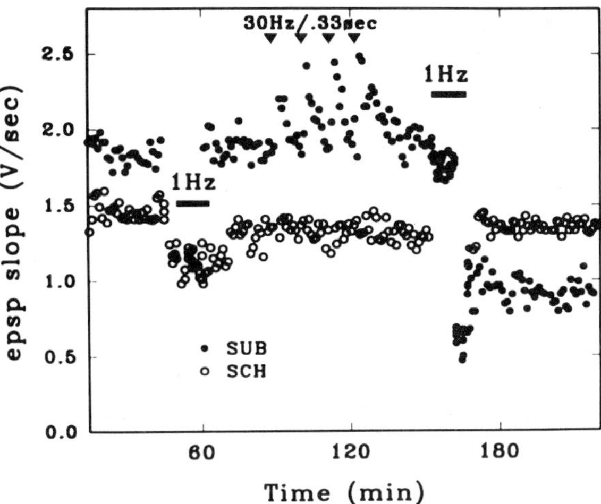

Figure 11.4 Repeated induction of short-term potentiation (STP) of synaptic transmission also lowers the threshold for LTD induction. LFS stimulation (1 Hz/15 min, solid bar) was applied to a control Schaffer collateral input in stratum radiatum of field CA1 (open circles), where it elicited an approximately 15% depression in EPSP slope (V/sec). STP was then induced four times in a second, test input (closed circles) by application of brief high-frequency bursts of stimuli (30 Hz/.33 sec each). 30 min after the last burst, a second LFS stimulation identical to the first was applied to the test pathway, where it now elicited about a 50% reduction in EPSP slopes. In 5 slices, LTD of the naive pathway synapses averaged $-12.9 \pm 3.5\%$, compared with $-32.7 \pm 5.6\%$ LTD following STP ($P < .05$, Student's t-test compared to unstimulated inputs).

conclude that other activity-dependent cellular events not triggered by NMDA-gated calcium influx alone are also necessary for lowering LTD threshold.

Given that metabotropic glutamate receptor activation has also been implicated in the induction of some forms of LTD, we were led to wonder whether second messenger systems that can be activated both by synaptic firing and metabotropic receptors might not be needed to prime synapses for enhanced LTD. Naturally, since metabotropic receptors activate production of both inositol triphosphate (IP_3), which releases Ca^{2+} from intracellular stores, and diacylglycerol (DAG), a logical candidate messenger system is the protein kinase C (PKC) family. We found that transient bath application of the general PKC activator phorbol 12,13-diacetate (PDA) 1 hr prior to low-frequency stimulation does, indeed, prime synapses to exhibit significantly greater homosynaptic LTD of synaptic potentials (controls $= -14.6 \pm 5.8\%$; PDA-pretreated $= -46.6 \pm 6.3\%$ LTD of EPSP slope $n = 5$; $P < .05$, Student's t-test). However, this does not prove that PKC activation is a required mediator of stimulus-dependent priming, simply that PKC stimulation is sufficient to shift the threshold for LTD to the left (see figure 11.2).

GLUTAMATE RECEPTOR SUBTYPES IMPLICATED IN INDUCTION OF LTD

Some of the studies cited earlier have tested the involvement of NMDA receptors and intracellular [Ca^{2+}] in the induction of LFS LTD. They found that NMDA receptor antagonists were able to block the induction of LFS LTD at Schaffer collateral–CA1 synapses in hippocampus (Dudek and Bear, 1992; Mulkey and Malenka, 1992). Furthermore, NMDA blockers have been found to prevent the induction of heterosynaptic LTD at perforant path–dentate granule cell synapses (Christie and Abraham, 1992), but the same group has also implicated voltage-dependent calcium channels in similar heterosynaptic LTD in field CA1 (Wickens and Abraham, 1991). Intracellular injection of the calcium chelator BAPTA prior to low-frequency stimulation also appears to be able to block LTD induction in both visual cortex (Kimura et al., 1990) and hippocampus (Mulkey and Malenka, 1992), suggesting that NMDA-gated influx of calcium is probably one, but not the only, contributor to LTD.

We have previously shown that, while an NMDA receptor blocker was not able to prevent induction of associative LTD by the Hebbian covariance paradigm (Stanton and Sejnowski, 1989; Chattarji et al., 1989), an antagonist of metabotropic glutamate receptors was able to block this form of hippocampal LTD (Stanton et al., 1991). Thus, the question of which receptor(s) can play a role in LTD induction is still open. We tested both the specific NMDA receptor antagonist 2-amino-5-phosphonopentanoic acid (AP5) and the metabotropic receptor blocker 2-amino-3-phosphonopropionic acid (AP3; Schoepp and Johnson, 1989) for their ability to block LFS-induced LTD at previously potentiated Schaffer collateral synapses.

In contrast to associative LTD, LFS-induced LTD appears to require activation of both NMDA and metabotropic glutamate receptors for its full expression. Figure 11.5 illustrates experiments where the NMDA receptor antagonist AP5 was bath-applied after the induction of LTP, but before application of LFS. When LFS (solid bar, 1 Hz/15 min) was given to one of two Schaffer collateral inputs in the presence of AP5 (10 μM, dotted line), we observed a transient homosynaptic depression of both EPSPs (closed circles) and population spikes (open squares) which did not persist as LTD. In contrast, when AP5 was washed out for 60 min and the same LFS given a second time, robust LTD lasting over 90 min resulted. Similar results with the metabotropic receptor antagonist AP3 (25 μM) are illustrated in figure 11.6, where only a transient depression of EPSP slopes (closed circles) and population spikes (open squares) was elicited by LFS (1 Hz/15 min) in AP3, while LTD was again produced after drug washout.

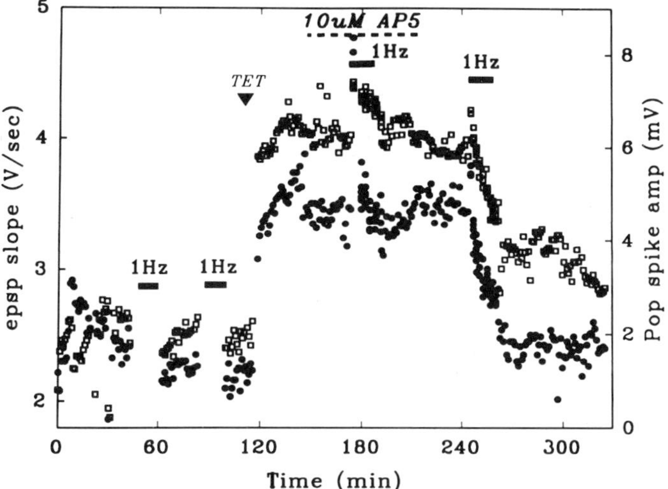

Figure 11.5 Induction of LTD by LFS requires activation of NMDA receptors. EPSP slope (closed circles) and population spike amplitude (open squares) was recorded in stratum radiatum of field CA1, and LFS (solid bar, 1 Hz/15 min) applied before and after induction of LTP. As in previous experiments, LFS did not elicit LTD at the naive input. LTP was then induced with high-frequency trains of stimuli (TET; 4 × 100 Hz/500 msec) to the Schaffer collaterals, and the NMDA receptor blocker 2-amino-5-phosphonopentanoic acid (AP5, 10 μM; dotted line) bath applied 30 min later. When a second LFS was given, no LTD was observed. The drug was washed out of the bath for 30 min, and a third identical LFS now elicited large LTD.

In untreated control slices, the average LTD (percent reduction) was: EPSP = $-21.24 \pm 8.6\%$, pop. spike = $-29.9 \pm 12.0\%$ (n = 8). Both AP5 and AP3 significantly reduced the amount of LTD elicited by LFS. In AP5, EPSP slope depressed only $-2.4 \pm 5.6\%$, and the pop. spike decreased only $-2.5 \pm 5.5\%$ (n = 5, $P < .05$, Student's t-test compared to untreated controls). Similarly, persistent depression in AP3-treated slices was only: EPSP = $-5.1 \pm 4.3\%$, pop. spike = $-9.3 \pm 1.6\%$ (n = 5, $P < .05$, Student's t-test). In these experiments, near-maximal LTP had been evoked 30 min prior to the start of bath application of antagonist, and LFS applied after a 30 min washin. In all experiments, after a 60-min drug-free washout, a second LFS consistently elicited robust LTD.

These findings correlate quite interestingly with those of Christie and Abraham (1992) comparing heterosynaptic and associative forms of LTD in the dentate gyrus. They found that, while heterosynaptic LTD was blocked by an NMDA receptor antagonist, the additional associative LTD they observed was not. Furthermore, the priming of associative LTD induction they induced with theta rhthym stimuli was, itself, NMDA receptor–dependent. Thus, NMDA, and metabotropic glutamate receptors and voltage-dependent calcium channels may all play a role in the induction of

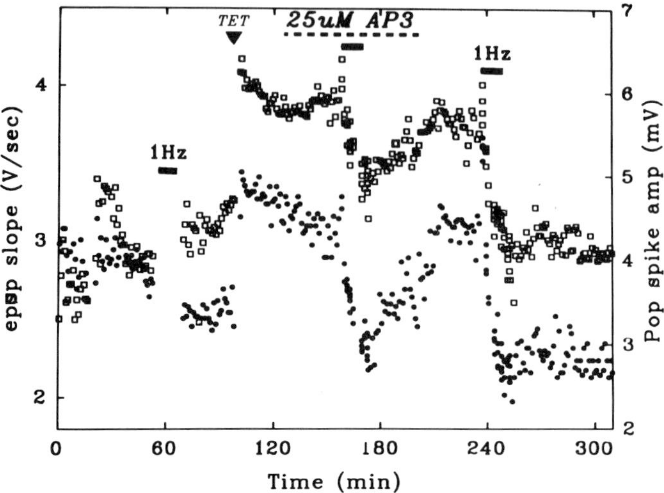

Figure 11.6 Induction of LTD also requires activation of metabotropic glutamate receptors. Using the same experimental paradigm as figure 11.5, EPSPs (closed circles) and pop spikes (open squares) were recorded in field CA1 (stratum radiatum) and LFS (solid bar, 1 Hz/15 min) applied before and after LTP. In this experiment, LFS in the unpotentiated slice elicited some LTD of the EPSP (14%). 30 min later, we bath applied the metabotropic receptor antagonist 2-amino-3-phosphonopropionic acid (AP3, 25 μM), waited an additional 30 min, and applied a second identical LFS. Although LFS elicited a large short-term depression, synaptic potentials completely recovered to baseline amplitudes by 30 min later. After a 30 min drug-free wash, a third identical LFS now elicited robust and persistent LTD.

LTD of synaptic transmission, perhaps depending on the neurons studied and input firing patterns. One thing all three receptors have in common is the ability to raise intracellular [Ca^{2+}], suggesting that increases in [Ca^{2+}]$_i$ independent of route, that fall within an appropriate window may elicit LTD.

LTD IN THE IMMATURE HIPPOCAMPUS

Taken together, the data described in the previous sections support the hypothesis that both NMDA and non-NMDA routes of calcium entry/release can contribute to the induction of LTD. However, a recent report directly tested the possibility of non-NMDA induced LTD at Schaffer collateral/commissural synapses in hippocampal slices, with negative results (Goldman et al., 1990). These investigators reported the failure to induce LTD with high-frequency stimulation applied in the presence of an NMDA receptor blocker in slices from adult rats. Our and other experience (Dudek and Bear, 1993) has suggested that using slices from younger animals may increase the amount of LTD observed. Moreover, an upregulation of LTD in immature hippocampus would be consistent with similar

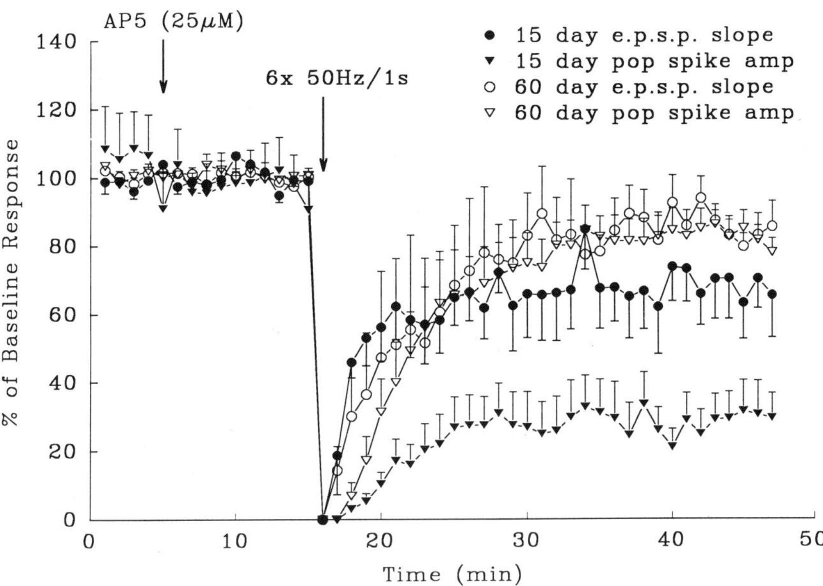

Figure 11.7 Under conditions of NMDA receptor blockade, high-frequency stimulation elicits homosynaptic LTD, which is largest in slices from immature (15-day-old) rats. EPSP slopes (circles) and pop spike amplitude (triangles) are plotted in slices from 15-day-old (closed symbols) and 60-day-old (open symbols) rats. 25 μM AP5 was bath applied (first arrow), prior to high-frequency Schaffer collateral stimulation (6 × 50 Hz/1 sec). Adult slices (60 day; n = 8) showed significant LTD of synaptically evoked EPSP (80.8 ± 0.5%; $P < .05$; paired t-test compared to prestimulus baseline) and pop spikes (80.5 ± 3.0%; $P < .05$; paired t-test compared to baseline). However, slices from 15-day-olds (n = 8) exhibited even larger LTD of both EPSP (69.5 ± 8.7%; $P < .05$; paired t-test) and pop spike (26.8 ± 6.1%; $P < .05$; paired t-test).

developmental observations that have already been made for LTP (Harris and Teyler, 1984).

We have recently conducted a comparison between slices from rats at 15, 30, and 60 days of age, to determine the relative presence of non-NMDA mediated LTD that can be evoked (Velíšek et al., 1993). In these studies, high-frequency Schaffer collateral/commissural stimulation (6 × 50 Hz/1 sec) was given while bath applying the NMDA receptor antagonist AP5 (25 μM). Figure 11.7 summarizes these experiments for the 15-and 60-day-old age groups, when AP5 was bath-applied 15 min prior to high-frequency stimulation. In adult (60 day) slices, we observed statistically significant homosynaptic LTD to approximately 80% of prestimulus baseline, in both EPSP slope (open circles) and population spike amplitude (open triangles). Furthermore, slices from 15-day-old rats exhibited significantly larger LTD of both EPSPs (closed circles; ~70% of baseline) and population spikes (closed triangles; ~25% of baseline). Note, especially, the apparent additional LTD of EPSP-spike coupling that appears to be prominent in the immature Schaffer collateral synapses. These findings indicate that (1) non-NMDA mediated LTD appears to be present in adult hippocampus, but is much more robust in developing animals, and (2) there may be separate mechanisms mediating LTD of synaptic transmission and of the coupling of postsynaptic potentials to action potential generation, with different developmental time courses. We can speculate (the joy of an absence of data) that greater LTD during early development may be an important balancer of the robust LTP that has also been observed, as well as possibly protecting neurons from excitotoxic damage from a variety of sources.

LTD IN NEOCORTEX

In addition to the hippocampus, there are some regions of neocortex that appear to have mechanisms for LTD of synaptic strength as well. A functional study of pairing of optical stimulation with either single neuron firing or inactivity in primary visual cortex suggested that covariance-linked LTP and LTD might occur (Frégnac et al., 1988). Artola et al. (1990) confirmed the presence, at synapses on visual cortical pyramidal neurons, of an LTD mechanism with properties very similar to hippocampus. However, unlike our hippocampal work, they found that small postsynaptic *depolarizations* coupled with presynaptic firing yielded LTD, while large depolarizations plus presynaptic stimulation elicited LTP. A number of factors could explain the apparent difference. The mean resting potential of cortical pyramidal neurons in slices is 10–15 mV more hyperpolarized than CA1 pyramidal cells to start with, and differences in levels of circuit inhibition may also contribute to modulating the effect of synaptic stimulation on intracellular depolarization and calcium. In any event, Artola (personal communication) has since observed that raising extracellular [Ca^{2+}] can cause stimuli that did not persistently alter synaptic strength to induce LTD, further supporting the notion that some critical level of [Ca^{2+}] above basal, but below that needed for LTP, is necessary for induction of LTD.

Recent work in our laboratory has given some indications that associative LTD may also occur at synapses in cingulate cortex. Cingulate cortex (area 29c), by virtue of reciprocal connections with hippocampus, plays an important role in processing hippocampal output, and is also probably receives oscillatory excitation at theta rhythm frequencies via the subiculum at the same time sensory input arrives from the thalamus. Therefore, we wondered whether associative LTP and LTD can be elicited at synapses in this neocortical area.

We examined the effects of stimulating converging subicular and contralateral cingulate (callosal) projections in cingulate slices from adult rats.

Since the terminal fields of these on projections are segregated on the apical and basal dendrites of deep lamina pyramids (Vogt, 1985), this area is well-suited for study of both homosynaptic and heterosynaptic plasticity at these synapses. Trains of high-frequency subicular conditioning stimuli (100 Hz/2 sec) reliably yields homosynaptic LTP of synaptic transmission, while trains at near-theta frequencies (2–10 Hz/2 sec, 15-sec intertrain) to the same input yielded a frequency-tuned LTD. In contrast, theta-frequency stimulation of callosal afferents alone did not alter synaptic strength. Instead, the stimulus paradigm previously found to elicit associative LTD in hippocampus (see figure 11.1b) also evoked associative LTD, when callosal synapses received low-frequency (theta) stimulation out of phase with subicular bursts. Thus, both associative and LFS LTD may be present in this and other neocortical areas.

CONCLUDING REMARKS

Although lagging behind our knowledge of LTP, it is becoming increasingly evident that mechanisms also exist in hippocampus and cortex, as well as cerebellum, which can presistently reduce the strength of synaptic connections. However, a bewildering collection of patterns of neuronal activity have been found to elicit LTD in multiple forms. While heterosynaptic LTD as a consequence of LTP is one form (Lynch et al., 1977), LTD can also be homosynaptic when induced by prolonged low-frequency synaptic activity (Dudek and Bear, 1992; Mulkey and Malenka, 1992; Wexler and Stanton, 1992, 1993). Another interesting form is associative LTD, which, combined with associative LTP, permits neurons to compute and store the temporal covariance between coactive inputs (Stanton and Sejnowski, 1989; Chattarji et al., 1989). While low-frequency LTD data support the "pre not post" rule for LTD induction, at least one study has also offered evidence that "post not pre" may also induce LTD (Pockett et al., 1990).

Are all these forms of LTD the same underneath? That is, is the consistent finding of a calcium dependence for LTD induction (Kimura et al., 1990; Mulkey and Malenka, 1992) telling us that the only requirement for inducing LTD is to reach the necessary window of intracellular $[Ca^{2+}]$? Alternatively, the involvement of metabotropic glutamate receptors in associative and low-frequency LTD (Stanton et al., 1991; Wexler and Stanton, 1992, 1993), presence of a non-NMDA form of LTD (Velíšek et al., 1993), and the priming effect of synaptic activity and phorbol esters, may indicate that synergistic actions of calcium, second messengers such as IP_3 and DAG, and protein kinases can all affect the magnitude of LTD. The entire question of the cellular expression of LTD is unexplored, including even rudimentary hints of a presynaptic versus postsynaptic locus. Such parameters as glutamate sensitivity, release studies, quantal analysis, and pharmacologic blockade all remain to be assessed during LTD, although one report has indicated that isolated NMDA conductances can exhibit LTD (Xie et al., 1993). One has the sinking feeling that the tendency toward multiple mechanisms with differing time courses emerging for LTP will be upheld for LTD as well.

There are, however, some hints indicating a postsynaptic decision point for LTD induction. We have previously shown that pairing of postsynaptic hyperpolarization (−20 mV from rest) with brief, low-frequency presynaptic stimulation induces LTD in hippocampus (Stanton and Sejnowski, 1989). Mulkey and Malenka (1992) found that much stronger postsynaptic hyperpolarization (−40 mV from rest) blocked the induction of low-frequency LTD. In fact, the study of Pockett and coworkers (1990) suggests that postsynaptic firing alone can be sufficient to induce LTD. Having learned from experience with LTP, we will hasten to add that none of these observations preclude the presynaptic terminal as a target for the expression of LTD, via retrograde messenger interactions.

The experimental evidence indicating that the magnitude of both LTP and LTD induced is under constant dynamic regulation may be extremely important for stability properties of neural networks. The data add important support for the "sliding threshold" functions that a number of theoretical models have required to appropriately mimic neuronal plasticity and learning (Bienenstock et al., 1982). We propose that this dynamic regulation is necessary for maximal storage and retrieval of information. In addition, it may play a role in preventing pathologies related to hyperexcitability such as seizures or stroke-related neuronal damage. The cellular mechanisms regulating plasticity threshold offer both a novel site of computation in neural networks and a potential locus for therapeutic intervention.

ACKNOWLEDGMENTS

This work was supported by NIMH grant no. 45752 and an Office of Naval Research Young Investigator Award to P.K.S., an Epilepsy Foundation of America Fellowship to L.V., an American Heart Association fellowship to E.M.W., and NIH training grant no. 5T32NS07183.

REFERENCES

Artola, A., Brocher, S., and Singer, W. (1990) Different voltage-dependent thresholds for inducing long-term depression and long-term potentiation in slices of rat visual cortex. *Nature* 347:69–72.

Bienenstock, E., Cooper, L., Munro, P. (1982) Theory for the development of neuron selectivity: Orientation specificity and binocular interaction in visual cortex. *J. Neurosci.* 2:32–48.

Bliss, T. V. P., and Lømo, T. (1973) Long-lasting potentiation of synaptic transmission in the dentate area of the anaesthetised rabbit following stimulation of the perforant path. *J. Physiol. (Lond.)* 232:331–356.

Chattarji, S., Stanton, P. K., and Sejnowski, T. J. (1989) Commissural synapses, but not mossy fiber synapses, in hippocampal field CA3 exhibit associative long-term potentiation and depression. *Brain Res.* 495:145–150.

Christie, B. R., and Abraham, W. C. (1992) Priming of associative long-term depression in the dentate gyrus by θ frequency synaptic activity. *Neuron* 9:79–84.

Dudek, S. M., and Bear, M. F. (1992) Homosynaptic long-term depression in area CA1 of hippocampus and effects of *N*-methyl-D-aspartate receptor blockade. *Proc. Natl. Acad. Sci. USA* 89:4363–4367.

Dudek, S. M., and Bear, M. F. (1993) Bidirectional long-term modification of synaptic effectiveness in the adult and immature hippocampus. *J. Neurosci.* 13:2910–2918.

Frégnac, Y., Schultz, D., Thorpe, S., and Bienenstock, E. (1988) A cellular analogue of visual cortical plasticity. *Nature* 333:367–370.

Fujii, S., Saito, K., Miyakawa, H., Ito, K., and Kato, H. (1991) Reversal of long-term potentiation (depotentiation) induced by tetanus stimulation of the input to CA1 neurons of guinea pig hippocampal slices. *Brain Res.* 555:112–122.

Goldman, R. S., Chavez-Noriega, L. E., and Stevens, C. F. (1990) Failure to reverse long-term potentiation by coupling sustained presynaptic activity and *N*-methyl-D-aspartate receptor blockade. *Proc. Natl. Acad. Sci. USA* 87:7165–7169.

Gustaffson, B., Wigström, H., Abraham, W. C., and Huang, Y. Y. (1987) Long-term potentiation in the hippocampus using depolarizing current pulses as the conditioning stimulus to single volley synaptic potentials. *J. Neurosci.* 7:774–780.

Harris, K. H., and Teyler, T. J. (1984) Developmental onset of long-term potentiation in area CA1 of the rat hippocampus *J. Physiol. (Lond.)* 346:27–48.

Huang, Y., Colino, A., Selig, D. K., and Malenka, R. C. (1992) The influence of prior synaptic activity on the induction of long-term potentiation. *Science* 255:730–733.

James, W. (1890) *Psychology (Briefer Course).* New York: Holt, pp. 253–279.

Kimura, F., Tsumoto, T., Nighigori, A., and Yoshimura, Y. (1990) Long-term depression but not potentiation is induced in calcium-chelated visual cortex neurons. *NeuroReport* 1:65–68.

Larson, J., Wong, D., and Lynch, G. (1986) Patterned stimulation at the theta frequency is optimal for the induction of hippocampal long-term potentiation. *Brain Res.* 368:347–350.

Levy, W. B., and Steward, O. (1979) Synapses as associative memory elements in the hippocampal formation. *Brain Res.* 175:233–245.

Lynch, G. S, Dunwiddie, T., and Gribkoff, V. (1977) Heterosynaptic depression: A post-synaptic correlate of long-term potentiation. *Nature* 266:737–739.

Mulkey, R. M., and Malenka, R. C. (1992) Mechanisms underlying induction of homosynaptic long-term depression in area CA1 of the hippocampus. *Neuron* 9:967–975.

Pockett, S., Brookes, N. H., and Bindman, L. J. (1990) Long-term depression at synapses in slices of rat hippocampus can be induced by bursts of postsynaptic activity. *Exp. Brain Res.* 80:196–200.

Schoepp, D. D., and Johnson, B. G. (1989) Inhibition of excitatory amino acid-stimulated phosphoinositide hydrolysis in the neonatal rat hippocampus by 2-amino-phosphonopropionate. *J. Neurochem.* 53:1865–1870.

Sejnowski, T. J. (1977) Storing covariance with nonlinearly interacting neurons. *J. Math. Biol.* 4:303–321.

Stanton, P. K., Chattarji, S., and Sejnowski, T. J. (1991) 2-Amino-3-phosphonopropionic acid, an inhibitor of glutamate-stimulated phosphoinositide turnover, blocks induction of homosynapic long-term depression, but not potentiation, in rat hippocampus. *Neurosci. Lett.* 127:61–66.

Stanton, P. K., and Sejnowski, T. J. (1989) Associative long-term depression in the hippocampus induced by hebbian covariance. *Nature* 339:215–218.

Staubli, U., and Lynch, G. (1990) Stable depression of potentiated synaptic responses in the hippocampus with 1–5 Hz stimulation. *Brain Res.* 513:113–118.

Tanzi, E. (1893) Facts and inductions in current histology of the nervous system. *Rivista Sperimentale di Freniatriae Medicina Legale delle Mentali Alienazioni* 19:419–472.

Velíšek, L., Moshé, S. L., and Stanton, P. K. (1993) Age-dependence of homosynaptic long-term depression in field CA1 of rat hippocampal slices under conditions of *N*-methyl-D-aspartate blockade. *Dev. Brain Res.,* 75:253–260.

Vogt, B. A. (1985) Cingulate cortex. In *Cerebral Cortex,* A. Peters and E. G. Jones (eds.), New York: Plenum Press.

von der Malsburg, C. (1973) Self-organization of orientation selective cells in the striate cortex. *Kybernetik* 14:85–100.

Wexler, E. M., and Stanton, P. K. (1992) Prior synaptic activity enhances the induction of long-term depression (LTD) in hippocampus. *Soc. Neurosci. Abstr.* 18:1351.

Wexler, E. M., and Stanton, P. K. (1993) Priming of homosynaptic long-term depression in hippocampus by previous synaptic activity. *NeuroReport* 4:591–594.

Wickens, J. R., and Abraham, W. C. (1991) The involvement of L-type calcium channels in heterosynaptic long-term depression in the hippocampus. *Neurosci. Letters* 130:128–132.

Wilshaw, D., and Dayan, P. (1990) Optimal plasticity from matrix memories: What goes up must come down. *Neural Computing* 2:85–93.

Xie, X., Berger, T. W., and Barrionuevo, G. (1993) Isolated NMDA receptor–mediated synaptic responses express both LTP and LTD. *J. Neurophysiol.* 67:1009–1013.

12 Postsynaptic Induction of Long-Term Depression of Synaptic Transmission in the Hippocampal Slice

Lynn J. Bindman, Geri Christofi, Stephen R. Bolsover, Alex V. Nowicky, and Michael F. Barry

Long-term depression (LTD) of synaptic transmission is thought to be as important for learning as is long-term potentiation (LTP) and has been incorporated into computational models of efficient learning. LTD may be the neuronal basis for the behavioral responses of extinction and forgetting, enabling an animal to change its behavioral response when environmental conditions change.

Induction of LTP conforms to a Hebbian model of learning (Hebb, 1949) in which synaptic inputs are strengthened when they are associated in time with sufficient depolarization and firing of the postsynaptic neuron. The obverse phenomenon of LTD can be induced at synapses in one pathway by three separate paradigms: (1) When presynaptic activity in the afferent pathway produces postsynaptic depolarization that is above a threshold value but is inadequate to give rise to LTP, then a synapse-specific LTD may result (Artola et al., 1990; Dudek and Bear, 1992; Mulkey and Malenka, 1992); presynaptic and postsynaptic depolarizations are associated in time, and the LTD is homosynaptic. (2) When postsynaptic depolarization and firing are produced via a separate converging input while presynaptic activity is absent in the test pathway (Lynch et al., 1977; Levy and Steward, 1979, 1983), a nonassociative, heterosynaptic LTD can be produced. (3) An associative LTD has been described in which presynaptic impulses in the test pathway at 5 Hz were repeatedly evoked at a time when the postsynaptic neuron was hyperpolarized, as a result of preceding short, high-frequency trains at 5 Hz in a second input pathway or hyperpolarizing current (Stanton and Sjenowski, 1989; Chatterji et al., 1989; Stanton et al., 1991). Thus the presynaptic depolarization in the test pathway was out of phase with the postsynaptic depolarization, but occurred associated with the subsequent after-hyperpolarization. Either input on its own did not give rise to LTD in the same or the other pathway.

LTD has been best studied in cerebellum (Ito, 1989; Crepel and Jaillard, 1991). LTD has also been observed in neocortex and hippocampus, where, until recently, it has been an elusive phenomenon. Heterosynaptic LTD in the hippocampal area CA1 was first described by Lynch et al. (1977) but, using the same experimental procedure, Andersen et al. (1977) did not obtain LTD. Dunwiddie and Lynch (1978) found that trains of antidromic stimuli applied to the CA1 cell population (exciting axons but also interneurons) resulted in depression lasting at least 15 min in test field synaptic potentials in some, but not all slices. More recently, associative LTD in CA1 and CA3 in vitro was described, in which the test input was paired with hyperpolarization of the postsynaptic neurone (Stanton and Sejnowski, 1989). However, others were unable to repeat these findings (see Christie and Abraham, 1992, for references). In the dentate region of hippocampus in anesthetized rats, Christie and Abraham (1992) found that the application of priming stimulation to the lateral afferent pathway at theta frequency (5 Hz) permitted the subsequent appearance of associative LTD in the pathway, whereas in the absence of priming, no LTD was produced.

Homosynaptic LTD has been induced in neocortex using apparently identical stimulus parameters to those eliciting LTP (Bindman et al., 1988; Hirsch and Crepel, 1990; see Teyler et al., 1990, for review). Associative LTD following postsynaptic depolarization paired with test shocks has also been described, using the same paradigm as that giving rise to associative LTP (Bindman et al., 1988).

The biochemical and pharmacological mechanisms underlying induction and maintenance of LTD have been studied in far less detail than those involved in LTP. In 1991 Bear suggested that "progress in this area has been hampered by the fact that extant models of synaptic weakening in cortex have not as yet proven to be very reliable in the hands of others." However, two robust paradigms have recently been developed for the induction of LTD in CA1, in isolated slices from rat brain. One is low-frequency (e.g., 1 Hz) induction of homosynaptic LTD (Dudek and Bear, 1992; Mulkey and Malenka, 1992). The second is postsynaptic induction of LTD, which has been described in two laboratories (Pockett and Lippold, 1986; Pockett et al., 1990; Christofi et al., 1991, 1992, 1993a,b). Postsynaptic induction is an experimental maneuver that can be used to investigate the mechanism underlying the induction of heterosynaptic, nonassociative LTD, while eliminating the presynaptic limb of the conditioning stimulus. Although the postsynaptic induction of LTD of excitatory postsynaptic potentials (EPSPs) onto pyramidal neurons was unpredictable when the slice was bathed in artificial cerebrospinal fluid (CSF) (Pockett et al., 1990, and results presented in this chapter) we have developed three experimental protocols in which pharmacological intervention enables nonassociative, postsynaptic induction of LTD to be produced more predictably. Surprisingly, LTD was reliably induced when post synaptic conditioning was applied during transient pharmacological block or severe reduction of synaptic transmission (Pockett et al. 1990; Christofi et al., 1991, 1992, 1993a). We also found a significant nonassociative LTD could be induced in some CA1 neurons when slices were bathed in a $GABA_A$ antagonist, bicuculline methiodide (Christofi et al., 1993b).

EXPERIMENTAL STRATEGY AND METHODS

This chapter describes experiments that fall into two sections: first, those in which postsynaptic induction of LTD was brought about with *antidromic* conditioning during the temporary block of synaptic transmission with a raised Mg^{2+} concentration in the bathing medium. This paradigm was also used to investigate the need for extracellular Ca^{2+} in the induction process. The second paradigm used *postsynaptic injection of current* to excite the neuron; this conditioning procedure was carried out (1) in the presence of transient synaptic block with raised [Mg^{2+}], or (2) in the presence of ionotropic glutamate receptor antagonists. In either situation, the bathing medium with raised [Mg^{2+}] or drugs was used to perfuse the slice for 5 min, and then washed out. (3) We also applied intracellular current for conditioning when the slice was bathed in CSF alone, or in the continuous presence of a $GABA_A$ receptor antagonist, bicuculline methiodide.

Ca^{2+} entry into the soma during intracellular conditioning in CSF, in raised [Mg^{2+}], and in CSF with no added Ca^{2+} was measured in some experiments using the fluorescent, ratiometric, cytosolic [Ca^{2+}] indicator FURA-2 injected into the cell.

Full details of methods are given in Christofi et al. (1993a). Transverse hippocampal slices, 200 μm thick, were isolated from 160–180 g male Sprague-Dawley rats. During recording, slices were submerged in artificial CSF at 31–32°C. High Mg^{2+} bathing solutions contained 25 mM $MgCl_2$ and 2 mM $CaCl_2$, or 15 mM $MgCl_2$ with no added $CaCl_2$; NaCl was reduced to maintain isosmolarity with CSF (at 285 mOsm$\cdot l^{-1}$). Drugs used in specified experiments were D-AP5 from Sigma at 5–20 μM, CNQX (Tocris Neuramin) at 2–10 μM, and bicuculline methiodide (Sigma) at 1–2 μM diluted in CSF and applied by bath perfusion.

For intracellular calcium measurements the slice was placed in a modified perfusion chamber on the stage of a Zeiss IM microscope. FURA-2 (Molecular Probes) was introduced into the cell through an intracellular micropipette by iontophoresis. The slice was illuminated with ultraviolet light to excite the FURA-2, and emitted light was imaged by an intensified CCD camera. The fluorescence of dye in the cell body was estimated by subtracting the background fluorescence recorded from regions of CA1 adjacent to the impaled cell from the fluorescence in the cell body. For resting cytosolic [Ca^{2+}] measurements, we illuminated the slice successively with

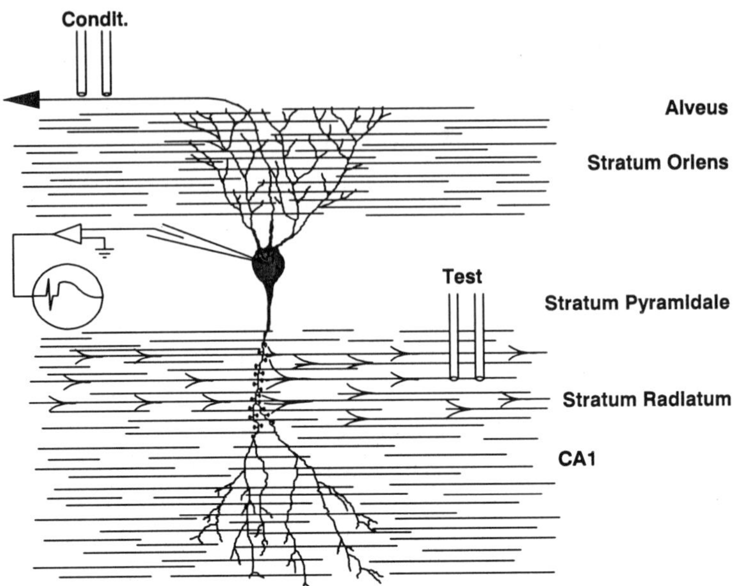

Figure 12.1 Diagram of pyramidal neuron in CA1 of hippocampus, showing position of test stimulating electrodes on stratum radiatum afferent fibers, and antidromic conditioning electrodes on axons in the alveus. In some experiments, the conditioning stimuli were intracellular (i.e., depolarizing current pulses injected through the recording electrode).

light at 350 and 380 nm, using narrow-band filters and the $[Ca^{2+}]$ was calculated (see Grynkiewicz et al., 1985). To measure cytosolic $[Ca^{2+}]$ during electrical stimulation we illuminated the slice with steady 380-nm excitation light for 2 min; cytosolic $[Ca^{2+}]$ was calculated from the fluorescence at this single wavelength as described by Silver et al. (1992).

Standard intracellular recording techniques were used to record from CA1 neurons. Stimulating electrodes were positioned in the stratum radiatum to evoke orthodromic (test) EPSPs at 0.1 Hz. Antidromic conditioning stimuli were applied to the alveus (condit. in figure 12.1): six trains at 100 Hz for 0.5 sec every 10 sec were used. Intracellular conditioning consisted of current pulses (six to nine 0.5-sec depolarizing pulses of 1.5–3.5 nA at 10-sec intervals).

EPSP slopes were measured over the first 1 to 1.5 msec from averages of eight consecutive responses. Statistical comparisons were made (unpaired Student's t-test, two-tailed) between measurements over the last 5 min in the control period and the mean EPSP slope of measurements taken from 30 to 35 min (high Mg^{2+} experiments, CSF and bicuculline methiodide experiments) or 40–45 min (CNQX with or without AP5 experiments) after conditioning stimulation or the reperfusion of the slice with CSF when no conditioning stimuli were applied.

RESULTS

Intracellular recordings were made from 83 cells with the following electrophysiological characteristics: membrane potential (V_m) at end of recording, -72.4 ± 5.6 mV (mean $\pm$ 1 SD); total spike amplitude, 97.8 ± 8.9 mV; firing threshold, 15.8 ± 3.9 mV.

LTD Induced with Antidromic Conditioning in High [Mg^{2+}] Solutions

Antidromic Conditioning When antidromic conditioning trains were applied during synaptic block of both the PSP evoked from the stratum radiatum pathway and that evoked from the alveus, LTD was produced in all the experiments ($n = 8$). The EPSP slope remained depressed more than 30 min after reperfusion with CSF, by which time full recovery from the 25 mM [Mg^{2+}] in CSF would be expected.

Control Experiments with No Antidromic Conditioning We know the time of the recovery from the 25 mM [Mg^{2+}] in CSF because in the absence of antidromic conditioning, a 5-min period of bathing the slices in 25 mM [Mg^{2+}] in CSF abolished EPSPs throughout the slices within 3–5 min, but recovery occurred by about 20–25 min after reperfusing the slices with CSF. By 30–35 min, the mean EPSP slope was $109 \pm 8\%$ (mean $\pm$ 1 SE, $n = 5$) of its control. In contrast, in the cells subjected to antidromic conditioning after the EPSP had been blocked by a high [Mg^{2+}], the mean EPSP slope was $52.6 \pm 11.4\%$ of its control at 30–35 min after conditioning and reperfusion with CSF ($n = 5$). The difference between the conditioned and control groups was significant at $P < .01$ (figure 12.2, columns 1 and 2). In the eighth cell given antidromic conditioning, only the peak amplitude was measured; there was a significant depression compared with its control.

LTP Induction Once LTD had been established, LTP of more than 30 min was produced in two cells by tetanic stimulation of stratum radiatum afferents, showing that synaptic transmission could be modulated in both directions. In three other cells that were not exposed to high Mg^{2+} solutions, antidromic conditioning trains induced LTD in one cell, but LTP in the other two (see Pockett et al., 1990).

In this section of results, we have confirmed that an LTD of EPSP amplitude occurs following antidromic conditioning (Pockett et al., 1990) and have established that the initial EPSP slope is depressed.

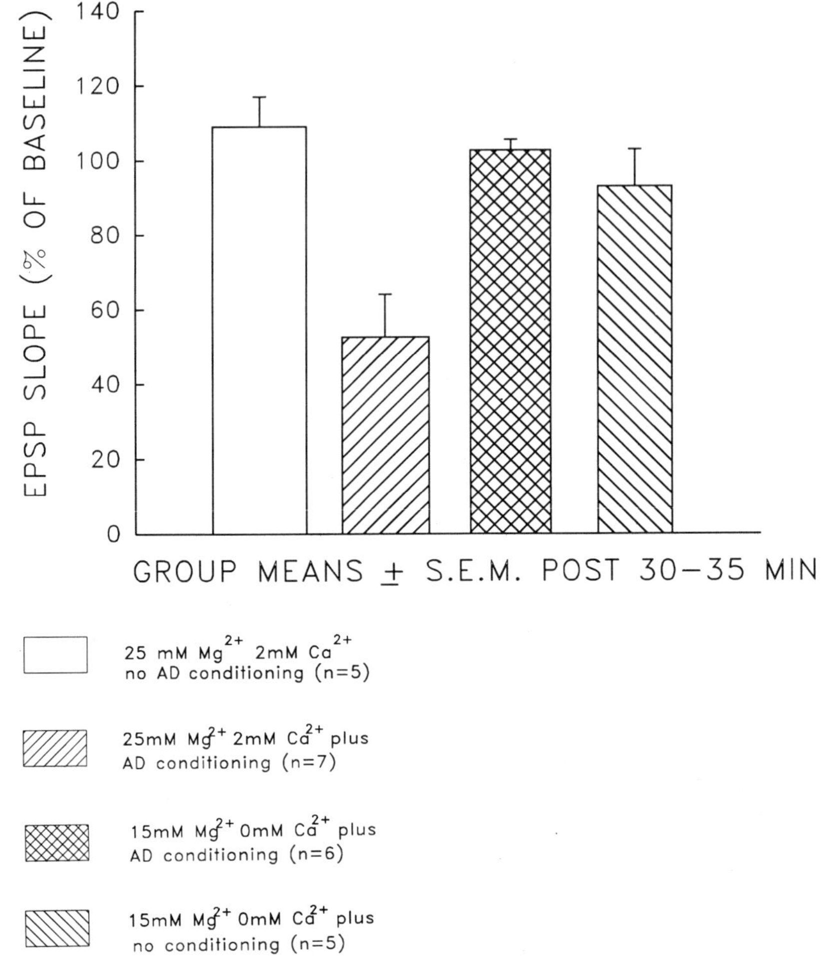

Figure 12.2 Bar chart summarizing results of nonassociative antidromic conditioning on the mean EPSP slope when conditioning was applied after bathing the slices for 5 min in solutions that transiently blocked synaptic transmission, with (column 2) or without (column 3) added Ca^{2+}. Control experiments in which the slices were exposed to the experimental bathing media but no conditioning was applied are shown in columns 1 and 4. The EPSP slope measurements were made at 30–35 min after conditioning (expressed as a % of baseline).

Failure to Induce LTD in High Mg^{2+} with No Added Ca^{2+}

Control Experiments without Antidromic Conditioning When 25 mM Mg^{2+} with no added Ca^{2+} was bath applied for 5 min (n = 2), the time course of recovery of the EPSP on reperfusion with CSF was complete, but slower (30–40 min) than for 25 mM Mg^{2+} and 2 mM Ca^{2+}. Thereafter, we bathed slices in CSF with 15 mM Mg^{2+} but no added Ca^{2+}, which blocked synaptic transmission within 3–5 min; there was complete recovery when no antidromic conditioning stimulation was applied, by 30–35 min post reperfusion with CSF (mean EPSP slope 96.7% ± 7.7% of its control, n = 5) (figure 11.2 column 4).

Experiments with Antidromic Conditioning In contrast to the LTD produced by antidromic conditioning during synaptic block but with Ca^{2+} in the bathing medium, LTD was not produced by antidromic conditioning applied during perfusion with a bathing medium containing no added Ca^{2+}. The EPSP slope at 30–35 min post conditioning was not significantly different from its control (mean EPSP slope 102.7% ± 2.8% of its control, n = 6) (figure 12.2, column 3). Similarly, no LTD was produced in a seventh neuron in which PSP amplitude but not slope was analyzed.

LTD following Intracellular Conditioning in High Mg^{2+}

We repeated the experiments of conditioning during block of synaptic transmission, but used discrete postsynaptic stimulation instead of antidromic excitation. Repeated application of depolarizing current pulses produced firing of the neurons and induced a pronounced LTD. The mean EPSP slope at 30–35 min post conditioning was 57.2% ± 11.6% of its control (n = 7). The depression was significantly different from the control (figure 12.3, columns 1 and 2) at $P < .01$.

LTD following Intracellular Current Conditioning with $GABA_A$ Recentor Antagonists

Intracellular Conditioning in CSF We attempted to induce LTD in CSF by intracellular current pulses, but were unsuccessful in five experiments (figure 12.3, column 4). The mean EPSP slope at 30–35 min after conditioning was 103 ± 1% of its control.

Intracellular Conditioning in Bicuculline We then investigated the effect of intracellular conditioning when the slice was bathed in 1 μM bicuculline methiodide, since Abraham and Wickens (1991) had shown

POSTSYNAPTIC INDUCTION OF LTD

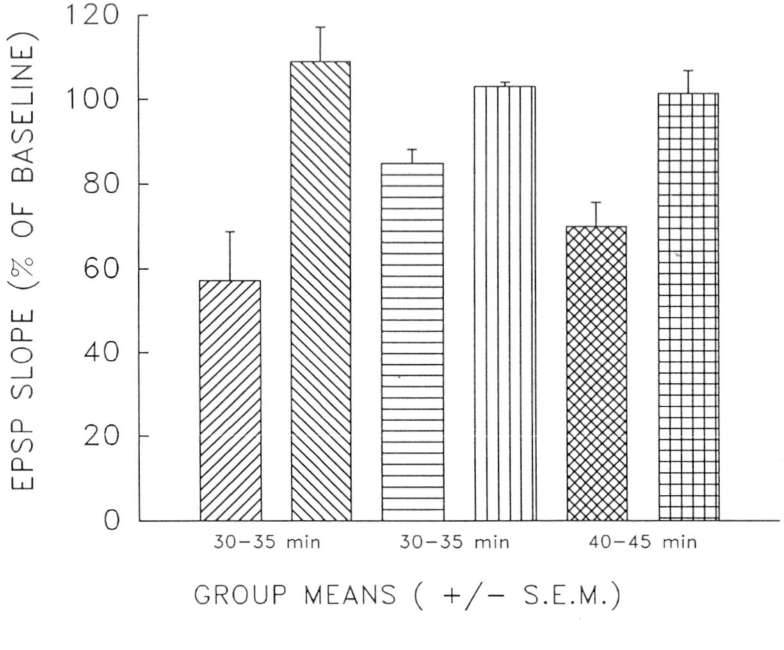

25 mM Mg^{2+} i.c conditioning (n=7)

25 mM Mg^{2+} control (n=6)

Bic meth i.c conditioning (n=9)

ACSF i.c conditioning (n=5)

CNQX +/– AP5 i.c conditioning (n=5)

CNQX +/– AP5 control (n=10)

Figure 12.3 Bar chart summarizing results of nonassociative intracellular conditioning on the mean EPSP slope (columns 1, 3, 4, 5) when conditioning was applied after bathing the slices for 5 min in solutions that transiently blocked synaptic transmission (column 1; compare with control experiments in column 2) or greatly reduced excitatory synaptic transmission (column 5; compare with control experiments in column 6). In columns 3 and 4, intracellular conditioning was applied in CSF with added bicuculline methiodide (1 μM present throughout the experiment) or in CSF alone, respectively. The EPSP slopes were measured at 30–35 min after conditioning (columns 1, 3, 4) or after replacement of bathing solution with CSF (column 2) or at 40–45 min after conditioning (column 5) or after replacement of drugs with CSF (column 6).

that induction of heterosynaptic LTD was facilitated by blockade of $GABA_A$ inhibition. Out of ten experiments, a significant LTD was produced by intracellular conditioning in four cells, but there was no effect of similar conditioning in six cells. An example of one of the LTD experiments is shown in figure 12.4, and the pooled results for a group of nine are represented in column 3 in figure 12.3. The mean EPSP slope at 30–35 min was $80.75 \pm 1.4\%$ ($n = 10$) of the mean in the control (Christofi et al., 1993b). The LTD of the mean EPSP slope in bicuculline methiodide is significantly different at $P < .001$ from the mean control EPSP slope (paired t-test) and also from the mean EPSP slope after conditioning in CSF alone ($n = 6$) at $P < .001$ (unpaired t-test) (Christofi et al., 1993b).

Sample averaged EPSPs are shown in figure 12.4A, obtained during the control and, below, at 34 min after conditioning. The time course of the experiment is shown in B. Since the bicuculline was present throughout the experiment, we are able to see that the depression of the EPSP slope was present by the time that the conditioning trains ended, that is, by about 2–6 min after the first conditioning train. The longest we monitored an LTD produced in bicuculline was 45 min; in all of the four experiments the magnitude of the LTD remained the same after induction throughout the recording period (from 35 to 45 min).

It is clear that a marked LTD can be produced, albeit less reliably, with discrete intracellular conditioning without the block of transmitter release brought about with the high $[Mg^{2+}]$ solutions.

LTD following Intracellular Current Conditioning with Glutamate Receptor Antagonists

Instead of attempting to block both $GABA_A$ and $GABA_B$ receptors, which would be likely to lead to a hyperexcitable slice, we used CNQX, an antagonist of ionotropic glutamate receptors. The action of CNQX was to block both excitation of inhibitory interneurons, and spontaneous and evoked release of glutamate on ionotropic receptors on the test neurons during the intracellular conditioning procedure. We do not know of a monosynaptic recurrent excitatory pathway, but conceivably polysynaptic recurrent excitation could occur (Lacaille et al., 1987).

Control Experiments in CNQX with or without AP5 Preliminary experiments were carried out to test the action of CNQX, bath-applied for 5 min. Using 5 μM CNQX in CSF gave rise to an approximately 80% reduction in EPSP slope. A residual small EPSP seen after 10 μM CNQX in six cells was resistant to bicuculline methiodide at 1 μM ($n = 3$) but was abolished by subsequent application of 10 μM D-AP5 ($n = 6$).

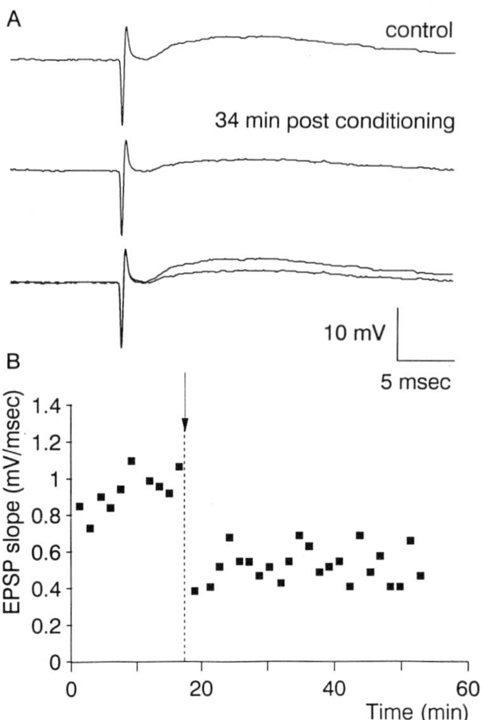

Figure 12.4 The postsynaptic induction of LTD by intracellular conditioning pulses in slices bathed with the $GABA_A$ receptor antagonist, bicuculline methiodide (1 μM). (*A*) Representative averaged waveforms (n = 8 successive EPSPs) from the experiment. From the top down: control; 34 min post conditioning; first and second waveforms superimposed. (*B*) Plot of initial EPSP slope versus time for the experiment shown in *A*. The conditioning paradigm (onset at arrow) consisted of depolarizing the membrane potential by 14 mV by injection of steady current through the recording electrode followed by the injection of 8 intracellular depolarizing current pulses (2 nA for 500 msec at 10 sec intervals) each pulse eliciting 22 spikes. Before LTD: mean R_{in} 47.6 MΩ (n = 16); mean spike threshold 15.8 mV (n = 8); mean spike amplitude (threshold to peak) 89 mV (n = 8). During LTD: mean spike threshold 15.2 mV (n = 8); mean spike amplitude (threshold to peak) 88.4 mV (n = 8); mean R_{in} 48 MΩ (n = 16). V_m − 72 mV.

The recovery of the PSPs after CNQX application (5 or 10 μM) with or without D-AP5 (at 5–20 μM) took longer following reperfusion with CSF than for high Mg^{2+} solutions. We therefore used 40–45 min after reperfusion with CSF as the time to measure EPSP slope. There was a complete recovery of PSPs from the effect of all the various drug doses by this time: the mean EPSP slope was 101.3% $\pm$ 5.5% of its control at 40–45 min after reperfusion with CSF (n = 10) (figure 12.3, column 6). Separating the five results with CNQX and AP5 from the five with CNQX but no AP5, the recovery of the initial slope was complete by 40–45 min in both groups.

Intracellular Conditioning in CNQX with or without AP5 A significant LTD was produced by intracellular conditioning in the presence of CNQX, with or without added AP5. The mean EPSP slope was 70% $\pm$ 5.7% of its control at 40–45 min after intracellular conditioning (n = 5) (see legend to figure 6 in Christofi et al., 1993, for drug doses). The difference between the control and conditioned groups (figure 12.3, columns 5 and 6) was significant at $P < .01$.

Two experiments were carried out in which intracellular conditioning took place when the slice was bathed in 10 μM D-AP5 only; no effect of conditioning was seen.

The Activity Dependence of Induction of LTD

The firing rate and depolarization during spike trains showed that LTD could be induced by antidromic spiking where the steady baseline depolarization was only a few millivolts due to the Mg^{2+} perfusion (see Christofi et al., 1993) and there was little summation of depolarizing afterpotentials (DAPs). LTD could also be obtained using depolarizing current which elicited comparatively few action potentials per train (as summarized in table 1 of Christofi et al., 1993). The failure to induce LTD when no Ca^{2+} was added to the bathing medium was due neither to a lack of postsynaptic depolarization nor to too few antidromic spikes.

Measurements of Intracellular [Ca^{2+}]

The implication of the experiments in which LTD could not be induced in the absence of extracellular calcium is that entry of Ca^{2+} into the neurons during the conditioning trains is required for the induction of LTD. Direct evidence that cytosolic Ca^{2+} could increase as a result of action potentials elicited in a high [Mg^{2+}] bathing solution was obtained using intracellular FURA-2. EPSPs were monitored, and were blocked within 3 min of perfusion of high Mg^{2+} solutions within the imaging apparatus, as in the previous experiments.

Resting [Ca^{2+}] was 55 ± 7 nM (n = 10) with the slice bathed in CSF and did not change when the bathing medium was switched to 25 mM Mg^{2+} 2 mM Ca^{2+} (n = 4).

We estimated the rise in [Ca^{2+}] per action potential by dividing the calcium change at the end of an action potential train by the number of action potentials in that train: the mean was 13 ± 2 nM per spike with the slice bathed in CSF (n = 11). The mean rise in [Ca^{2+}] was reduced to 6 ± 0.8 nM per spike during exposure of the slice to 25 mM Mg^{2+} 2 mM Ca^{2+} (n = 7). A higher depolarizing current was used in high [Mg^{2+}] because of the raised threshold for firing.

Measurements of cytosolic Ca^{2+} were also carried out in nine somata when slices were bathed in 15 mM Mg^{2+} but with no added Ca^{2+}. Resting Ca^{2+} levels did not change significantly during the perfusion. The mean rise in [Ca^{2+}] was only 4 ± 0.5 nM per spike when the bathing medium was switched from CSF to 15 mM Mg^{2+} with no added Ca^{2+} (n = 9). The difference of 2 nM in the rise of [Ca^{2+}] per action potential in the two experimental bathing media (25 mM Mg^{2+} 2 Ca^{2+}, and 15 mM Mg^{2+} 0 Ca^{2+}) was significant at $P < .05$ (two tailed t-test).

Input Resistance and Spike Threshold during LTD

There were no long-lasting changes in input resistance either as a result of drug perfusions or accompanying LTD.

There was no change in mean spike threshold: a comparison of the mean value during the control and during LTD indicated that there was no net change in membrane potential (for details, see Christofi et al., 1993a).

DISCUSSION

There is now a wealth of experimental evidence showing that the efficacy of synaptic transmission in many regions of the CNS can be modified as a function of the previous firing history of the presynaptic and/or postsynaptic elements of the synapses. Experiments have shown that there is a postsynaptic depolarization threshold for the induction of either LTP or LTD (Artola et al., 1990; Stanton and Sejnowski, 1989; see also Frégnac et al., 1990).

In the experiments we have reported in this chapter and in Christofi et al. (1993a,b) we experimentally increased postsynaptic activity during the conditioning procedule while the test input pathway was not stimulated. This is analogous, although over a shorter time course, to the situation of monocular deprivation (Wiesel and Hubel, 1963). The input from the open eye drives the postsynaptic cell (conforming to the increase in mean firing rate due to our nonassociative postsynaptic conditioning stimulation), but the closed eye has no patterned input (akin to the absence of test shocks during conditioning). There is synaptic disconnection from the deprived eye following monocular deprivation and, in our experiments, LTD of the test input. Our results, however, do not conform with those of Tamura et al. (1992). They found heterosynaptic LTD in visual cortex in vivo following tetanic stimulation of one optic nerve, which could not be produced when tetrodotoxin (TTX) silenced the spontaneous inputs.

Frégnac (1991) has set out various models of the dependency of plasticity thresholds for homosynaptic LTP and LTD on the postsynaptic activity or depolarization. In some models there is a continuous transition of the dependence of synaptic plasticity—from depression to potentiation—on the postsynaptic activity or membrane potential (e.g., in the covariance hypothesis, with fixed threshold, Sejnowski, 1977; and with floating threshold, Bienenstock et al., 1982). The dependence of the direction of synaptic plasticity on membrane potential conforms to the experimental results of Artola et al. (1990) in visual cortex. In other models there is a range of postsynaptic membrane potential over which no change occurs; at the lower edge of this range is the threshold for depression and above it is the threshold for potentiation. Frégnac (1991) considers this model fits best with the predictions from the electrophysiological evidence of Frégnac et al. (1990). The postsynaptic, heterosynaptic LTD we have induced in hippocampus cannot be described by these models of homosynaptic synaptic change because the presynaptic firing is zero, but our electrophysiological and imaging data (Christofi et al., 1993a) suggest two different ranges of activity-dependent increases in calcium concentration are required for induction of LTP and LTD. They may be separate by a range in which no effect is induced.

What is the intracellular messenger determining the direction of the persistent change in synaptic efficacy? Many experiments using intracellular Ca^{2+} chelator drugs (Kimura et al., 1990; Yoshimura et al., 1991; Brocher et al., 1992; Hirsch and Crepel, 1992; Mulkey and Malenka, 1992) in neocortex and hippocampus, and caged Ca^{2+} (Malenka et al., 1988) in hippocampus, point to the crucial role played by the $[Ca^{2+}]$ in the induction of LTP when it is at a high level, and of LTD when at a lower level, but above a certain threshold. Wickens and Abraham (1991) showed an involvement of L-type Ca^{2+} channels in the induction of heterosynaptic LTD in hippocampus. Our measurements of activity-dependent changes in somatic cytosolic $[Ca^{2+}]$ provide evidence for different levels of $[Ca^{2+}]$ in

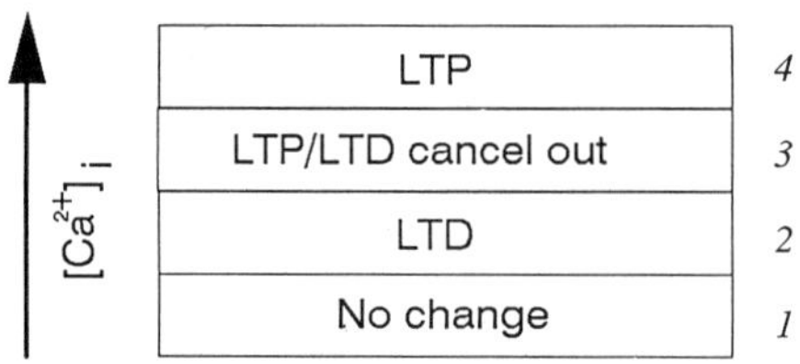

Figure 12.5 A model summarizing four levels of cytosolic [Ca^{2+}] achieved by conditioning stimulation in various experimental procedures, and the resulting synaptic plasticity. Levels 1, 2, and 3 correspond to the soma cytosolic changes in [Ca^{2+}] we measured following conditioning in CSF with 15 mM Mg^{2+} but no added calcium (level 1), in CSF with 25 mM Mg^{2+} (level 2) and in CSF (level 3). The descriptions in the boxes refer to the results of nonassociative conditioning when the hippocampal slices were bathed in these media, but note that the statement that LTP/LTD cancel out in level 3 is an *interpretation* of the lack of change we observed with intracellular conditioning in CSF, and the variable results obtained with AD conditioning in CSF (see text; Pockett et al., 1990). Level 4 is hypothetical, based on other experiments. We have not measured cytosolic [Ca^{2+}] in experiments in which LTP was produced. Our attempts to elicit LTP in imaging experiments failed, presumably due to the chelating action of FURA-2.

three circumstances in which LTD or no effect result from intracellular conditioning. The changes in somatic [Ca^{2+}] are less than but are correlated with the apical dendritic [Ca^{2+}] (Jaffe et al., 1992 and our own unpublished observations).

With intracellular conditioning, we were able to elicit a marked LTD using a variety of pharmacological manipulations. The common factor was the reduction in both excitatory and inhibitory transmitter action on postsynaptic receptors, either because of blocked evoked transmitter release or block of ionotropic receptors during the conditioning stimulation.

We argue that there are two possible (but conflicting) explanations of how the reduction of synaptic transmission could enhance the probability of obtaining LTD in our experiments. The hypotheses can best be understood with reference to figure 12.5. The diagram indicates four ranges of intracellular [Ca^{2+}]. We propose that when the level of intradendritic Ca^{2+} remains in the lowest of the four ranges shown during conditioning, no synaptic plasticity can be produced. This range would reflect the small activity-dependent increases that we measured in the soma when the bathing medium contained no added Ca^{2+}. The mean soma rise in [Ca^{2+}] was about 4 nM per action potential. The rise in [Ca^{2+}] could be due to a residual entry of calcium from the extracellular medium, or to a Na^+-induced release of intracellular Ca^{2+}.

We propose that when the intradendritic [Ca^{2+}] rises during conditioning to level 2, LTD is produced at some test synapses. This would be the situation when our block or reduction of synaptic transmission during conditioning produces LTD in a reliable fashion. For example, in 25 mM

Mg^{2+} we measured a mean activity-dependent rise of soma [Ca^{2+}] of about 6 nM per action potential.

We suggest that when the [Ca^{2+}] during conditioning is further increased to level 3, there could be LTP at some synapses, but LTD at others, since there is a temporal and spatial gradient of the rise of [Ca^{2+}] within the pyramidal cell. The net effect on the EPSP recorded in the soma would be no change, although it is in fact made up of potentiated EPSPs at some synapses and reduced EPSPs at others. Alternatively, it could be the biochemical processes required to initiate LTP or LTD that cancel one another's effectiveness.

We suspect this cancelling out of different processes is occurring when the intracellular conditioning trains are applied in CSF, when the mean rise in soma [Ca^{2+}] was about 13 nM per action potential. Only when the rise in dendritic [Ca^{2+}] reaches level 4 would the conditions for LTP be produced at a sufficient number of synapses for the net change to be registered as potentiation. This is presumably the level reached when an apparently nonassociative LTP was produced in some of our experiments by antidromic conditioning in CSF (Pockett et al., 1990; Christofi et al., 1993a; see also Alonso et al., 1990) but activation of recurrent excitatory inputs cannot be ruled out in these conditions.

The mutually exclusive hypotheses explaining how a pharmacological reduction in synaptic transmission in our experiments could facilitate the induction of LTD are as follows:

1. The conditions for, or the expression of, LTP and LTD will cancel out when conditioning is in CSF; the reduction of synaptic transmission during the conditioning stimulation by raised [Mg^{2+}] or drugs prevented intracellular [Ca^{2+}] from rising in some of the dendrites to a level that would have induced LTP, and hence a net LTD is unmasked (i.e., the drugs reduce the activity-dependent rise in dendritic [Ca^{2+}] from level 3 during conditioning in CSF to level 2).

2. The loss or reduction of inhibitory transmitter release from recurrent collaterals activated during the conditioning stimulation could reduce the shunting effect of GABA on the soma and apical dendrite, thereby allowing more current to spread into distal *dendrites* and permit sufficient Ca^{2+} to enter to reach the level needed to initiate LTD (i.e., to rise from level 1 to level 2 in the dendrites, even though in the *soma* and apical dendrite the [Ca^{2+}] is reduced from level 3 to level 2.

Recent preliminary evidence obtained from measurements of activity-dependent changes in dendritic [Ca^{2+}] suggest the first hypothesis is more likely. All the pharmacological agents we used to reduce synaptic transmission also reduced the activity-dependent rise in [Ca^{2+}] within the neuron (soma and apical dendrite) compared with the rise produced in CSF. The raised [Mg^{2+}] would do so by virtue of partial block of voltage-sensitive Ca^{2+} channels (Almers and Palade, 1981) and bicuculline and CNQX by removing the depolarizing action of GABA released from recurrent collaterals onto the apical dendrite (Langmoen et al., 1978; Andersen et al., 1980; Alger and Nicoll, 1982; Grover et al., 1993). CNQX would also remove the depolarizing action of spontaneously released glutamate at synapses on the dendrites. The reliable induction of LTD in our experiments was presumably serendipitous in that we happened to achieve the right range of increased [Ca^{2+}] during conditioning. The hypothesis is now open to further experimental testing by manipulation of stimulation parameters and bathing media.

SUMMARY

We have found robust methods for inducing nonassociative LTD of stratum radiatum EPSPs, using postsynaptic conditioning stimuli. The conditioning stimuli consisted of either brief trains of high-frequency antidromic stimuli, or pulses of intracellular depolarizing current sufficient to elicit about 14 spikes per pulse. In each method for the induction of LTD, either evoked release of transmitter was blocked, or a selective antagonist of postsynaptic glutamate receptors, or of $GABA_A$ receptors was used during the conditioning procedure. In this chapter we describe these experimental maneuvers leading to LTD, each successively less pharmacologically interventionist but also finally less effective for LTD induction. The familiar experimental situation used for the induction of LTP, namely intracellular conditioning pulses applied in slices bathed in a $GABA_A$ antagonist—but without the paired presentation of the test shock—was used successfully to induce nonassociative LTD. However, under these conditions a significant LTD was not produced in every experiment.

One of the more reliable procedures for eliciting LTD, involving transient block of evoked transmitter release during conditioning, was used to show the need for extracellular calcium ions during induction. Also, measurements of cytosolic calcium ion concentration were made of the activity-dependent rise in intracellular calcium in different bathing media. We found that a small activity-dependent rise in cytosolic calcium concentration occurred when the slice was bathed in a medium containing no added calcium, but no LTD was evoked. Under conditions in which LTD was reliably produced, the activity-dependent rise in somatic, cytosolic calcium concentration was greater than that in the absence of Ca^{2+} added to the bathing medium, but less than that when the slice was bathed in CSF alone, and LTD could not be produced.

ACKNOWLEDGMENTS

We thank the Wellcome Trust for salary (A.V.N.) and support of the research. G.C. was an MRC scholar. Yves Frégnac made helpful comments on the manuscript.

REFERENCES

Abraham, W. C., and Wickens, J. R. (1991) Heterosynaptic long-term depression is facilitated by blockade of inhibition in area CA1 of the hippocampus. *Brain Res.* 546:336–340.

Alger, B. E., and Nicoll, R. A. (1982) Pharmacological evidence for two kinds of GABA receptor on rat hippocampal pyramidal cells studied in vitro. *J. Physiol.* 328:125–141.

Almers, W., and Palade, P. T. (1981) Slow calcium and potassium currents across frog muscle membrane: Measurements with a vaseline-gap technique. *J. Physiol. (Lond.)* 312: 159–176.

Alonso, A., de Curtis, M., and Llinás R. (1990) Postsynaptic Hebbian and non-Hebbian long-term potentiation of synaptic efficacy in the entorhinal cortex in slices and in the isolated adult guinea pig brain. *Proc. Natl. Acad. Sci. USA* 87:9280–9284.

Andersen, P., Sundberg, S. H., Sveen, O., and Wigström, H. (1977) Specific long-lasting potentiation of synaptic transmission in hippocampal slices. *Nature* 266:736–737.

Andersen, P., Dingledine, R., Gjerstad, L., Langmoen, I. A., and Mosfeldt Laursen, A. (1980) Two different responses of hippocampal pyramidal cells to application of gamma-aminobutyric acid. *J. Physiol. (Lond.)* 305:279–296.

Artola, A., Brocher, S., and Singer, W. (1990) Different voltage-dependent thresholds for inducing long-term depression and long-term potentiation in slices of rat visual cortex. *Nature* 347:69–72.

Bear, M. (1991) Use of developing visual cortex as a model to study the mechanism of experience-dependent synaptic plasticity. *Brain Res. Rev.* 16:198–200.

Bienenstock, E., Cooper, L. N., and Munro, P. (1982) Theory for the development of neurone selectivity: Orientation specificity and binocular interaction in visual cortex. *J. Neurosci.* 2:23–48.

Bindman, L. J., Murphy, K. P. S. J., and Pockett, S. (1988) Postsynaptic control of the induction of long-term changes in efficacy of transmission at neocortical synapses in slices of rat brain. *J. Neurophysiol.* 60:1053–1065.

Brocher, S., Artola, A., and Singer, W. (1992) Intracellular injection of Ca^{2+} chelators blocks induction of long-term depression in rat visual cortex. *Proc. Natl. Acad. Sci. USA* 89:123–127.

Chattarji, S., Stanton, P. K., and Sejnowski, T. J. (1989) Commissural synapses, but not mossy fibre synapses, in the hippocampal field CA3 exhibit associative long-term potentiation and depression. *Brain Res.* 495:145–150.

Christie, B. R., and Abraham, W. C. (1992) Priming of associative long-term depression in the dentate gyrus by θ frequency synaptic activity. *Neuron* 9:79–84.

Christofi, G., Nowicky, A. V., and Bindman, L. J. (1991) The postsynaptic induction of long-term depression (LTD) of synaptic transmission in isolated rat hippocampal slices requires extracellular calcium. *J. Physiol. (Lond.)* 438:257P.

Christofi, G., Nowicky, A. V., Ray, J., Khan, H., and Bindman, L. J. (1992) Long-term depression (LTD) of synaptic transmission in isolated rat hippocampal slices can be induced postsynaptically in an anti-Hebbian paradigm by depolarizing current pulses. *J. Physiol. (Lond.)* 452:33P.

Christofi, G., Nowicky, A. V., Bolsover, S. R., and Bindman, L. J. (1993a) The postsynaptic induction of non-associative long-term depression (LTD) of excitatory synaptic transmission in rat hippocampal slices. *J. Neurophysiol.* 69:219–229.

Christofi, G., Barry, M. F., and Bindman, L. J., (1993b) Heterosynaptic long-term depression of synaptic transmission in isolated slices of rat hippocampus can be induced postsynaptically in the presence of a $GABA_A$ antagonist. *J. Physiol. (Lond.)* in press.

Crepel, F., and Jaillard, D. (1991) Pairing of pre- and postsynaptic activities in cerebellar Purkinje cells induces long-term changes in synaptic efficacy in vitro. *J. Physiol. (Lond.)* 432:123–141.

Dudek, S. M., and Bear, M. F. (1992) Homosynaptic long-term depression in area CA1 of hippocampus and effects of *N*-methyl-D-aspartate receptor blockade. *Proc. Natl. Acad. Sci. USA* 89:4363–4367.

Dunwiddie, T., and Lynch, G. (1978) Long-term potentiation and depression of synaptic responses in the rat hippocampus: Localization and frequency dependency. *J. Physiol. (Lond.)* 276:353–367.

Frégnac, Y. (1991) Computational approaches to network processing and plasticity. In *Long-term Potentiation: A Debate of Current Issue,* M. Baudry and J. L. Davis (eds.). Cambridge, Mass.: pp. 425–435.

Frégnac, Y., Smith, D., and Friedlander, M. (1990) Postsynaptic membrane potential regulates synaptic potentiation and depression in visual cortical neurones. *Soc. Neurosci. Abstr.* 16:798.

Grover, L. M., Lambert, N. A., Schwartzkroin, P. A., and Tyler, T. J. (1993) Role of HCO_3^- ions in depolarizing $GABA_A$ receptor-mediated responses in pyramidal cells of rat hippocampus. *J. Neurophysiol.* 69:1541–1555.

Grynkiewicz, G., Poenie, M., and Tsien, R. Y. (1985) A new generation of Ca^{2+} indicators with greatly improved fluorescence properties. *J. Biol. Chem.* 260:3440–3450.

Hebb, D. O. (1949) *The Organization of Behavior.* New York: Wiley.

Hirsch, J. C., and Crepel, F. (1990) Use-dependent changes in synaptic efficacy in rat prefrontal neurones in vitro. *J. Physiol. (Lond.)* 427:31–49.

Hirsch, J. C., and Crepel, F. (1992) Postsynaptic calcium is necessary for the induction of LTP and LTD in prefrontal neurones. An in vitro study in the rat. *Synapse* 10:173–175.

Ito, M. (1989) Long-term depression. *Annu. Rev. Neurosci.* 12:85–102.

Jaffe, D. B., Johnston, D., Lasser-Ross, N., Lisman, J. E., Miyakawa, H., and Ross, W. N. (1992) The spread of Na^+ spikes determines the pattern of dendritic entry into hippocampal neurons. *Nature* 357:244–246.

Kimura, F., Tsumoto, T., Nishigori, A., and Yoshimura, Y. (1990) Long-term depression but not potentiation is induced in Ca^{++}-chelated visual cortex neurons. *NeuroReport* 1:65–68.

Lacaille, J.-C., Mueller, A. L., Kunkel, A. D., and Schwartzkroin, P. A. (1987) Local circuit interactions between oriens/alveus neurons and CA1 pyramidal cells in hippocampal slices: Electrophysiology and morphology. *J. Neurosci.* 7:1979–1993.

Langmoen, I. A., Andersen, P., Gjerstad, L., Laursen, A. M., and Ganes, T. (1978) Two separate effects of GABA on hippocampal pyramidal cells in vitro. *Acta Physiol. Scand.* 102:28A–29A.

Levy, W. B., and Steward, O. (1979) Synapses as associative memory elements in the hippocampal formation. *Brain Res.* 175:233–245.

Levy, W. B., and Steward, O. (1983) Temporal contiguity requirements for long-term associative potentiation/depression in the hippocampus. *Neuroscience* 8:791–797.

Lynch, G. S., Dunwiddie, T., and Grybkoff, V. (1977) Heterosynaptic depression: A postsynaptic correlate of long-term potentiation. *Nature* 266:737–739.

Malenka, R. C., Kauer, J. A., Zucker, R. S., and Nicoll, R. A. (1988) Postsynaptic calcium is sufficient for potentiation of hippocampal synaptic transmission. *Science* 242:81–87.

Mulkey, R. M., and Malenka, R. C. (1992) Mechanisms underlying induction of homosynaptic long-term depression in area CA1 of the hippocampus. *Neuron* 9:967–975.

Pockett, S., and Lippold, O. C. J. (1986) Long-term potentiation and depression in hippocampal slices. *Exp. Neurol.* 91:481–487.

Pockett, S., Brookes, N. H., and Bindman, L. J. (1990) Long-term depression at synapses in slices of rat hippocampus can be induced by bursts of postsynaptic activity. *Exp. Brain Res.* 80:196–200.

Sejnowski, T. (1977) Storing covariance with non-linearly interacting neurons. *J. Math. Biol.* 4:303–321.

Silver, R. A., Bolsover, S. R., and Whitaker, M. (1992) Intracellular ion imaging using fluorescent dyes: Artefacts and limitations to resolution. *Pflugers Arch.* 420:595–602.

Stanton, P. K., and Sejnowski, T. J. (1989) Associative long-term depression in the hippocampus induced by Hebbian covariance. *Nature* 339:215–218.

Stanton, P. K., Chattarji, S., and Sejnowski, T. J. (1991) 2-Amino-3-phosphonopropionic acid, and inhibitor of glutamate-stimulated phosphoinositide turnover, blocks induction of homosynaptic long-term depression, but not potentiation, in rat hippocampus. *Neurosci. Lett.* 127:61–66.

Tamura, H., Hata, Y., and Tsumoto, T. (1992) Activity-dependent potentiation and depression of visual cortical responses to optic nerve stimulation in kittens. *J. Neurophysiol.* 68:1603–1612.

Teyler, T. J., Aroniadou, V., Berry, R. L. Borroni, A., DiScenna, P., Grover, L., and Lambert, N. (1990) LTP in neocortex. *Semin. Neurosci.* 2:365–383.

Wickens, J., and Abraham, W. C. (1991) The involvement of L-type calcium channels in heterosynaptic long-term depression in the hippocampus. *Neurosci. Lett.* 130:128–132.

Wiesel, T. N., and Hubel, D. H. (1963) Single-cell responses in the striate cortex of kittens deprived of vision in one eye. *J. Neurophysiol.* 26:1003–1017.

Yoshimura, Y., Tsumoto, T., and Nishigori, A. (1991) Input-specific induction of long-term depression in Ca^{2+}-chelated visual cortex neurons. *NeuroReport* 2:393–396.

13 Homosynaptic Long-Term Depression and Its Relationship to Long-Term Potentiation in the Rat Neocortex In Vitro

Alain Artola

Use-dependent long-term changes of synaptic efficacy are thought to form a basis for learning and memory. Theoretical considerations suggest that these modifications should include both an increase and a decrease of synaptic strength, and, indeed, evidence is now available in several brain areas for both long-term potentiation (LTP) and long-term depression (LTD) of synaptic transmission. Since the discovery of LTP in the hippocampus (Bliss and Lømo, 1973; for review see Bliss and Collingridge, 1993), use-dependent enhancement of synaptic transmission has been observed in a variety of brain structures, including neocortex (for review see Tsumoto, 1992; Singer and Artola, in press). In the majority of cases this LTP is associative, that is, a synapse will potentiate only if it is active at the time when the respective dendrite is sufficiently depolarized. But it has been claimed that there is also a nonassociative form of LTP which requires only strong presynaptic activation, such as LTP of hippocampal mossy fibers (Zalutsky and Nicoll, 1990; but see Jonston et al., 1992) and LTP of parallel fibers in the cerebellum (Sakurai, 1987, 1990; Hirano, 1990).

LTD can also occur under different conditions, depending mainly on whether the input needs to be activated to undergo LTD. There is evidence both from hippocampus (Lynch et al., 1977; Dunwiddie and Lynch, 1978) and neocortex (Hirsch et al., 1992) that tetanic stimulation of one excitatory input can cause depression of another, nonstimulated excitatory input to the same neuron (for review see Artola and Singer, 1993). Since in this case depression occurs at inactive synapses and is induced by other synaptic inputs, this form of LTD is referred to as "heterosynaptic" LTD. But there are also cases in which depression is induced by activity of the modified inputs themselves. This form of LTD is called "homosynaptic." By definition, homosynaptic LTD is input specific. The first evidence for homosynaptic LTD came from studies in the cerebellum (Ito and Kano, 1982; Ito et al., 1982; Kano and Kato, 1987; Sakurai, 1987). Later, homosynaptic LTD was also demonstrated in the hippocampus (Chattarji et al., 1989; Stanton and Sejnowski, 1989). However, in both cases induction of

LTD required that the stimulation of the modifiable pathway be combined with activation of other inputs. In the cerebellum, the synaptic response of Purkinje cells (PCs) to parallel fiber (PF) stimulation was only depressed when PFs were coactivated with climbing fibers (CFs), which also converge onto PCs. In the hippocampus, tetanic stimulation of subicular afferents led to homosynaptic LTD, but only when the Schaffer collaterals were additionally activated in counter phase. That a synaptic connection can undergo LTD following its own tetanic activation alone has only been demonstrated very recently, first in the neocortex (Artola et al., 1990; Hirsch and Crépel, 1990; see also Bindman et al., 1988) and then in the hippocampus (Dudek and Bear, 1992; Mulkey and Malenka, 1992; see also Dunwiddie and Lynch, 1978), the striatum (Calabresi et al., 1992a), and the nucleus accumbens (Pennartz et al., 1993). In all preparations except the hippocampus (see Discussion) the frequencies of the tetanic stimuli suitable for LTD induction were similar to those used for the induction of LTP. The fact that LTD and LTP of excitatory transmission are induced by apparently similar conditions raises a pivotal question: How is an excitatory synapse able to decide whether it should depress or potentiate?

EXPERIMENTAL PROCEDURE

Experiments were performed in slices of the rat visual cortex. Slices (350 μm) were prepared by standard techniques (Artola and Singer, 1990) and were allowed to recover at room temperature. A single slice was then transferred to the recording chamber where it was held completely submerged. The slice was maintained at 28–30°C and continuously perfused with a solution containing 124 mM NaCl, 5 mM KCl, 1.25 mM $NaHPO_4$, 2 mM $MgSO_4$, 2 mM $CaCl_2$, 26 mM $NaHCO_3$, and 10 mM D-glucose, saturated with 95% O_2, 5% CO_2. During exposure to high Ca^{2+}, only the concentration of Ca^{2+} was changed and raised to 4 mM. In some experiments, the $GABA_A$ receptor antagonist bicuculline (0.1–0.5 μM), the NMDA receptor antagonist AP5 (Sigma; 25 to 100 μM) and the AMPA receptor antagonist CNQX (Ferrosan; 10 μM) were bath-applied.

Intracellular recordings were obtained with 3 M potassium acetate-filled electrodes (80–120 MΩ) from regular spiking cells in layer III (McCormick et al., 1985). To examine the role of postsynaptic Ca^{2+} concentration on the induction of LTD, we injected neurons with high concentrations of one of the two Ca^{2+} chelators EGTA or BAPTA. Intracellular electrodes containing either 500 mM EGTA or 100 mM BAPTA and 3 M potassium acetate were used. For injection, hyperpolarizing current pulses (1.5 nA, 0.5 sec, 0.5 Hz) were applied for 10–20 min; an additional 30 min was allowed to elapse between injection and LTD induction to ensure diffusion

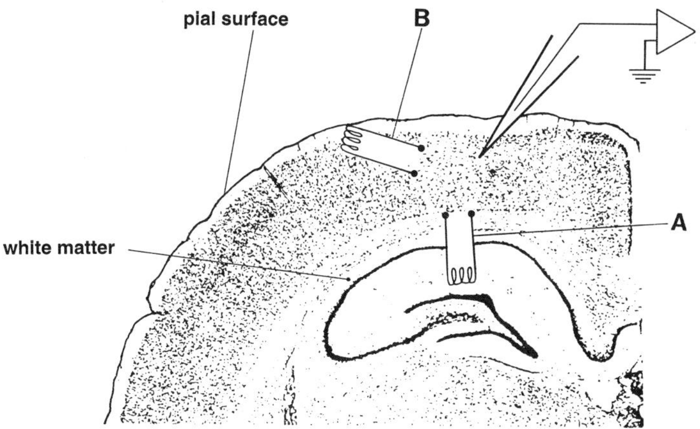

Figure 13.1 Positions of recording and stimulating electrodes in the slice preparation.

of the chelators to the dendrites. The efficacy of the Ca^{2+} chelators was assessed by measuring the amplitude changes of the afterhyperpolarizing potentials following current-induced depolarizing steps (500 msec) (see figure 13.4).

Synaptic responses were elicited by electrical stimulation through bipolar tungsten electrodes that were placed in the underlying white matter (w.m.) and intracortically (i.c.) in layers I–II adjacent to the recorded neuron (figure 13.1). The distances between the recording electrode and the two stimulating electrodes were roughly the same. The tetanus was applied through the w.m. electrode and consisted of five 2-sec long stimulus trains (50 Hz) delivered at 10-sec intervals. In most experiments, the stimulation intensity, during tetanus, was raised to suprathreshold levels as indicated in the text and in the legends to the figures.

Synaptic responses were digitized online (digitizing rate of 0.2 msec) and averaged over five to 10 successive trials. To assess post-tetanic modifications (+ for potentiation, — for depression) of the monosynaptic response (see below), the amplitude of the initial part of the postsynaptic potential (PSP) and the initial slope were measured between 20 and 30 min after tetanus and expressed as percentages of control.

RESULTS

White matter (w.m.) stimulation elicits in layer III pyramidal cells of the rat visual cortex a characteristic pattern of PSPs, namely, an excitatory PSP (EPSP) followed by two inhibitory PSPs (IPSPs), an initial, $GABA_A$ receptor–mediated IPSP (iIPSP), and a late, $GABA_B$ receptor–mediated IPSP (lIPSP). Two EPSP components can be dissociated on the basis of their susceptibility to the selective antagonists of glutamate receptors: an

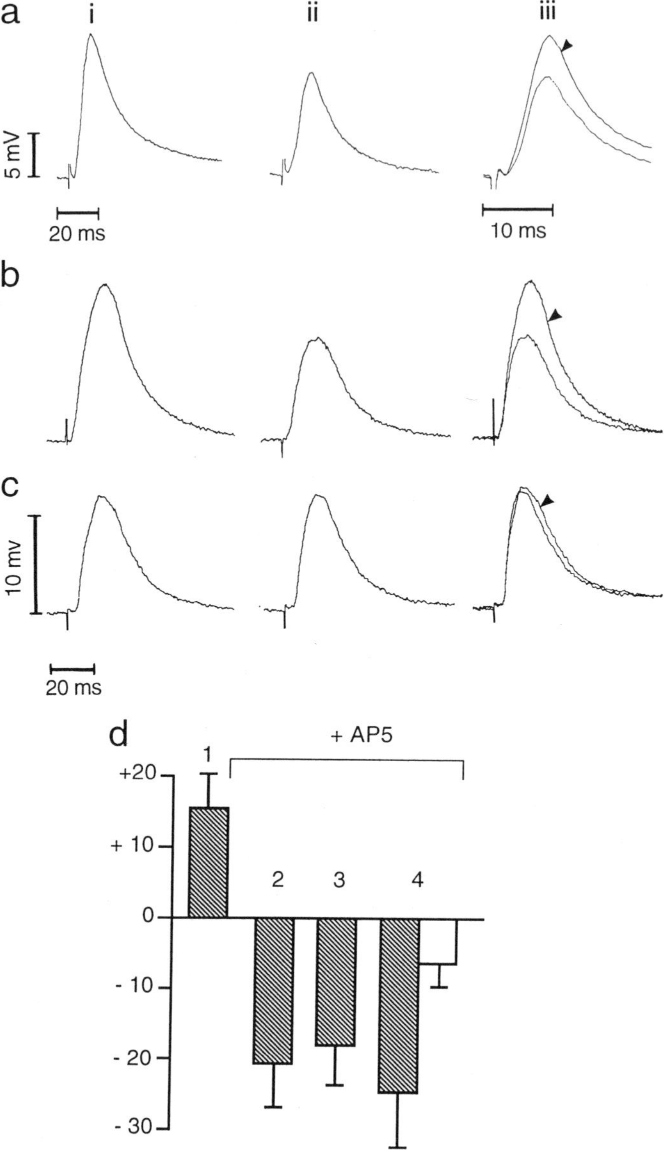

Figure 13.2 Effect of w.m. tetanic stimulation on synaptic responses in slices treated with either 0.3 μM bicuculline (*a, d*) or 0.3 μM bicuculline and 25–100 μM AP5 (*b, c, d*). (*a*) Synaptic responses to w.m. stimulation in a cell treated with 0.3 μM bicuculline. $V_{mr} = -71$ mV. (*b, c*) Synaptic responses to w.m. (*b*) and i.c. (*c*) stimulation in a cell treated with both 0.3 μM bicuculline and 25 μM AP5. $V_{mr} = -70$ mV. Responses i and ii are recorded before and 20 min after tetanus, respectively, and are superimposed at expended time scale in iii. Arrows indicate the pretetanic response. Note that only the response to w.m. (tetanized pathway) is modified. (*d*) Amplitude changes observed in responses to w.m. stimulation 20–30 min after w.m. tetanus in slices treated with 0.3 μM bicuculline (group 1: n = 20), 0.3 μM bicuculline and 25 μM AP5 (group 2: n = 10) and 0.3 μM bicuculline and 100 μM AP5 (group 3: n = 3). In 7 (group 4) out of the 11 cells of group

AMPA receptor–mediated EPSP of large amplitude and short peak latency (8–10 msec) and an NMDA receptor–mediated EPSP of smaller amplitude and longer peak latency (20–25 msec). The two EPSPs begin at the same time atter the stimulation. The fact that the onset latency of the synaptic response remains stable during high-frequency stimulation (50–100 Hz) indicates that at least the initial part of the excitatory response is monosynaptic. This monosynaptic response is mediated by both non-NMDA and NMDA glutamate receptors (for details see Artola and Singer, 1990; Artola et al., 1990b).

Induction of LTD and LTP in the Neocortex

LTP is very easily produced in hippocampus by the application of brief trains of high-frequency stimulation to any of the main excitatory pathways (Bliss and Lømo, 1973; Alger and Teyler, 1976). To test whether the monosynaptic excitatory afferents to layer III neurons were also able to undergo use-dependent long-term synaptic modifications, a similar high-frequency stimulation was applied to w.m.

Reduction of synaptic inhibition facilitates the induction of LTP in hippocampus (Wigström and Gustafsson, 1983). Thus, in a first series of experiments, we investigated the effect on synaptic strength of a constant w.m. high-frequency stimulation in slices treated with various concentrations (0.1–0.3 μM) of the $GABA_A$ receptor antagonist bicuculline. The stimulation intensity during the conditioning tetanus was raised to two times the test intensity. The long-term effect of this high-frequency stimulation on the amplitude of the initial part of the synaptic response turned out to vary with the degree of blockade of the $GABA_A$ receptor-mediated synaptic inhibition. In cells bathed in normal Ringer's solution, tetanic stimulation failed to induce any long-term modification of the synaptic response ($-2.5 \pm 1.7\%$, n = 12; mean $\pm$ SEM). In cells in which bicuculline (0.1–0.2 μM) had only slightly reduced synaptic inhibition, the tetanus induced LTD ($-18.3 \pm 4.5\%$, n = 7) and in cells in which inhibition was strongly decreased by 0.3 μM bicuculline, it led to LTP ($+15.5 \pm 4.6\%$, n = 20; figure 13.2a,d; for details see Artola et al., 1990a; Artola and Singer, 1990).

Disinhibition enhances the depolarizing response of the recorded neuron by two mechanisms: First, it reduces Cl^- conductances on the recorded neuron and, second, by increasing activity in polysynaptic pathways, it

2, responses to w.m. (hatched bar) and to i.c. (empty bar) were recorded. The amplitude change in the tetanized pathway differs significantly ($P < .05$) from that in the nontetanized pathway. Stimulus intensity during tetanus was twice the test intensity. (Adapted from Artola et al., 1990)

enhances the excitatory input. Thus, according to its concentration, bicuculline allows the neuron to be more or less depolarized. As bicuculline treatment should not alter the number of monosynaptically activated synapses, the occurrence of long-term modifications of the monosynaptic input appears to depend on the level of the postsynaptic depolarization. By analogy with the mechanisms for LTP induction, this suggests that the mechanisms underlying LTD induction also involve a voltage-dependent threshold in the postsynaptic neuron which is, however, lower than that for LTP induction.

To test this hypothesis, slices were treated with 0.3 μM bicuculline to raise excitability, and at the same time NMDA receptors were blocked by AP5 (25 μM, n = 11 and 100 μM, n = 3). In slices of the neocortex, NMDA receptors participate in monosynaptic (Aram et al., 1987; Artola and Singer, 1987, 1990; Jones and Baughman, 1988; Hirsch and Crépel, 1990) and polysynaptic (Sutor and Hablitz, 1989) transmission. In addition, NMDA receptor-gated conductances are known to be highly permeable to Ca^{2+} (MacDermott et al., 1986) as well as to monovalent cations and therefore activation of these conductances together with postsynaptic depolarization, which removes the voltage-dependent Mg^{2+} block of the channel (Mayer et al., 1984; Nowak et al., 1984), will lead to an influx of Ca^{2+} into the postsynaptic spine. Thus, blockade of NMDA receptors decreases both postysnaptic depolarization and $[Ca^{2+}]_i$ rise. In these conditions of raised excitability and blocked NMDA receptors tetanic stimulation induced a strong LTD (figure 13.2b,d). Induction of LTD was selective for the stimulated pathway. Responses evoked from the intracortical stimulation were not modified (figure 13.2c,d). LTD was reversible. After AP5 has been washed out, a second tetanic stimulation induced LTP (n = 8). As expected, the same tetanic stimulation had no effect when only the NMDA receptor antagonist AP5 was bath-applied.

In a second series of experiments, we attempted to reach the cooperativity threshold for the induction of LTD by raising the intensity of the tetanic stimulus and, thus, increasing the number of excitatory afferents that were activated during tetanus. When stimulus intensity was high enough (see legend to figure 13.3), LTD was consistently obtained in slices kept in normal medium (figures 13.3 and 13.4c). As before, this LTD was confined to the tetanized afferents (figure 13.3). As a direct test of the hypothesis that LTD and LTP have different voltage-dependent thresholds, current was injected throughout tetanic stimulation to produce hyper- and depolarizing shifts of membrane potential (V_m) of -40 and $+20$ mV, respectively. Hyperpolarizing currents consistently prevented the occurence of LTD (figure 13.3), but application of depolarizing currents led to LTP in 9 of 10 cells ($+27.2 \pm 4.2\%$, n = 9). Both LTD and LTP

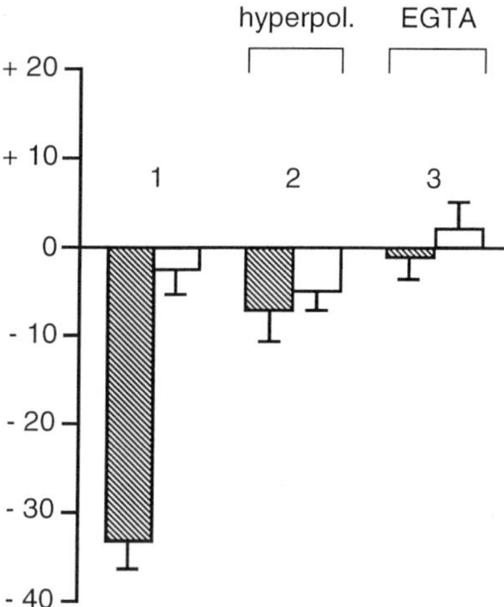

Figure 13.3 Amplitude changes observed in responses to w.m. (hatched bars) and i.c. (empty bars) stimulation 20–30 min after high intensity w.m. tetanus in control cells (group 1: n = 13), in cells hyperpolarized by −40 mV below V_{mr} during tetanus (group 2: n = 6) and in cells injected with EGTA (group 3: n = 8). The amplitude change in the tetanized pathway of group 1 (LTD) differs signĩicantly ($P < .001$) from (a) that in the nontetanized pathway of the same group and from (b) those in the tetanized pathways of groups 2 and 3. This indicates that homosynaptic LTD is input specific and is prevented by hyperpolarizing the postsynaptic cell during tetanus (group 2) or by injecting the Ca^{2+} chelator EGTA (group 3). The intensity required to induce LTD in normal Ringer's solution depends on the tip separation of the bipolar stimulation electrode, i.e. on the ability to recruit the number of excitatory afferents required to reach the cooperativity threshold for the induction of LTD. In this set of experiments, the tip separation of the stimulating electrodes was 350 μm. The minimum stimulation intensity required for reliable LTD induction with this electrode corresponded to around 3 times the threshold intensity for spike generation. Stimulus intensity during tetanus was thus set to 4 times the threshold intensity. (Adapted from Artola et al., 1990 and Bröcher et al., 1992)

could be reversed. As in the previous set of experiments, potentiation but not depression was prevented when AP5 was present in the bath. Thus LTD and LTP can be induced in alternation in the same pathway and with the same stimulation parameters if different levels of postsynaptic depolarization are maintained during tetanic stimulation.

Altogether, these results suggest that the mechanisms of induction of LTD and LTP are both postsynaptic. LTD is obtained if postsynaptic depolarization exceeds a critical level but remains below a threshold whereas LTP is induced if this second threshold is reached.

Intracellular Injection of Ca^{2+} Chelators Blocks the Induction of Homosynaptic LTD

Recent evidence in cerebellum suggests that LTD induction depends on the presence of Ca^{2+} in the postsynaptic neuron (Sakurai, 1990). As a

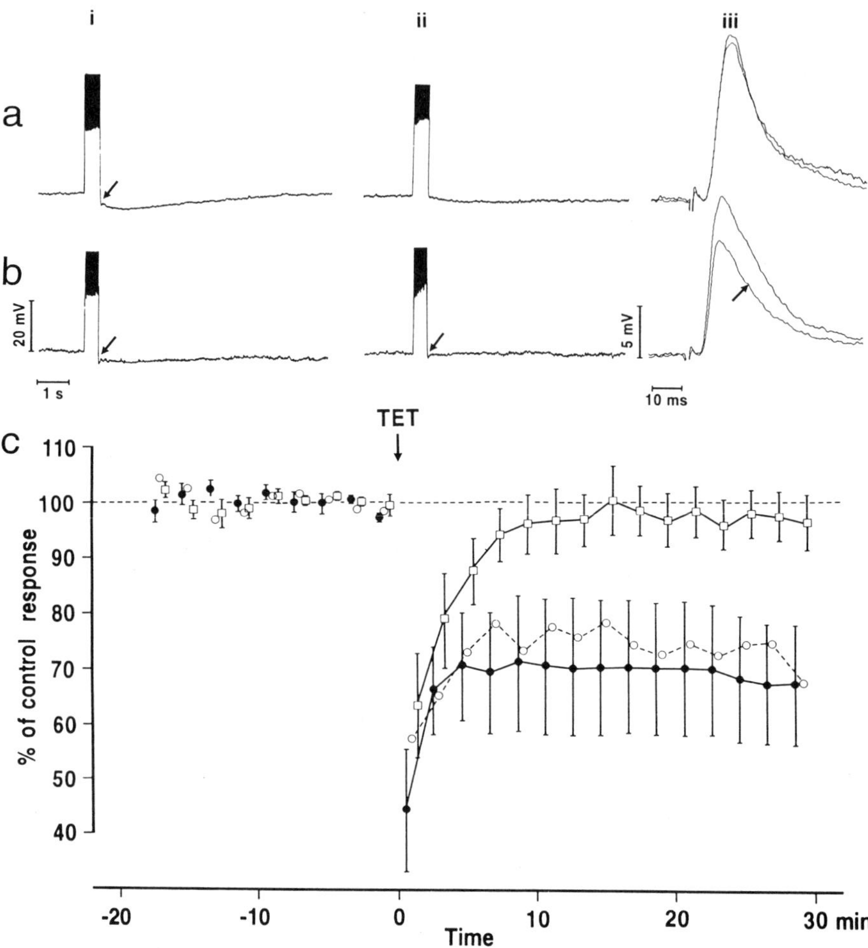

Figure 13.4 Effect of intracellular injection of BAPTA on induction of LTD in rat visual cortex. (*a* and *b*) Effect of BAPTA injection on induction of LTD in two cells. (*a*, i and ii and *b*, i and ii) responses to brief (500 msec) depolarizing current pulses (*a*, +0.7 nA; *b*, +0.8 nA) before (i) and after (ii) BAPTA injection. In the cell in (*a*), blockade of early afterhyperpolarizing potentials (AHPs) is complete except for the slow Ca^{2+}-independent AHP. Comparison of PSPs recorded before and 20 min after the tetanus shows no evidence of LTD (*a*, iii). In the cell in *b*, BAPTA injection has been interrupted *before* complete disappearance of the fast, Ca^{2+}-dependent AHP (arrow in *b*, ii). Comparison of PSPs recorded before and 20 min after (arrow) tetanus reveals that LTD was still elicitable in this cell (*b*, iii). For the AHP measurements in *a* and *b*, cells have been depolarized to −49 (A) and −50 (B) mV by current injection (Vm, −79 mV in *a* and −81 mV in *b*). (*c*) Time course of post-tetanic changes of the amplitude of PSPs elicited from white matter in control cells (solid circles; n = 4) and in cells injected with BAPTA (squares; n = 5). Open circles refer to the cell shown in (*b*). In this set of experiments, the tip separation of the stimulating electrodes was 270 μm. The minimum stimulation intensity required for reliable LTD induction corresponded to around 7.5 times the threshold intensity for spike generation. Stimulus intensity during tetanus was thus set to 10 times the threshold intensity (18.0 ± 2.0 V in control cells and 16.3 ± 1.4 V in injected cells; mean ± SEM). (From Bröcher et al., 1992)

direct test for this hypothesis, we examined whether LTD induction in slices of the rat visual cortex is influenced by intracellular injection of the Ca^{2+} chelators EGTA and BAPTA. As shown above, homosynaptic LTD is consistently induced by high-intensity tetanic stimulation of w.m. By contrast, no depression was observed in cells in which either EGTA or BAPTA had led to a complete blockade of the early afterhyperpolarizing potentials (AHPs) (figures 13.3 and 13.4a,c; for details see Bröcher et al., 1992).

These results indicate that effective buffering of intracellular free Ca^{2+} prevents LTD induction. As two recent studies (Kimura et al., 1990; Yoshimura et al., 1991) report that LTD can be successfully induced in cortical cells injected with EGTA or BAPTA, we performed two additional experiments attempting to resolve this discrepancy. In the studies that showed LTD in the presence of Ca^{2+} chelators, the effectiveness of the injection procedure was assessed by measuring the level of spike adaptation during a depolarizing current pulse and the tetanus was applied before complete suppression of spike adaptation (see figure 2 of Kimura et al., 1990). This suggested the possibility that, in these experiments, buffering of intracellular Ca^{2+} may have been less complete than in our study. To test this hypothesis, tetanic stimulation was applied in two BAPTA-treated cells before early AHPs were completely blocked. Despite the fact that BAPTA had already caused a near complete blockade of the early AHPs, the tetanus induced a marked LTD that was indistinguishable from that obtained in non-treated cells (figure 13.4b,c).

Calcium-induced LTD

These results confirm that a minimum $[Ca^{2+}]_i$ in the postsynaptic neuron is required for the induction of LTD. To test whether LTD induction actually depends on a surge of $[Ca^{2+}]_i$ in the postsynaptic neuron, we examined the influence of raising $[Ca^{2+}]_o$.

In a first series of experiments we examined the effect of a brief exposure (10 min) to high extracellular $[Ca^{2+}]$ (4 mM) alone (i.e., without synaptic activation). In all cells whose resting membrane potential (V_{mr}) was less negative than -79 mV (-79 mV $< V_{mr} <$ -71 mV; -71 mV was the V_{mr} of the most depolarized of the recorded cells; $n = 8$), raising $[Ca^{2+}]_o$ transiently from 2 to 4 mM for 10 min produced, after return to 2 mM $[Ca^{2+}]_o$, a marked ($> -10\%$ and up to -35%: $-17.2 \pm 2.7\%$) and lasting (>90 min) depression of w.m. and i.c. EPSPs. This Ca^{2+}-induced LTD was equally pronounced for both w.m. and i.c. responses. In more polarized cells (-96 mV $< V_{mr} <$ -79 mV; -96 mV was the V_{mr} of the most polarized of the recorded cells) transiently raising $[Ca^{2+}]_o$ had no lasting effect on synaptic transmission ($+1.4 \pm 8.6\%$, $n = 5$). Further support for the voltage-dependency of Ca^{2+}-induced LTD comes from experiments which show that LTD in the first group of cells (-79 mV $< V_{mr} <$ -67 mV) was prevented by hyperpolarizing by -20 mV the postsynaptic neuron during bath application of high $[Ca^{2+}]_o$. This indicates that Ca^{2+}-induced LTD also involves postysnaptic, voltage-dependent mechanisms and suggests similarities with tetanus-induced LTD.

In a second series of experiments, we thus examined the effect of high-frequency stimulation combined with high Ca^{2+}. The stimulus intensity during tetanus was the same as for the test stimuli. Such a tetanus, which would have been ineffective in normal $[Ca^{2+}]_o$ (Artola and Singer, 1990; Artola et al., 1990a), now induced LTD ($-27.4 \pm 1.1\%$, $n = 5$) in hyperpolarized cells (-92 mV $< V_{mr} < -79$ mV) when applied during high Ca^{2+}. This depression was selective for the tetanized pathway and could be blocked by loading the cell with BAPTA before stimulation ($n = 3$). In cells with V_{mr} below -79 mV (-79 mV $< V_{mr} < -67$ mV), in which already high Ca^{2+} alone induced LTD, tetanic stimuli applied in conjunction with the Ca^{2+} pulse caused no further enhancement of LTD: depression was of similar amplitude for the tetanized and non-tetanized input. This indicates that the two forms of LTD, Ca^{2+}- and tetanus-induced LTD, occlude one another.

These results suggest that both Ca^{2+}- and tetanus-induced LTD depend for their induction on the same postsynaptic process. The trigger signal in both cases appears to be a surge of intracellular Ca^{2+}, which results at least in part from Ca^{2+} entry through voltage-gated Ca^{2+} conductances (Artola et al., 1992).

DISCUSSION AND CONCLUSION

These results demonstrate that, in the visual cortex, the same monosynaptic excitatory afferents from w.m. to layer III cells can undergo either LTD or LTP. Homosynaptic LTD is input specific and exhibits an associative property. It thus closely ressembles associative LTP. A major difference is that LTP induction requires a stronger postsynaptic depolarization. LTD is obtained if postsynaptic depolarization exceeds a critical level but remains below a threshold, whereas LTP is induced if this second threshold is reached. Since homosynaptic LTD can be obtained in the presence of the NMDA receptor antagonist AP5, activation of NMDA receptors is not an absolute requirement for the induction of LTD in the neocortex. Postsynaptic processes responsible for LTD seem to be triggered by a rise of $[Ca^{2+}]_i$. The Ca^{2+} responsible for this rise appears to enter at least partially through voltage-gated Ca^{2+} channels.

Long-term increases and decreases of the same initial part of the synaptic response have been obtained in normal Ringer's solution as well as in the presence of 0.3 μM bicuculline, which abolishes almost completely the initial IPSP (see figure 4 of Artola and Singer, 1990). Thus, these modifications represent actual changes of synaptic efficacy in excitatory afferents. Several observations suggest that the modified synapses are located on the recorded neuron. First, the direction of the long-term synaptic modification can be reversed by injecting current into the postsynaptic neuron. Second, LTD is prevented if the recorded neuron is either hyperpolarized or filled with a Ca^{2+} chelator. In addition, neither the initial slope nor any other part of the PSP is modified in these latter cases (see figure 13.4a,iii). It is thus very likely that most of the EPSP recorded in control conditions is monosynaptically activated from w.m. Injections of Lucifer yellow in neocortical neurons have revealed that the regular spiking cells are spiny pyramidal neurons (McCormick et al., 1985; Artola, unpublished observations). However, the actual origin of the excitatory afferents to layer III pyramidal cells (geniculo-cortical, cortico-cortical, and so on) that we activate from w.m. cannot be specified.

Existence of Two Voltage-dependent Thresholds: The ABS Rule

These results have led to the notion of two voltage-dependent thresholds for the induction of LTD and LTP, respectively. If the first threshold (θ^-) is reached, the activated synapses depress, and if the second threshold (θ^+) is reached, which requires stronger depolarization, the activated synapses potentiate. Accordingly, there are two ranges of V_m in which the synapses will not display any long-term modification. First, when V_m remains below the threshold for LTD (θ^-) and, second, when V_m is around the second threshold (θ^+), which separates the two ranges of V_m leading to LTD and LTP, respectively. This evidence, which is no longer compatible with the classic Hebb rule (1949), led to the formulation of a new rule for synaptic modifications, the ABS rule (Artola et al., 1990a; figure 13.5). It is pertinent to consider here that the existence of a threshold θ^+, that separates LTP from LTD has already been postulated by Sejnowski (1977) in the context of the covariance hypothesis and by Bienenstock, Cooper, and Monroe (1982; BCM) to account for experience-dependent modifications in the developing visual cortex (figure 13.5, inset).

Evidence is now available that the ABS rule applies also to use-dependent synaptic modifications in other brain areas such as the prefrontal cortex, the striatum, and the hippocampus. In slices of the prefrontal cortex a tetanic stimulus that would normally elicit LTP induces LTD when the depolarizing response to the tetanus is reduced by blocking NMDA recep-

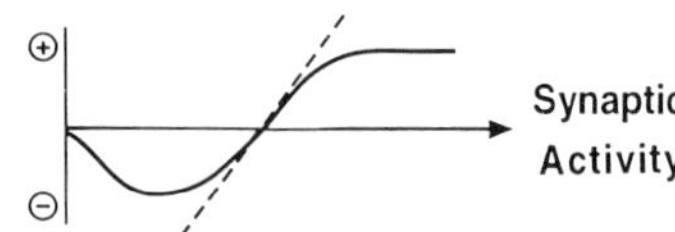

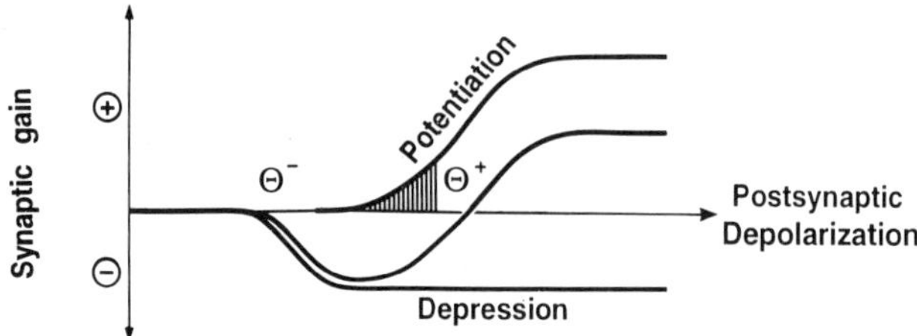

Figure 13.5 Shematic representation of the two different voltage-dependent thresholds (Θ^- and Θ^+) for the induction of homosynaptic LTD and LTP. The ordinate indicates the direction of the synaptic gain change and the abscissa the *membrane potential level of the postsynaptic neuron* that is maintained during presynaptic activation. If the first threshold Θ^- is reached, a mechanism is activated that leads to a long-lasting depression of synaptic efficacy, and if the second threshold Θ^+ is reached, another process is triggered that leads to the potentiation of the synapse. *Inset*: relationship between the direction of the synaptic gain and *synaptic activity* for the BCM rule (solid line) and the covariance rule (dotted line). Note the absence of Θ^- in these two curves.

tors (Hirsch and Crépel, 1991). In the striatum, the same tetanus, which produces LTP in Mg^{2+} free medium, induces LTD in the presence of 1.2 mM $[Mg^{2+}]_o$ and no modification when the postsynaptic cell is hyperpolarized during the conditioning tetanus (Calabresi et al., 1992ab). Thus the induction of LTD and LTP in the striatum shows the same voltage-dependence as in the visual cortex. In area CA1 of the hippocampus, low-frequency (ca. 1 Hz) stimulation induces LTD while high-frequency stimulation ($>$ 10 Hz) induces LTP (Dunwiddie and Lynch, 1978; Dudek and Bear, 1992). This is also consistent with different thresholds for the induction of LTD and LTP as postsynaptic depolarization increases with stimulation frequency. Furthermore, it is also possible in CA1 neurons to change the direction of the synaptic modification by modifying V_m with current injection during application of a tetanus of intermediate frequency. A 5–10 Hz tetanus which alone has no long-term effect on synaptic transmission (Dunwiddie and Lynch, 1978; Dudek and Bear, 1992) induces LTD if it is paired with hyperpolarizing current injection and LTP if it is paired with depolarizing current injection (Stanton and Sejnowski, 1989; Xie et al., 1992). This result can be accounted for if one assumes that stimulation with 5–10 Hz, which is intermediate between the frequencies suitable for the induction of LTD and LTP, produces a postsynaptic depolarization around the θ^+ theshold. Finally, there is also evidence for the θ^- threshold in the hippocampus, as low-frequency tetani which normally induce LTD (see above) are no longer effective if cells are hyperpolarized during stimulation (Mulkey and Malenka, 1992).

The Role of Ca^{2+} in the Induction of LTD and LTP

The conclusion that LTD induction actually requires a surge of $[Ca^{2+}]_i$ above the normal resting level comes from two observations. First, in the visual cortex, as in the cerebellum (Sakurai, 1990), the frontal cortex (Hirsch and Crépel, 1992), and the hippocampus (Mulkey and Malenka, 1992), LTD induction is prevented by intracellular injection of Ca^{2+} chelators. Second, it is facilitated by an increase of extracellular $[Ca^{2+}]_o$. Furthermore, the fact that a brief exposure to high $[Ca^{2+}]_o$ can induce LTD suggests that an appropriate rise of $[Ca^{2+}]_i$ is sufficient to trigger LTD. Additional evidence for this hypothesis comes from experiments in CA1 which showed that (1) a depression of nonactivated synaptic inputs can be obtained if the postsynaptic neuron is strongly depolarized either by antidromic activation or by current injection (Pockett and Lippold, 1986; Pockett et al., 1990; Christofi et al., 1993) and that (2) this depression is blocked if $[Ca^{2+}]_o$ is low (Christofi et al., 1993). There is also evidence both in hippocampus (Lynch et al., 1983) and in neocortex (Kimura et al., 1990; Hirsch and Crépel, 1991) that an increase in postsynaptic $[Ca^{2+}]_i$ is necessary to induce LTP and is sufficient to potentiate synaptic transmission (Malenka et al., 1988). Thus, how are membrane depolarization and the consecutive rise of $[Ca^{2+}]_i$ differentiated to produce either LTD or LTP?

There are several ways to account for the fact that different levels of postsynaptic depolarization produce opposite synaptic modifications. One way is to assume that the critical variable determining the sign of the change is the origin of the Ca^{2+} surge following tetanic stimulation. Early evidence favored the hypothesis that Ca^{2+} entering through NMDA receptor–gated channels mediates the induction of LTP, while Ca^{2+} liberated from intracellular stores following activation of metabotropic glutamate receptors (mGluR) leads to LTD. This view equated the LTP threshold θ^+ with the activation threshold of NMDA receptor–gated conductances, that is, the voltage-dependent removal of the Mg^{2+} block (Mayer et al., 1984; Nowak et al., 1984). The evidence that, in hippocampus, neocortex, and striatum, blockade of NMDA receptors prevents the induction of LTP (Collingridge et al., 1983; Artola and Singer, 1987) but not of LTD (Stanton and Sejnowski, 1989; Artola et al., 1990a; Hirsch and Crépel, 1991; Calabresi et al., 1992a; but see below), and the fact that cerebellar PCs lack NMDA receptors (Llano et al., 1988; Audinat et al., 1990; Llano et al., 1991b) altogether appeared compatible with this interpretation. Moreover, there is evidence that activation of mGlu receptors facilitates LTD induction. In cultured PCs (Linden et al., 1991), in cerebellar wedges (Ito and Karachot, 1990), and in slices of area CA1 (Stanton, P., personal communication), combined application of AMPA and *trans*-ACPD, an agonist of the mGlu receptor, induces a persistent desensitization of AMPA receptors. Conversely, blockade of mGlu receptors by AP3 can prevent the induction of LTD in hippocampus (Stanton et al., 1991), neocortex (Crépel, personal communication; Artola, unpublished observations), and cultured PCs (Linden et al., 1991). However, more recent evidence indicates clearly that neither LTP nor LTD induction depends selectively on activation of NMDA or mGlu receptors. There is clear evidence that LTP can be induced without activation of NMDA receptors. Under certain conditions such as stimulation at very high frequencies (Grover and Teyler, 1991) or blockade of K^+ channels (Aniksztejn and Ben-Ari, 1991), associative LTP can be induced in hippocampal CA1 after blockade of NMDA receptors. Likewise, NMDA receptor–independent LTP has been obtained in the visual cortex of young rats and kittens (Komatsu et al., 1991). These conditions have in common to induce a very strong postsynaptic depolarization and to presumably cause a massive Ca^{2+} entry through voltage-gated Ca^{2+} channels (Grover and Teyler, 1991; Komatsu and Iwakiri, 1992). Potentiation of synaptic transmission has also been obtained in the absence of synaptic activation in CA1 and the dorsolateral septal nucleus by bath application of *trans*-ACPD (Zheng and Gallagher, 1992; Bortolotto and Collingridge, 1992) or by exposure to high $[Ca^{2+}]_o$ (Turner et al., 1982; Grover and Teyler, 1990), conditions in which NMDA receptors are not activated. On the other hand, there is evidence that activation of NMDA receptor–gated conductances can actually contribute to LTD induction. This has been demonstrated in the hippocampus (Dudek and Bear, 1992; Mulkey and Malenka, 1992). In this case the tetanic stimuli had very low trequency (1 Hz) and presumably produced only weak postsynaptic depolarization. Finally, an influx of Ca^{2+} through voltage-gated Ca^{2+} channels can also contribute to the induction of LTD. In the cerebellum, PF activation (Crépel and Krupa, 1988; Crépel and Jaillard, 1991) or bath application of *trans*-ACPD (Daniel et al., 1992) induces LTD of PF-PC synapses if PCs are in addition depolarized by current injection which is likely to activate voltage-gated Ca^{2+} channels. Thus, there is no stringent correlation between the activation of particular Ca^{2+} sources such as the NMDA and mGlu receptors or voltage-gated Ca^{2+} channels and the induction of LTP and LTD, respectively.

The evidence for a Ca^{2+}-dependence of LTD suggests the possibility that the postsynaptic depolarization above a critical threshold is only required to achieve an appropriate surge of $[Ca^{2+}]_i$ in the postsynaptic neuron. It is conceivable, therefore, that the amplitude of the Ca^{2+} surge rather than its cause determines the sign of the synaptic modification, the induction of LTP requiring a greater surge of intracellular Ca^{2+} than the induction of LTD. This is supported by several observations. First, in neocortex, incomplete buffering of intracellular Ca^{2+} by EGTA or BAPTA readily prevents the induction of LTP but still allows for the induction of LTD (Bröcher et al., 1992; see also Kimura et al., 1990 and Yoshimura et al., 1991). Second, in CA1 a 20-Hz tetanus normally induces LTP, but when $[Ca^{2+}]_o$ is lowered from 2.5 to 0.5 mM it leads to LTD (Mulkey and Malenka, 1992). The interpretation that high $[Ca^{2+}]_i$ leads to LTP while intermediate $[Ca^{2+}]_i$ induces LTD agrees also with numerous other observations. It accounts well for the different voltage-dependent thresholds of LTP and LTD as the influx of Ca^{2+} through NMDA receptor- and voltage-gated Ca^{2+} channels increases with depolarization. In addition, it fits with the notion that coactivation of NMDA receptors favors LTP induction, as NMDA receptor activation leads to a particularly strong surge of Ca^{2+}. It is also compatible with the evidence that any source of Ca^{2+} can contribute to either LTP or LTD induction and that the critical determinants are the stimulation protocols, activation conditions causing strong increases of $[Ca^{2+}]_i$ being more likely to induce LTP than LTD (see above). However, there must be additional variables that influence the induceability of LTP and LTD. It appears rather unlikely, for instance, that the sole reason for the noninduceability of LTP in the cerebellum is an insufficient increase of $[Ca^{2+}]_i$. Although PCs lack NMDA receptors, influx of Ca^{2+} through voltage-gated Ca^{2+} channels and release from intracellular stores has been shown to cause very substantial increases of $[Ca^{2+}]_i$ (Ross and Werman, 1987; Llano et al., 1991a). Still, one might argue that Ca^{2+} from these sources does not reach the concentration required for LTP at sites where this is required, but the possibility should not be dismissed that PCs lack mechanisms required for LTP induction that are downstream of the Ca^{2+} signal, such as certain Ca^{2+}-activated enzyme systems. There is evidence that different concentrations of $[Ca^{2+}]_i$ can lead to differential activation of Ca^{2+}-dependent enzymes because of different Ca^{2+} affinities of the proteins. Based on this notion, Lisman (1989) postulated that low levels of $[Ca^{2+}]_i$ should lead to preferential activation of phosphatases, while high levels should activate predominantly kinases, and he related this differential effect to the induction of LTD and LTP, respectively. Recently, evidence has become available that the efficacy of glutamate receptor–gated ion channels can be changed in opposite directions by dephosphorylation and phosphorylation (Greengard et al., 1991; Wang et al., 1991). Thus, it is tempting to speculate that the Ca^{2+}-dependent depression and potentiation of excitatory transmission at glutamatergic synapses, such as occurs with LTP and LTD, is indeed brought about by differential activation of Ca^{2+}-regulated enzymes with different Ca^{2+} affinity.

The data suggest that (1) a rise of postsynaptic free Ca^{2+} is necessary and sufficient to trigger both LTD and LTP, (2) the increase of $[Ca^{2+}]_i$ required to induce LTP has to be significantly larger than that necessary for LTD induction, and (3) the amplitude of the Ca^{2+} increase rather than the source of Ca^{2+} appears to be the relevant variable for the induction of synaptic modifications. It follows that it should not matter whether the depolarization required for Ca^{2+} entry is caused by the very synapses that are going to be modified or by other synaptic inputs. Thus, it should be possible to generalize these results to heterosynaptic modifications (Artola and Singer, 1993).

REFERENCES

Aniksztejn, L., and Ben-Ari, Y. (1991) Novel form of long-term potentiation produced by K^+-channel blocker in the hippocampus. *Nature* 349:67–69.

Alger, B. E., and Teyler, T. J. (1976) Long-term and short-term plasticity in the CA1, CA3 and dentate regions of the rat hippocampal slice. *Brain Res.* 110:463–480.

Aram, J. A., Bindman, L. J., Lodge, D., and Murphy, K. P. S. J. (1987) Glutamate receptor-mediated EPSPs in layer V neurones in rat neocortical slices. *J. Physiol.* 394:117P.

Artola, A., and Singer, W. (1987) Long-term potentiation and NMDA receptors in rat visual cortex. *Nature* 330:649–652.

Artola, A., and Singer, W. (1990) The involvement of *N*-methyl-D-aspartate receptors in induction and maintenance of long-term potentiation in rat visual cortex. *Eur. J. Neurosci.* 2:254–269.

Artola, A., and Singer, W. (1993) Long-term depression of excitatory synaptic transmission and its relationship to long-term potentiation. *Trends Neurosci.*, 16:480–487

Artola, A., Bröcher, S., and Singer, W. (1990a) Different voltage-dependent thresholds for inducing long-term depression and long-term potentiation in slices of rat visual cortex. *Nature* 347:69–72.

Artola, A., Bröcher, S., and Singer, W. (1990b) An in vitro model of plasticity in neocortical slices. In *The Biology of Memory*, L. Squire and E. Lindenlaub, (eds.) Stuttgart: Schattauer Medical Publishers, pp. 939–952.

Artola, A., Hensch, T., and Singer, W. (1992) A rise of $[Ca^{++}]_i$ in the postsynaptic cell is necessary and sufficient for the induction of long-term depression (LTD) in neocortex. *Soc. Neurosci.* 18:567.30.

Audinat, E. Knöpfel, T., and Gäwhiler, B. H. (1990) Responses to excitatory amino acids of Purkinje cells and neurones of the deep nuclei in cerebellar slice cultures. *J. Physiol.* 430: 297–313.

Bienenstock, E. L., Cooper, L. N., and Munro, P. W. (1982) Theory for the development of neuron selectivity: Orientation specificity and binocular interaction in visual cortex. *J. Neurosci.* 2:32–48.

Bindman, L. J., Murphy, K. P. S. J., and Pockett, S. (1988) Postsynaptic control of the induction of long-term changes in efficacy of transmission at neocortical synapses in slices of rat brain. *J. Neurophysiol.* 60:1053–1065.

Bliss, T. V. P., and Collingridge, G. (1993) A synaptic model of memory: Long-term potentiation in hippocampus. *Nature* 361:31–39.

Bliss, T. V. P., and Lømo, T. (1973) Long-lasting potentiation of synaptic transmission in the dentate area of the anaesthetized rabbit following stimulation of the perforant path. *J. Physiol.* 232:331–356.

Bortolotto, Z. A., and Collingridge, G. L. (1992) Activation of glutamate metabotropic receptors induces long-term potentiation. *Eur. J. Pharmacol.* 214:297–298.

Bröcher, S., Artola, A., and Singer, W. (1992) Intracellular injection of Ca^{++} chelators blocks induction of long-term depression in rat visual cortex. *Proc. Natl. Acad. Sci. USA* 89:123–127.

Calabresi, P., Maj, R., Pisani, A., Mercuri, N. B., and Bernardi, G. (1992a) Long-term synaptic depression in the striatum: Physiological and pharmacological characterization. *J. Neurosci.* 12:4224–4233.

Calabresi, P., Pisani, A., Mercuri, N. B., and Bernardi, G. (1992b) Long-term potentiation in the striatum is unmasked by removing the voltage-dependent magnesium block of NMDA recceptor channels. *Eur. J. Neurosci.* 4:929–935.

Chattarji, S., Stanton, P. K., and Sejnowski, T. J. (1989) Commissural synapses, but not mossy fiber synapses, in hippocampal field CA3 exhibit associative long-term potentiation. *Brain Res.* 495:145–150.

Christofi, G., Nowicky, A. V., Bolsover, S. R., and Bindman, L. J. (1993) The postsynaptic induction of non-associative long-term depression of excitatory synaptic transmission in rat hippocampal slices. *J. Neurophysiol.* 69:219–229.

Collingridge, G. L., Kehl, S. J., and McLennan, H. (1983) Excitatory amino acids in synaptic transmission in the Schaffer collateral-commissural pathway of the rat hippocampus. *J. Physiol.* 334:33–46.

Crépel, F., and Jaillard, D. (1991) Pairing of pre- and postsynaptic activities in cerebellar Purkinje cells induces long-term changes in synaptic efficacy in vitro. *J. Physiol.* 432:123–141.

Crépel, F., and Krupa, M. (1988) Activation of protein kinase C induces a long-term depression of glutamate sensitivity of cerebellar Purkinje cells. *Brain Res.* 458:397–401.

Daniel, H., Hemart, N., Jaillard, D., and Crépel, F. (1992) Coactivation of metabotropic glutamate receptors and of voltage-gated calcium channels induces long-term depression in cerebellar Purkinje cells in vitro. *Exp. Brain Res.* 90:327–331.

Dunwiddie, T., and Lynch, G. (1978) Long-term potentiation and depression of synaptic responses in the rat hippocampus: Localization and frequency dependency. *J. Physiol.* 276: 353–367.

Dudek, S. M., and Bear, M. F. (1992) Homosynaptic depression in area CA1 of hippocampus and effects of *N*-methyl-D-aspartate receptor blockade. *Proc. Natl. Acad. Sci. USA* 89:4363–4367.

Greengard, P., Jen, J., Nairn, A. C., and Stevens, C. F. (1991) Enhancement of the glutamate response by cAMP-dependent protein kinase in hippocampal neurons. *Science* 253:1135–1137.

Grover, L. M., and Teyler, T. J. (1990) Effects of extracellular potassium concentration and postsynaptic membrane potential on calcium-induced potentiation in area CA1 of rat hippocampus. *Brain Res.* 506:53–61.

Grover, L. M., and Teyler, T. J. (1991) Two components of long-term potentiation induced by different patterns of afferent activation. *Nature* 347:477–479.

Hebb, D. O. (1949) *The Organization of Behavior* New York: Wiley.

Hirano, T. (1990) Depression and potentiation of the synaptic transmission between a granule cell and a Purkinje cell in rat cerebellar culture. *Neurosci. Lett.* 119:141–144.

Hirsch, J. C., and Crépel, F. (1990) Use-dependent changes in synaptic efficacy in rat prefrontal neurons in vitro. *J. Physiol.* 427:31–49.

Hirsch, J. C., and Crépel, F. (1991) Blockade of NMDA receptors unmasks a long-term depression in synaptic efficacy in rat prefrontal neurons in vitro. *Exp. Brain Res.* 85:621–624.

Hirsch, J. C., and Crépel, F. (1992) Postsynaptic calcium is necessary for the induction of LTP and LTD of monosynaptic EPSPs in prefrontal neurons. An in vitro study in the rat. *Synapse* 10:173–175.

Hirsch, J. C., Barrionuevo, G., and Crépel. F. (1992) Homo- and heterosynaptic changes in efficacy are expressed in prefrontal neurons: An in vitro study in the rat. *Synapse,* 12:82–85.

Ito, M., and Kano, M. (1982) Long-lasting depression of parallel fiber-Purkinje cell transmission induced by conjunctive stimulation of parallel fibers and climbing fibers in the cerebellar cortex. *Neurosci. Lett.* 33:253–258.

Ito, M., and Karachot, L. (1990) Receptor subtype involved in, and time course of, the long-term desensitization of glutamate receptors in cerebellar Purkinje cells. *NeuroReport* 1:129–132.

Ito, M., Sakurai, M., and Tongroach, P. (1982) Climbing fibre induced depression of both mossy fibre responsiveness and glutamate sensitivity of cerebellar Purkinje cells. *J. Physiol.* 324:113–134.

Johnston, D., Williams, S., Jaffe, D., and Gray, R. (1992) NMDA-receptor–independent long-term potentiation. *Annu. Rev. Physiol.* 54:489–505.

Jones, K. A., and Baughman, R. W. (1988) NMDA- and non-NMDA-receptor components of excitatory synaptic potentials recorded from cells in layer V of rat visual cortex. *J. Neurosci.* 8:3522–3534.

Kano, M., and Kato, M. (1987) Quisqualate receptors are specifically involved in cerebellar synaptic plasticity. *Nature* 325:276–279.

Kimura, F., Tsumoto, T., Nishigori, A., and Yoshimura, Y. (1990) Long-term depression but not potentiation is induced in Ca^{2+}-chelated visual cortex neurons. *NeuroReport* 1:65–68.

Komatsu, Y., and Iwakiri, M. (1992) Low-threshold Ca^{2+} channels mediate induction of long-term potentiation in kitten visual cortex. *J. Neurophysiol.* 67:401–410.

Komatsu, Y., Fujii, K., Maeda, J., Sakaguchi, H., and Toyama, K. (1988) Long-term potentiation of synaptic transmission in kitten visual cortex. *J. Neurophysiol.* 59:124–141.

Komatsu, Y., Narajima, S., and Toyama, K. (1991) Induction of long-term potentiation without the participation of *N*-methyl-D-aspartate receptors in kitten visual cortex. *J. Neurophysiol.* 65:20–32.

Linden, D. J., Dickinson, M. H., Smeyne, M., and Connor, J. A. (1991) A long-term depression of AMPA currents in cultured cerebellar Purkinje neurons. *Neuron* 7:81–89.

Lisman, J. (1989) A mechanism for the Hebb and the anti-Hebb processes underlying learning and memory. *Proc. Natl. Acad. Sci. USA* 86:9574–9578.

Llano, I., Marty, A., Johnson, J. W., Ascher, P., and Gähwiler, B. H. (1988) Patch recordings of amino acid-activated responses in "organotypic" slice cultures. *Proc. Natl. Acad. Sci. USA* 85:3221–3225.

Llano, I., Dreesen, J., Kano, M., and Konnerth, A. (1991a) Intradendritic release of calcium induced by glutamate in cerebellar Purkinje cells. *Neuron* 7:577–583.

Llano, I., Marty, A., Armstrong, C. M., and Konnerth, A. (1991b) Synaptic and agonists-induced excitatory currents in Purkinje cells in rat cerebellar slices. *J. Physiol.* 434:183–213.

Lynch, G. S., Gribkoff, V. K., and Deadwyler, S. A. (1977) Long-term potentiation is accompanied by a reduction in dendritic responsiveness to glutamic acid. *Nature* 263:151–153.

Lynch, G., Larson, J., Kelso, S., Barrionuevo, G., and Schottler, F. (1983) Intracellular injections of EGTA block induction of hippocampal long-term potentiation. *Nature* 305: 719–721.

MacDermott, A. B., Mayer, M. L., Westbrook, G. L., Smith, S. J., and Barker, J. L. (1986) NMDA-receptor activation increases cytoplasmic calcium concentration in cultured spinal cord neurones. *Nature* 321:519–522.

Malenka, R. C., Kauer, J. A., Zucker, R. S., and Nicoll, R. A. (1988) Postsynaptic calcium is sufficient for potentiation of hippocampal synaptic transmission. *Science* 242:81–84.

Mayer, M. L., Westbrook, G. L., and Guthrie, P. B. (1984) Voltage-dependent block by Mg^{2+} of NMDA responses in spinal cord neurones. *Nature* 309:261–263.

McCormick, D. A., Connors, B. W., Lighthall, J. W., and Prince, D. A. (1985) Comparative electrophysiology of pyramidal and sparsely spiny stellate neurons of the neocortex. *J. Neurophysiol.* 54:782–806.

Mulkey, R. M., and Malenka, R. C. (1992) Mechanisms underlying induction of homosynaptic long-term depression in area CA1 of the hippocampus. *Neuron* 9:967–975.

Nowak, L., Bregestovski, P., Ascher, P., Herbet, A., and Prochiantz, A. (1984) Magnesium gates glutamate-activated channels in mouse central neurons. *Nature* 307:462–465.

Pennartz, C. M. A., Ameerun, R. F., Groenewegen, H. J., and Lopes da Silva, F. H. (1993) Synaptic plasticity in an in vitro slice preparation of the rat nucleus accumbens. *Eur. J. Neurosci.* 5:107–117.

Pockett, S., and Lippold, O. C. J. (1986) Long-term potentiation and depression in hippocampal slices. *Exp. Neurol.* 91:481–487.

Pockett, S., Brookes, N. H., and Bindman, L. J. (1990) Long-term depression at synapses in slices of rat hippocampus can be induced by bursts of postsynaptic activity. *Exp. Brain Res.* 80:196–200.

Ross, W. N., and Werman, R. (1987) Mapping calcium transients in the dendrites of Purkinje cells from the guinea-pig cerebellum in vitro. *J. Physiol.* 389:319–336.

Sakurai, M. (1987) Synaptic modification of parallele fibre-Purkinje cell transmission in in vitro guinea-pig cerebellar slices. *J. Physiol.* 394:319–336.

Sakurai, M. (1990) Calcium is an intracellular mediator of the climbing fiber in induction of cerebellar long-term depression. *Proc. Natl. Acad. Sci. USA* 87:3383–3385.

Sejnowski, T. J. (1977) Storing covariance with nonlinearly interacting neurons. *J. Math. Biol.* 4:303–321.

Singer, W., and Artola, A. (1993) Plasticity of the mature cortex. In *Cellular and Molecular Mechanisms Underlying Higher Neural Functions,* A. Selverston and P. Ascher, (eds.) New York: Wiley, in press.

Stanton, P. K., and Sejnowski, T. J. (1989) Associative long-term depression in the hippocampus induced by hebbian covariance. *Nature* 339:215–218.

Stanton, P. K., Chattarji, S., and Sejnowski, T. J. (1991) 2-Amino-3-phosphonopropionic acid, an inhibitor of glutamate-stimulated phosphonoinositide turnover, blocks induction of homosynaptic long-term depression, but not potentiation, in rat hippocampus. *Neurosci. Lett.* 127:61–66.

Sutor, B., and Hablitz, J. J. (1989) Long-term potentiation in frontal cortex: Role of NMDA-modulated polysynaptic excitatory pathways. *Neurosci. Lett.* 97:111–117.

Tsumoto, T. (1992) Long-term potentiation and long-term depression in neocortex. *Prog. Neurobiol.* 39:209–228.

Turner, R. W., Baimbridge, K. G., and Miller, J. (1982) Calcium-induced long-term potentiation in the hippocampus. *Neuroscience* 7:1411–1416.

Wang, L. Y., Salter, M. W., and MacDonald, J. F. (1991) Regulation of kainate receptors by cAMP-dependent protein kinase and phosphatase. *Science* 253:1132–1135.

Wigström, H., and Gustafsson, B. (1983) Facilitated induction of hippocampal long-lasting potentiation during blockade of inhibition. *Nature* 301:603–604.

Xie, X., Berger, T. J., and Barrionuevo, G. (1992) Isolated NMDA receptor-mediated synaptic responses express both LTP and LTD. *J. Neurophysiol.* 67:1009–1013.

Yoshimura, Y., Tsumoto, T., and Nishigori, A. (1991) Input-specific induction of long-term depression in Ca^{2+}-chelated visual cortex neurons. *NeuroReport* 2:393–396.

Zalutsky, R. A., and Nicoll, R. A. (1990) Comparison of two forms of long-term potentiation in single hippocampal neurons. *Science* 248:1619–1624.

Zheng, F., and Gallagher, J. P. (1992) Metabotropic glutamate receptors are required for the induction of long-term potentiation. *Neuron* 9:163–172.

14 Does Postsynaptic Membrane Potential Regulate Functional Plasticity in Kitten Visual Cortex?

Yves Frégnac, Dominique Debanne, Daniel Shulz, and Attila Baranyi

VISUAL CORTEX, WHERE THEORY MEETS EXPERIMENTS

Visual pathways have long been a model of choice for the study of activity-dependent rules acting in the formation and regulation of functional connectivity. In a prenatal phase of development, intrinsic synchronous bursting activity which arises from the retina (Maffei and Galli-Resta, 1990; Meister et al., 1990) contributes to the development of the retinogeniculate pathway (Shatz, 1992). Spontaneous firing of ganglion cells participates in an activity-dependent regulation of the branching of retinal axons and cellular segregation in distinct laminae at the thalamic level (reviewed in Shatz, 1990). In a later phase of postnatal development, the major source of phasic activity is evoked by the interaction with the visual environment. It has been shown to exert a structuring role in the periodic ordering of bands of geniculate fibers subserving each eye in layer IV of visual cortex, in the pruning of their terminal fields, and in the spatial restriction of periodic patches of intracortical connectivity (Callaway and Katz, 1990; Löwel and Singer, 1992). The grouping and sorting out of fibers, the morphological complexification of the terminal boutons of intrinsic and extrinsic axons, the functional expression and possibly silencing of synapses could all be, at some stage of postnatal development, under the influence of temporal correlation between the activity of presynaptic fibers converging onto the same target, or between presynaptic and postsynaptic partners. While this essential role of coactivity in self-organization was foreseen by theoreticians (Von der Malsburg, 1973; Willshaw and Von der Malsburg, 1976; Linsker, 1986a,b,c), experimental evidence remains limited concerning its implication during normal development. Most support comes from the study of the neuroanatomical and physiological effects of forced patterns of activity, which *simulate* the functional effects of visual experience. A remarkable illustration is given by the induction of segregation of geniculate afferents in visual cortex in the absence of retinal activity (blocked by intraocular injection of TTX) following the imposition of different degrees of temporal correlation between electrical stimulation of the optic nerves (Stryker and Strickland, 1984; Stryker and Harris, 1986).

HEBBIAN SCHEMES OF SYNAPTIC PLASTICITY

These neural processes of self-organization appear to obey simple activity-dependent rules, reminiscent of a general principle of association, originally proposed in non-neuronal terms by James in 1890. This postulate was later reworded in synaptic terms by Hebb in 1949, in the context of a general theory of cell assembly formation. It is assumed that coactivity arising in a given assembly controls the coupling strength of the functional links established between each activated neuronal element. This hypothetical scheme of synaptic plasticity has been tested experimentally at a variety of sites in the central nervous system, ranging from molluscan neuronal ganglia to the mammalian forebrain (Wurtz et al., 1967; Carew et al., 1984; Kelso et al., 1986; Wigström, et al., 1986; reviewed in Brown et al., 1990).

Local Hebbian association rules predict that cells that tend to fire together, either because they share the same input or because they have the same functional preference, tend to reinforce their synaptic connections (reviewed in Frégnac and Shulz, 1993). Experimental validation has been gathered with some success in the description of activity-dependent sharpening processes in retinotopic map formation (Schmidt and Edwards, 1983; Schmidt and Eisele, 1985), and of the effects of visual experience on the development of cortical specificity (Cynader, 1979; Rauschecker and Singer, 1981). The demonstration of synaptic potentiation following imposed correlation between afferent activity and depolarization of the postsynaptic element under the cooperative influence of other inputs was first obtained in motor cortex (Baranyi and Feher, 1981) and hippocampus (Kelso et al., 1986; Wigström et al., 1986).

In contrast, activity-dependent decreases in synaptic efficacy have received less attention, apart for the notable exception of cerebellum (Albus, 1971; see Crépel et al., this volume). However, mechanisms for selective weakening of excitatory synaptic transmission should also exist in cerebral cortex based on theoretical considerations (Rochester et al., 1956; Marr, 1970; Stent, 1973; Von der Malsburg, 1973; Von der Malsburg and Bienenstock, 1986). Depression mechanisms are indeed needed to counteract an obvious limitation of a straight application of Hebb's principle to excitatory synapses: maintained temporal correlation between presynaptic and postsynaptic activity at an already transmitting synapse would lead to a continuous increase in the gain of synaptic transmission. This "divergence" problem was solved by modelers using various normalization hypotheses: Saturation values were imposed at a given synapse and forgetting mechanisms were introduced with disuse; complementary plasticity rules were assumed to operate simultaneously in active pathways (homosynaptic depression), and in neighboring inactive afferents (heterosynaptic depression). The resulting prediction was that spatial and temporal competition would occur between synapses that impinge on a common target cell.

Most algorithms that are currently used to model synaptic plasticity are surprisingly uniform, and based on coactivity of presynaptic and postsynaptic neurons. They may be summarized by the same general equation where the change of synaptic efficacy with time is equal to the product of a presynaptic term and a postsynaptic term (reviewed in Frégnac and Shulz, 1993). The so-called "covariance hypothesis" (detailed in figure 14.1b) introduced by Sejnowski in 1977 for cerebellum (Sejnowski, 1977a,b), and later by Bienenstock, Cooper and co-workers for visual cortex (Bienenstock et al., 1982; Bear et al., 1987), replaces the presynaptic and postsynaptic terms by the departure of instantaneous pre- and postsynaptic activities from their respective average values over a certain window of time. Since this covariance term can be positive or negative, it predicts both increases and decreases in synaptic efficacy, which avoids progressive saturation of the efficacy of all the synapses in the network.

Indirect experimental evidence in visual cortex supports the covariance scheme as a possible rule of regulation of synaptic efficiency operating during at least a critical period of postnatal life (Hubel and Wiesel, 1970). Comparisons of ocular dominance histograms of cells recorded in normally reared and dark-reared kittens (figure 14.2a,b) indicate that disuse per se (imposed for several weeks) is not sufficient to degrade synaptic transmission, although it does significantly affect some receptive field properties such as orientation selectivity (Buisseret and Imbert, 1976; Frégnac, 1979) and to a lesser extent visual responsiveness (Ramoa et al., 1987; Tsumoto and Freeman, 1987). Most cells recorded in dark-reared kittens remain visually activated and still respond to visual stimuli through each eye (figure 14.2b). It remains possible that the retinal dark discharge participates in the maintenance of binocular connections in the absence of phasic structured input.

The effect of unilateral closure of eyelids on cortical binocularity is more dramatic: most visual cortical cells recorded after only 1 week of monocular deprivation imposed at the peak of the critical period respond only through the remaining open eye (Hubel and Wiesel, 1970). This suggests that vision through the open eye results in the functional silencing (or complete disconnection) of the input coming from the closed eye (figure 14.2c). This apparent paradox between the effects of monocular

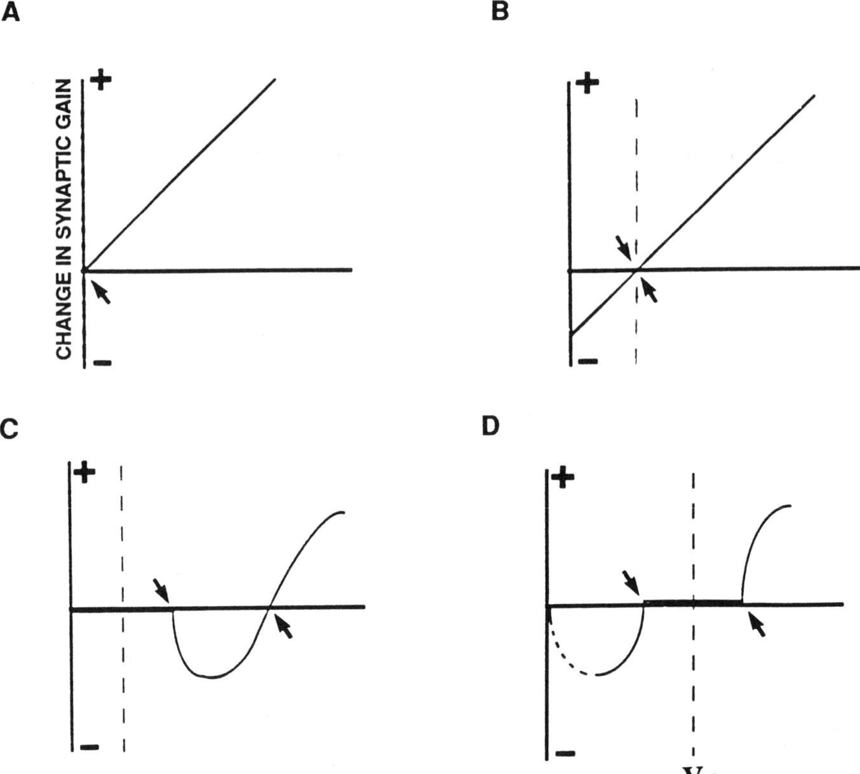

Figure 14.1 Comparison of the predictions of different rules of excitatory synaptic plasticity: changes in synaptic efficacy (+, potentiation; −, depression) in an active pathway as a function of concomitant levels of postsynaptic activity (*a, b*) or of postsynaptic membrane potential (*c, d*). (*a*) Hebb's postulate (1949) predicts only positive changes in synaptic efficacy, which increase linearly with postsynaptic activity (the slope being proportional to presynaptic activity). (*b*) The covariance hypothesis (Sejnowski, 1977*a,b*) predicts both increase and decrease in synaptic efficacy, depending on the level of postsynaptic activity at the time of synaptic activation. When the cell is functioning spontaneously or in a nonadaptative transmiting mode, limited random fluctuations in postsynaptic activity around its mean value (dotted line) should leave on average synaptic efficiency unchanged. Only one postsynaptic threshold (arrows) of synaptic plasticity is defined, above and below which, respectively, synaptic potentiation or depression occur. (*c*) Artola and colleagues (1990) proposed a variant in which synaptic change is made dependent on the membrane potential of the postsynaptic neuron. Progressive depolarizations from the resting membrane potential (dotted line) predicts the crossing of a low threshold above which depression of synaptic efficacy occurs (downward arrow), before a second higher threshold (upward arrow) is trespassed, above which synaptic potentiation is obtained. Two distinct ranges of membrane potential are proposed for which no synaptic changes are expected. (*d*) Frégnac and colleagues (1990) observed potentiation of compound postsynaptic potentials (PSPs) when paired with intracellular depolarization, and depression of the same PSP when paired with hyperpolarization. They propose the existence of a unique "read-only" state (around the dotted line) representing the admissible range of membrane potential between two plasticity thresholds (arrows), which still allows sensory neurons to transmit reliably visual information. V_f corresponds to the normal functioning state of the synapse, i.e., the local postsynaptic membrane potential reached at the time of activation of the test input.

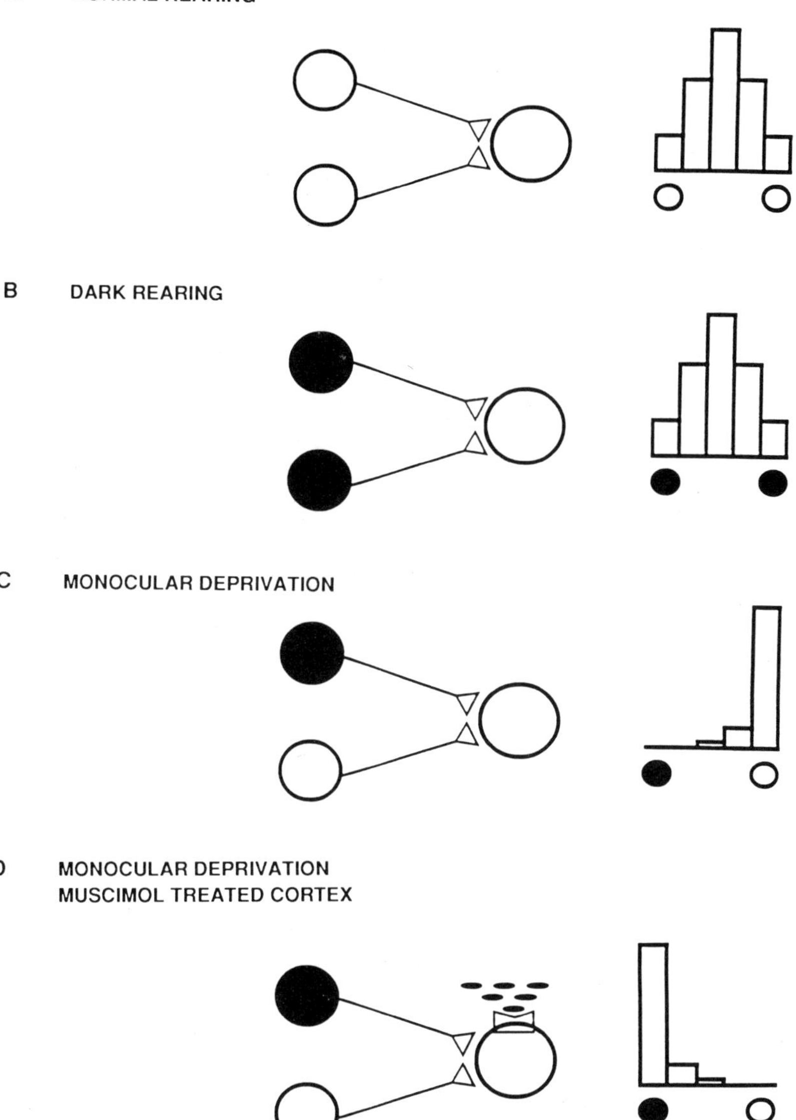

Figure 14.2 Comparison of the effects of visual experience during postnatal development. (*Left*) Schematized cortical neurons receiving separate inputs (through the LGNd, not represented) from each eye. (*Right*) Histograms of ocular dominance established on a five-class scale, ranging from monocular activation through one eye (class 1) to monocular activation through the other eye (class 5) via three intermediate binocular states (classes 2–4). Open circles symbolize experience of vision through one eye, whereas filled circles indicate total visual deprivation or eye lid closure. Four experimental situations and their effect on cortical binocularity are shown: (a) normal rearing, (b) dark rearing, (c) monocular deprivation, and (d) monocular deprivation performed conjointly with intracortical infusion of a $GABA_A$ agonist. See text for details.

and binocular deprivation has been taken as indicative of competitive processes between active and inactive synapses. It supports Stent's postulate of heterosynaptic plasticity, although the biophysical mechanism this author proposed appears to be very unlikely (Stent, 1973).

From these various experiments one may conclude that negative covariance between presynaptic and postsynaptic activity caused by a reduction of presynaptic activity associated with a high level of postsynaptic activity (under the influence of visual experience gained through the remaining open eye) could lead to synaptic depression. A recent ingenious rearing protocol explores the effects of a symmetrical situation, when presynaptic activity is associated with repetitive failure in synaptic transmission. For this purpose monocular deprivation was imposed concomitantly with a pharmacological blockade of postsynaptic activity (Reiter and Stryker, 1988). Intracortical infusion of a $GABA_A$ agonist (muscimol) is known to suppress cortical cell activity without affecting the activity of afferent geniculate fibers which are devoid of $GABA_A$ receptors. Such manipulations impose periods of negative covariance between presynaptic and postsynaptic activity at the active synaptic sites. The change in the ocular dominance distribution (figure 14.2d) agrees with the prediction made using the covariance model: recorded cells tended to respond exclusively to the previously closed eye, which suggests a depression in the efficacy of synapses subserved by the open eye. Similar findings have been reported following intracortical infusion of inactivating doses of *N*-methyl-D-aspartate (NMDA) (Bear et al., 1990). Thus, these four rearing procedures illustrate the validity of the covariance scheme in simulating correctly the phenomenology of functional changes occurring during epigenesis at the cortical level. Unfortunately these compelling data are based on population analysis techniques (i.e., comparison of statistical representation of receptive field properies across population of cells recorded in different animals of the same age, submitted to various rearing conditions and pharmacological manipulations). More direct experimental strategies needed to be developed to ascertain the validity of this theory at a cellular level.

STRATEGIES IN THE STUDY OF VISUAL CORTICAL PLASTICITY

A classic approach used in most studies of synaptic plasticity is to explain modifications of the integrative properties of neurons on the sole basis of long-term changes in the efficacy of synaptic transmission. This equation of functional plasticity with the activity-dependent expression of long-term potentiation (LTP) and depression (LTD) might have led experimenters to overemphasize their attention on long-lasting changes in monosynaptic links, and ignore reversible dynamics due to network influence. In the cortical network, the "effective" functional coupling between two neurons stems from the temporal integration of the synaptic influence of converging monosynaptic and polysynaptic pathways. Adaptative properties of visual cortical receptive fields such as those described in the previous section (for a more extensive review, see Frégnac and Imbert, 1984) are probably not the exclusive result of punctual modification processes occurring at the active synapse site, but could also be influenced by changes distributed across the network.

Several interpretations of a change in "effective" coupling can be made:

• It might be due to biochemical or structural modifications occurring primarily at the active synaptic site (homosynaptic plasticity).

• It could be the generalization of a change occuring primarily at a parent bouton of the active axon terminating on a neighboring cell (Bonhoeffer et al., 1990; Kossel et al., 1990) through the possible action of diffusible retrograde signals (i.e., arachidonic acid, see Williams et al., 1989; nitric oxide, see Gally et al., 1990; Montague et al., 1991; Bon and co-workers, and Madison and Schuman, this volume).

• It could be the result of changes occuring primarily in neighboring competing active synapses terminating on the same neuron (heterosynaptic plasticity, i.e., such as cross-depression in kitten visual cortex; Tsumoto and Suda, 1979; Tamura et al., 1993).

• In addition, part of the interaction between these different forms of synaptic plasticity at a given synaptic site will be construed by the spatial distribution of neighboring synaptic terminals in the recorded structure, and possibly by the glial compartment and the available extracellular volume of diffusion in the cortical network (Müller, 1992). We will refer to such processes using the term *network plasticity*.

In order to address the homosynaptic nature of the functional modifications more directly, the effects of pairing of presynaptic and postsynaptic activity on synaptic transmission should be examined ideally for only a homogeneous class of monosynaptic connections. This has been successfully achieved in other structures for excitatory synaptic connections, by recording simultaneously the pre- and postsynaptic partners, such as at the neuromuscular junction (Dan and Poo, 1992) or at the CA3-CA1 connections of the hippocampal slice (Friedlander et al., 1990; Sayer et al., 1990; Malinow, 1991). However, dual intracellular recordings in intact slices of visual cortex are difficult to master (but see Mason et al., 1991; Stern et al., 1992), and to date no attempt to apply Hebbian protocols has been yet reported.

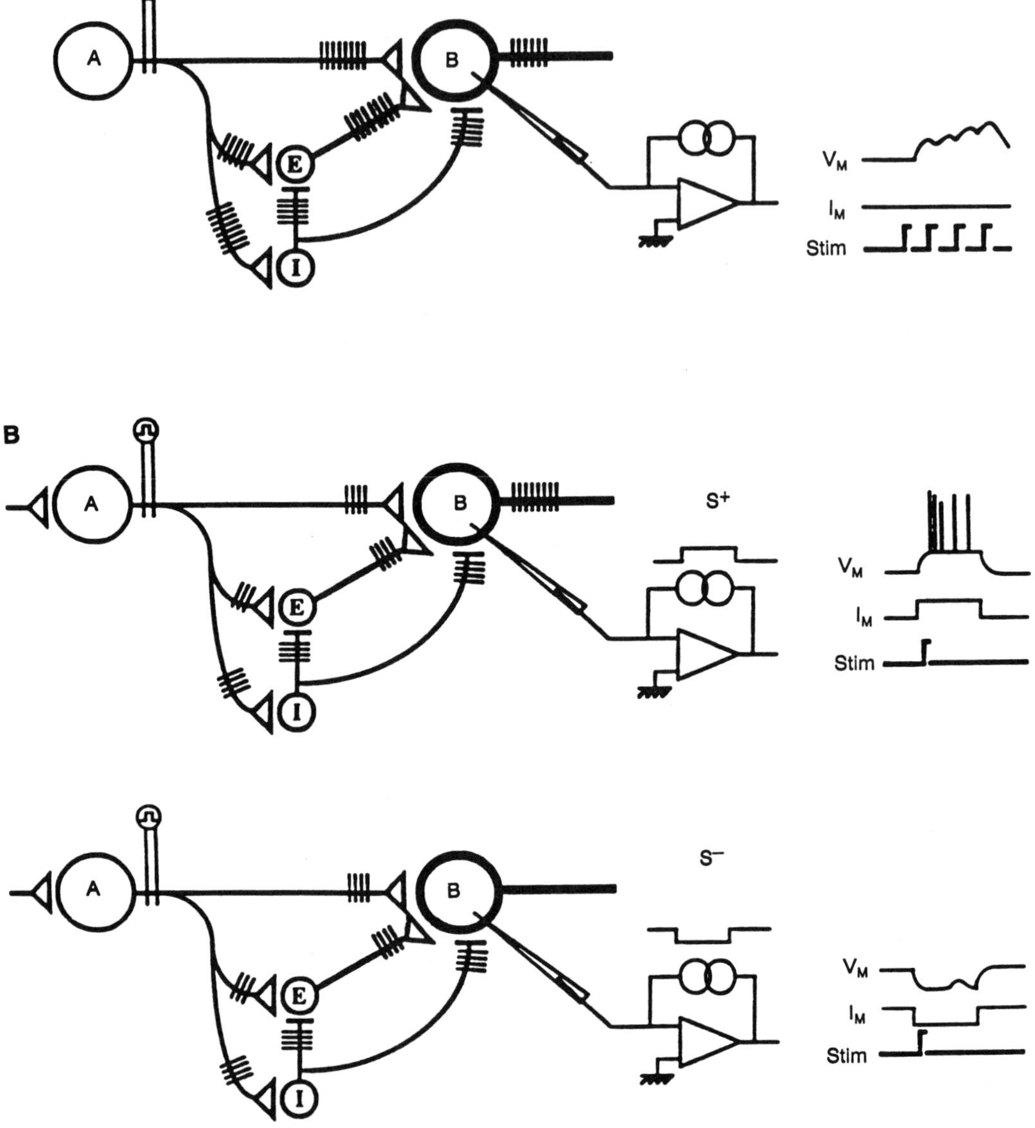

Figure 14.3 Conditioning protocols: high-frequency tetanization versus low-frequency pairing. The functional coupling between the input pathway (originating from presynaptic neuron A) and the postsynaptic cell B results from monosynaptic excitatory contacts and polysynaptic circuits involving both excitatory (E) and inhibitory (I) interneurons. Right, intracellular voltage deflection recorded in cell B in response to the visual or electrical stimulation of cell A (Stim). The firing frequency of the considered pathway is given by the density of the bars along the axon. Two protocols are classically used to induce activity-dependent transmission. (*a*) High-frequency tetanization: changes might potentially occur at every synaptic locus along the activated pathway. Analysis is often restricted to the early monosynaptic component of the response. (*b*) Low-frequency pairing: the temporal correlation between the afferent message and the postsynaptic state of the target cell (B) are directly controlled by injecting intracellular depolarizing (S+ pairing,

An ultimate problem is raised by the partial knowledge of the spatial distribution and temporal invariance of the control activity pattern evoked across the network. Theoretical simulations demonstrate that changes in "effective" connectivity measured at a synaptic site (which, for instance, could be interpreted as a direct effect of past activity) cannot be distinguished from a rapid gain control exerted by the network of "unseen" units, on the apparent efficacy of the tested pathway (Aertsen et al., 1988).

CONDITIONING PROTOCOLS: HIGH-FREQUENCY TRAINS OF PRESYNAPTIC ACTIVITY VERSUS LOW-FREQUENCY PAIRING BETWEEN PRESYNAPTIC AND POSTSYNAPTIC ACTIVITY

While no direct approach has yet been developed in mammalian visual cortex to understand functional adaptation in terms of synaptic changes, two types of protocols have been applied successfully at the cellular level to change synaptically evoked responses (figure 14.3). The first protocol is very much inspired from the literature of hippocampal plasticity (figure 14.3a): it advocates the use of high-frequency trains of stimulation of pathways afferent to the recorded neuron in order to induce activity-dependent changes in the efficacy of transmission of the next synapse (Artola and Singer, 1987; Komatsu et al., 1981, 1988). In the case of the study of intracellular postsynaptic potentials (PSPs) elicited by extracellular stimulation of an afferent pathway, assessment of monosynaptic excitatory changes would in theory require comparison measurements of the initial slope of the PSP before and after pairing during total blockade of inhibition. To a much greater extent than in hippocampus, such an experimental situation appears highly unstable in the visual cortical slice, and bicuculline application in the bath has been limited in previous pharmacological studies of visual cortical plasticity to a microdose level (less than 1.2–1.5 μM) in order to avoid epileptic seizures (Artola et al., 1990; Komatsu et al., 1991; Bear et al., 1992). It is not known in this latter case to what extent inhibition is reduced.

Interpretation of the change in the test PSP slope is further complicated by the fact that bath application of drugs (bicuculline, APV, AP3, raised

first row) or hyperpolarizing (S− pairing, second row) current pulses in the postsynaptic cell, concomitantly with the activation of the input neuron (A). The covariance hypothesis (figure 14.1b, d) predicts synaptic potentiation at the terminal level of the mono- *and* polysynaptic excitatory pathways, when depolarizing the target cell (S+). Depression is predicted at the same synapses impinging on the postsynaptic neuron (B), when it is kept silent by the intracellular injection of an hyperpolarizing current applied during synaptic activation.

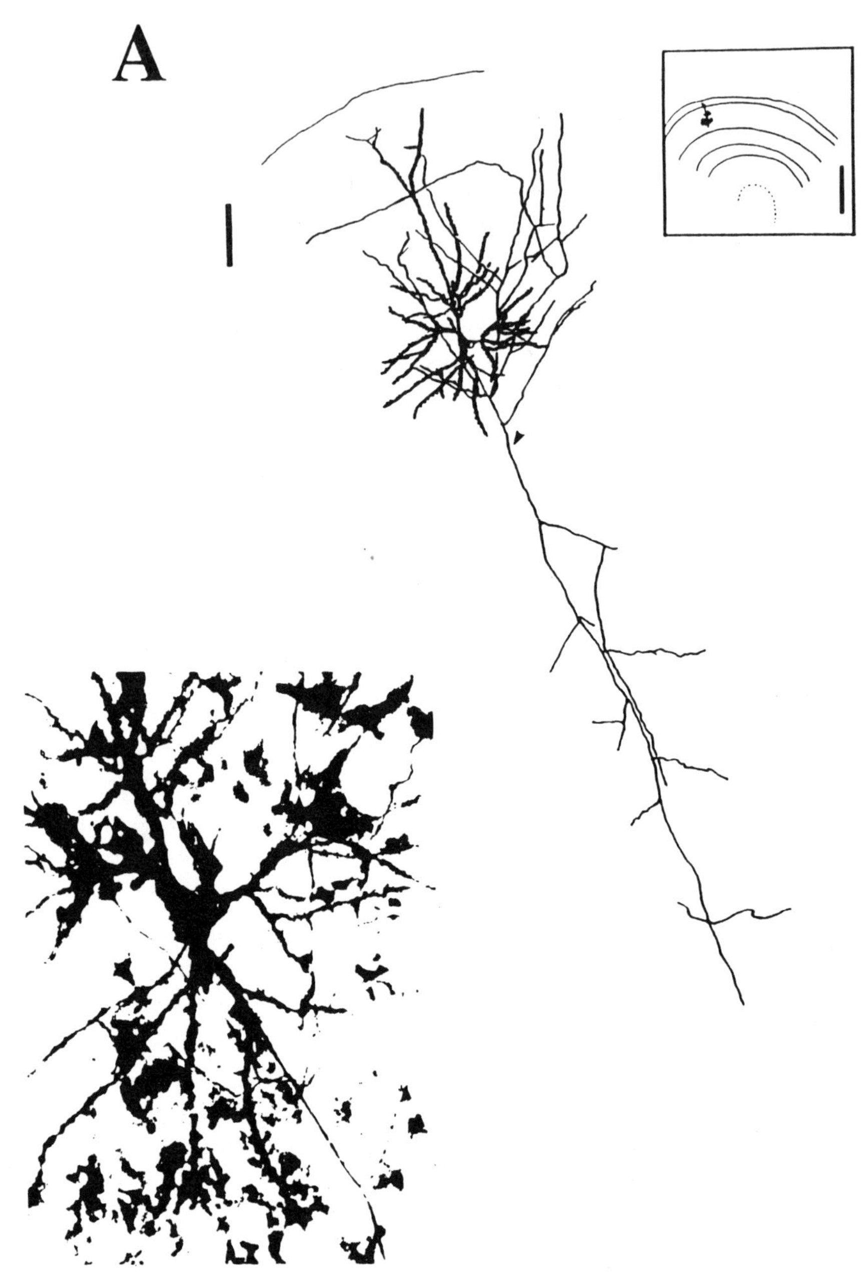
A

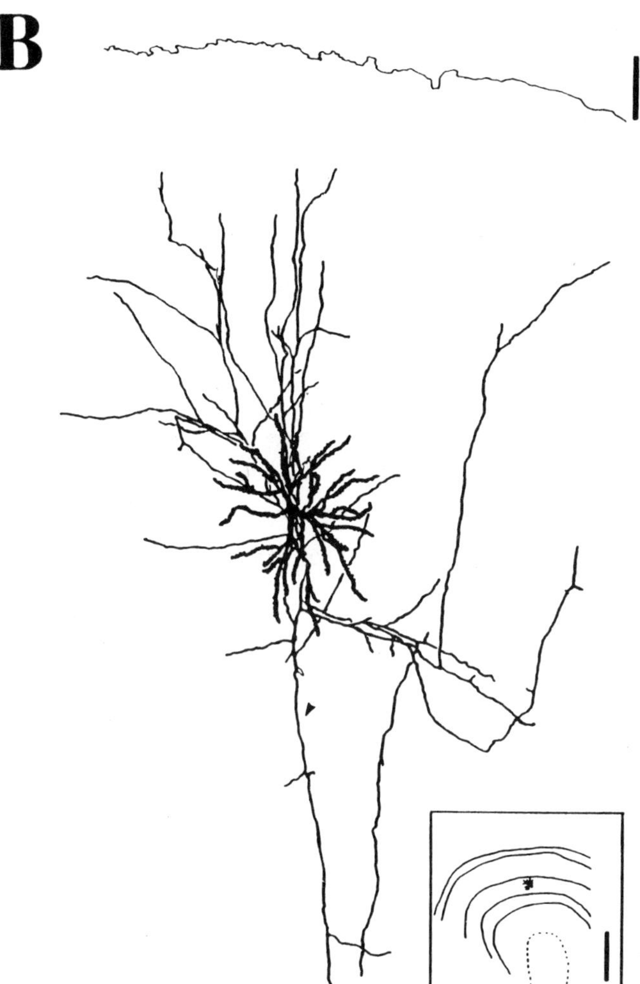

Figure 14.4 Morphological reconstruction of biocytin-injected neurons in kitten visual cortex. The reconstructions of dendritic and axonal (arrows) arborizations were made from several serial 80-μm sections of the fixed slice using a camera lucida drawing chamber, after intracellular recording and injection of biocytin (Frégnac et al., 1990; Grant et al., 1990). Calibration bars: 100 μm and 1 mm. (*A*) Typical spiny pyramidal cell recorded in layers II–III of area 17 in a 4-week-old kitten. The axon of the cell running on an axis perpendicular to the cortical surface (upper continuous line) could be followed in the same plane for more than 1.5 mm (not shown). Note the umbrella-like pattern of recurrent axonal branches originating 50–100 μm from the soma which remain restricted in the upper cortical layers inside the cortical volume defined by the apical dendritic tree. *Inset*: Original micrograph of the biocytin-injected cell. (*B*) Spiny stellate cell recorded in layer IV of area 17 in a 4-week-old kitten. Note the limited extension of the dendritic tree, which contrasts with the profusion of axonal branches outside layer IV, and their bifurcation pattern occurring often at right angles.

level of extracellular Ca^{2+}) is often used to control the permissive state of the network, and that the intensity of stimulation used during the conditioning protocol is often raised several times higher than that used in the test stimulation condition in order to induce a change (i.e., four times in Artola et al., 1990). Therefore, the synaptic modifications might involve consecutive effects produced by the activation during tetanus of a large number of "unseen" units which previously were not recruited by the test pathway. One cannot even exclude that some direct links might be functionally revealed or suppressed, affecting the rising slope of electrically evoked PSPs. Consequently, plasticity revealed by such protocols is probably also intrinsically network dependent.

An additional potential problem—that has not been clearly addressed in visual cortex—comes from the respective location sites of the stimulation and recording electrodes. Most intracellular recordings in visual cortex published so far have been made in spiny pyramidal cells of supragranular layers (figure 14.4A). The choice of these layers is justified by the fact that they are the origin and the termination zone of most "horizontal" connections that participate in the intrinsic cortical network (Gilbert and Wiesel, 1983; Kisvarday and Eysel, 1992). In addition, current source density studies have shown that it is where the most prominent changes in synaptic sinks are observed following prolonged low-frequency repetitive stimulation of optic radiations (Komatsu et al., 1981; Tsumoto and Suda, 1982). However, most supragranular layer cells are known to be di- or polysynaptically activated by stimulation of the white matter afferents (Bullier and Henry, 1979a,b; Bode-Greuel et al., 1987; Komatsu et al., 1988, 1991). Restriction of conditioning procedures to the stimulation of white matter while recording in layer IV, or to the stimulation of layer IV when recording in supragranular layer cells, could yield different conclusions, and be more adapted to the study of the plasticity of identified monosynaptic connections. However, the former situation might be more difficult to achieve, since an intracellular recording from neurons monosynaptically activated by white matter stimulation remains a rare event (see spiny stellate cell in figure 14.4B).

The experiments reviewed in this chapter are based on a second strategy (detailed in figure 14.3b), in which the analysis is not restricted to the early monosynaptic component, and where the conditioning protocol does not require a change in the frequency and in the intensity of the stimulation of the conditioned pathway. Typically, a test pathway will be stimulated at a low frequency (below 1 Hz, instead of 2–50 Hz used classically in LTP and LTD experiments), to avoid neuronal fatigue or complete suppression of normally acting inhibitory processes. Conditioning will rely primarily on the exogenous control of the postsynaptic state of the conditioned neuron, and on the temporal correlation between presynaptic and postsynaptic activity.

DESIGNING EXPERIMENTAL TESTS OF A POSTSYNAPTIC CONTROL OF FUNCTIONAL PLASTICITY

It is known from a large number of electrophysiological studies in the 1970s that repeated visual stimulation of a visual cortical cell does not induce long-lasting modifications in its level of firing in the anesthetized and paralyzed preparation (reviewed in Frégnac and Imbert, 1984). This freezing of visual cortical function, which made the success of the acute preparation in developmental studies of receptive field properties, is due largely to the absence of attentional signals and the suppression of eye movements and extraocular proprioceptive inflow. Exogenous activation is required to produce acute changes in receptive field properties (reviewed in Frégnac, 1987). The experimental tests of the covariance hypothesis we developed in collaboration with Elie Bienenstock fullfill this latter condition, and constitute typically supervised learning procedures (Frégnac et al., 1988). A differential pairing protocol is used to impose opposite changes in the temporal correlation between given parameters characterizing afferent activity and the output signal of the cell. Two pathways are stimulated alternately (using natural stimuli presentation), one being submitted to the pairing procedure, the other one being used as a control. Taking into account the possibility of nonassociative changes affecting cellular excitability and/or membrane potentials, the simultaneous comparison of visually evoked responses or of PSPs in response to two different inputs yields more reliable conclusions than the direct observation of absolute changes in responses to a single test stimulus (see also the work in *Aplysia* by Carew et al., 1984). The pairing protocol can be summarized in the following way: an external supervisor (i.e., the experimenter) helps the cell to respond to one input, and blocks the cell's response to another, different input.

This protocol was achieved the following way: various presynaptic inputs were selectively activated by changing the value of physical parameters characterizing the visual stimulus. The postsynaptic state was controlled either by extracellular iontophoretic techniques to regulate up and down the firing frequency of the recorded neuron (Frégnac and Debanne, 1993; Frégnac et al., 1988, 1992; Shulz and Frégnac, 1992; Shulz et al., 1993), or by intracellular current injection to directly manipulate the postsynaptic membrane potential (Baranyi et al., 1991a; Debanne et al., 1991).

A FUNCTIONAL APPROACH TO SYNAPTIC PLASTICITY: SEGREGATION OF ON- AND OFF-CENTER AFFERENTS

Using such Hebbian-like conditioning procedures, we could demonstrate that dynamic functional properties of visual cortical cells such as orientation selectivity, ocular dominance, and interocular orientation disparity could be changed during the time of recording of a single neuron (Frégnac et al., 1988, 1992; Shulz and Frégnac, 1992). These experimentally induced modifications of specificity of the visual response are cellular analogs of cortical plasticity occurring during development or following early manipulation of the visual environment (Wiesel and Hubel, 1963; Blakemore and Cooper, 1970; Shinkman and Bruce, 1977). However, it remains unknown to what extent differences in the physical parameters of the visual stimuli (between two orientations, between the two eyes, or between two interocular orientation disparities) correspond to an effective separation in terms of activation of distinct synaptic afferents. Since in a certain number of cases, the observed changes in dynamic responses could be ascribed to restricted portions of the receptive field (RF) (see, for instance, figure 4 in Shulz and Frégnac, 1992), and since modifications of the spatial organization of RFs have also been reported after dark rearing or manipulations of the visual environment (Singer and Tretter, 1976; Milleret et al., 1988), we decided to adapt our original protocol to the conditioning of ON and OFF responses.

Receptive field organization is usually defined on the basis of the spatial distribution of ON and OFF responses across the whole extent of the RF. The ON/OFF dichotomy is not simply an ad hoc classification: it corresponds to a structural parallel organization of the retino-geniculo-cortical pathway. The ON-center and OFF-center channels are activated asynchronously (by definition of the physical stimulus!) and arise from the selective visual processing of two specialized and segregated networks in the retina ((Wässle et al., 1981a,b; reviewed in Schiller, 1982). At the geniculate level, ON- and OFF-channels remain functionally separated, and ON- and OFF-center relay cells tend to occupy two distinct sublaminae within layers A and A1 (Bowling and Caverhill, 1989). We assume schematically that ON and OFF inputs to cortical cells are fed through distinct retino-geniculo-cortical pathways. This input separation applies to "simple" RFs, which represent 70% of the population of cells encountered in area 17, and to a lesser extent to "complex" RFs.

Once a cell had been functionally characterized, its receptive field was stimulated using automated protocols with a light bar flashed ON and OFF in different positions. One position was chosen for the pairing procedure and the other(s) as the control position(s). Two experimental paradigms were used to control the temporal correlation between the afferent message and postsynaptic activity:

• A first *extracellular method* was to impose various levels of postsynaptic activity by applying iontophoretic current pulses (less than ± 10 nA) through the extracellular recording electrode (1–3 M KCl electrode, 10–20 MΩ) in temporal conjunction with the presentation or extinction of the visual stimulus (Frégnac et al., 1989; Shulz et al., 1993; Debanne et al., in preparation).

• A second method relied on *intracellular recordings* (2M potassium methylsulfate electrode, 70 MΩ) and pulses (100–400 msec) of current (less than ± 4 nA) were directly injected in the recorded neuron (Baranyi et al., 1991a; Debanne et al., 1991).

Typically, one contrast transition of the visual stimulation (for example, ON) was associated with an artificial increase of responsiveness (S^+), induced by positive current applied through the recording electrode, which imposed a depolarizing action through ejection of K^+ in the extracellular medium or by direct intracellular current injection. The response to the antagonist characteristic (OFF, same position) was unchanged (S^0) or reduced (S^-) by negative current application, which imposed an hyperpolarizing action through a field effect (extracellular iontophoresis) or via direct current action (intracellular pairing). These elementary associations within one cycle of visual stimulation were repeated every 3–8 sec up to 50 or 100 times. The stimulus was then presented ON and OFF alternately in the paired and unpaired positions of the RF, without current injection through the electrode. To ensure that the receptive field position remained unchanged throughout the whole recording session, additional controls were performed before and after pairing by studying the stability of the peak of discharges in the responses evoked by light and dark bars swept across the RF. The temporal evolution of the relative ratio of complexity—given by ON/(ON + OFF)—in both the paired and unpaired positions was calculated using a moving average technique on two to four pairs of successive ON- and OFF-stimulations, and statistically compared before versus after pairing, using nonparametric tests (significance level of .05 using Kolmogorov-Smirnov and Mann-Whitney tests).

Using extracellular iontophoresis, we were able to induce significant long-lasting modifications of the ratio of ON- and OFF-responses in more than 40% of the paired cells (see example in figure 14.5a). Such changes in the evoked firing of the paired neurons were never observed spontaneously and the functional reorganization of the ON/OFF balance in the paired position was predicted in 80% of the cases by the pattern of activity imposed during the pairing procedure. In half the cases, the effect was

A

OFF/(ON+OFF)
RATIO
C
P
+ 5 mn
+ 45 mn
ON
OFF
0.0
1.0
S-
S+

B

CONTROL
-67 mV
S+ PAIRING
+ 1 mn
+ 3 mn
+ 15 mn
+ 35 mn
100 ms
ON
PSP (mV)
0
5
S+

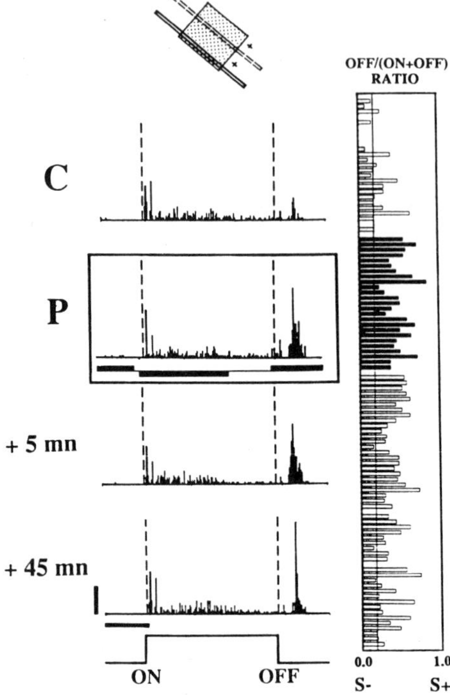

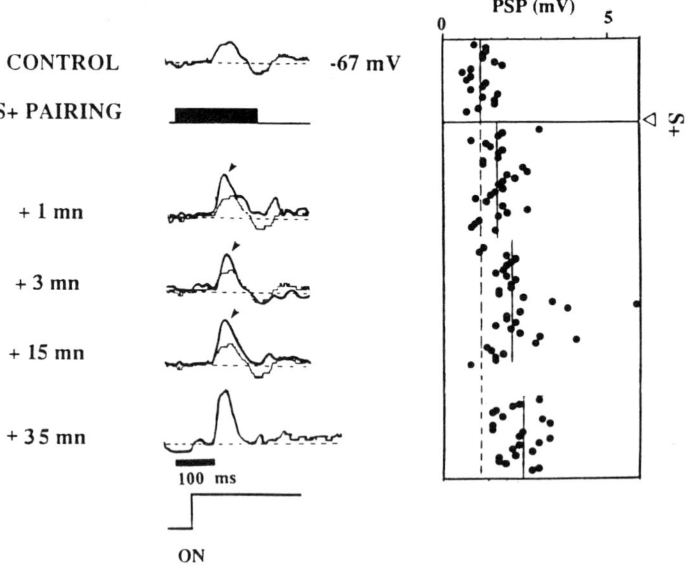

selective of the paired position. The relative preference between the two test stimulus characteristics (ON/OFF) was generally displaced toward that which had been paired with imposed increased visual responsiveness. The probability of inducing functional changes was comparable in the kitten from 4 to 15 weeks of age and in juvenile and adult cats (from 16 weeks to adulthood). However, as had been observed for the other functional properties of cortical cells, the largest effects were induced in the youngest animals at the peak of the critical period.

These in vivo extracellular results relied on application of iontophoretic pulses to control postsynaptic activity. While the effect of such manipulations on the suprathreshold activity of the postsynaptic neuron was assessed by measuring changes in firing rate, subthreshold changes in the postsynaptic neuron's membrane potential were not determined. To control the membrane potential of the target neuron during imposed coactivity protocols, and possibly test the synaptic nature of these effects, we achieved in collaboration with Attila Baranyi a series of in vivo intracellular recordings. Stability of recordings was improved by minimal craniotomy (1–2 mm), and eventually by bilateral pneumothorax and

Figure 14.5 Extracellular and intracellular demonstration of plasticity of ON- and OFF-responses in visual cortical neurons (*a*) Extracellular recordings in area 17 of a 5-week-old kitten. This cell exhibited initially a strong dominant ON response through the whole extent of the receptive field (shaded rectangle). The control visual response of one (solid bar) of the two positions which were quantitatively studied is represented in the first row (C). The pairing procedure consisted of 50 associations of a negative current pulse (−4 nA, −100 msec delay, 2000 msec duration) in phase with the presentation of the visual stimulus, and of a positive current (+4 nA, −100 msec delay, 2000 msec duration) in phase with the extinction of the visual stimulus. A significant increase of the OFF response was imposed during pairing whereas the negative current was ineffective in reducing the ON reponse. Five minutes after the end of the pairing, the OFF response was significantly potentiated (Kolmogorov-Smirnov, $P < .008$). This effect was still present 45 min after pairing and was selective for the paired position (the response was not modified in the control position, data not shown). The complexity index—given by the ratio of OFF/(OFF + ON) responses—was calculated in each position using a moving average technique (Frégnac et al., 1988), and plotted (with time) downward. Calibration bars: horizontal, 1 sec; vertical, 10 action potentials/second. (*b*) Intracellular recordings in area 17 of a 10-week-old kitten. The ON and OFF zones, symbolized respectively by empty and filled rectangles, were stimulated alternately. The ON presentation of the stimulus (solid bar) in the ON zone was paired 30 times with a concomitant intracellular depolarizing current pulse (+2 nA, 200 msec duration). (*Left*) Control PSPs (in the absence of current through the recording electrode), averaged on 20 successive trials. (*Right*) Trial-by-trial plot of the peak amplitude of the PSP (ordinate in mV, with time downward). The resting membrane potential (−67 mV) was unchanged by the pairing procedure. The size of the paired PSP doubled following pairing (arrows) and this significant change was maintained 35 min later (Kolmogorov-Smirnov, $P < .0175$). No significant change of the unpaired response (OFF response to stimulation of the OFF zone) was observed (data not shown).

cysterna drainage. Seventy-seven stable intracellular recordings (from 10 min up to 2 hr) were made in area 17 of 7- to 15-week-old kittens, using 2M potassium methylsulfate–filled glass micropipettes. The membrane potential (-63 ± 7 mV), spike amplitude (54 ± 11 mV), and membrane resistance (25 ± 9 MΩ) were similar to those reported in adult visual cortex (Douglas et al., 1991). However, many more bursting neurons (28% of cells) with low threshold Ca^{2+} spikes were recorded in the in vivo preparation than in in vitro slices of kittens of the same age (Frégnac et al., 1990).

Significant differences between the extracellular and intracellular protocols can be observed in the nature of the postsynaptic response paired with afferent activation. In the extracellular condition, the suprathreshold spiking response was maintained at a "high" or "low" level by iontophoretic action. In the intracellular recording situation, the test visual response was a subthreshold compound postsynaptic potential, and the polarization of the membrane potential during pairing was imposed through direct current injection restricted to the time course of the PSP. In these latter experiments the area (rather than the peak amplitude) of the evoked compound PSP was taken as a measure of the net efficacy of the population of synapses activated directly or indirectly by the test input. In spite of an unexpectedly high variability of subthreshold responses, which makes difficult to prove synaptic modifications on a given cell in the in vivo situation (but see figure 14.5b), the analysis of the relative changes in the visually evoked PSPs pooled from our sample of conditioned cells confirms our previous conclusions, based on extracellular conditioning. The PSPs evoked by the characteristic paired with a depolarizing pulse were on average significantly strengthened (average change = +35%, probability of increase = 78%, shown in the cumulative distribution of changes in figure 14.6b). Those paired with an hyperpolarizing pulse, which did not show a depolarizing rebound at the anodal break, tended to be significantly reduced (average change = −11%, probability of decrease = 69%).

In spite of their rather high technicality in visual cortex, in vivo intracellular pairings do not completely rule out unwanted effects of spurious correlations of membrane potential changes with spontaneous or extraretinal postsynaptic potentials. A parallel study was undertaken by one of us (Y. F.) in collaboration with Michael Friedlander, using intracellular recording in the kitten or adult guinea-pig visual cortical slices. The simplified in vitro neural network offers the advantage that numerous uncontrolled sources of evoked and spontaneous activity are eliminated. The tests of the covariance hypothesis we applied were similar to the para-

A

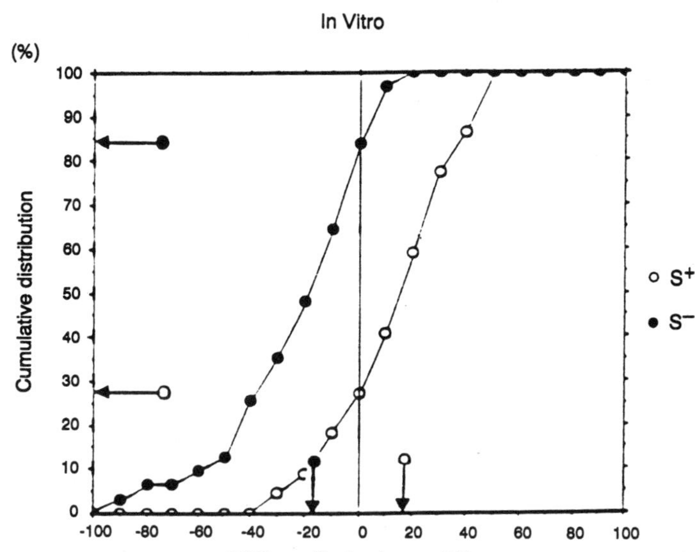

B

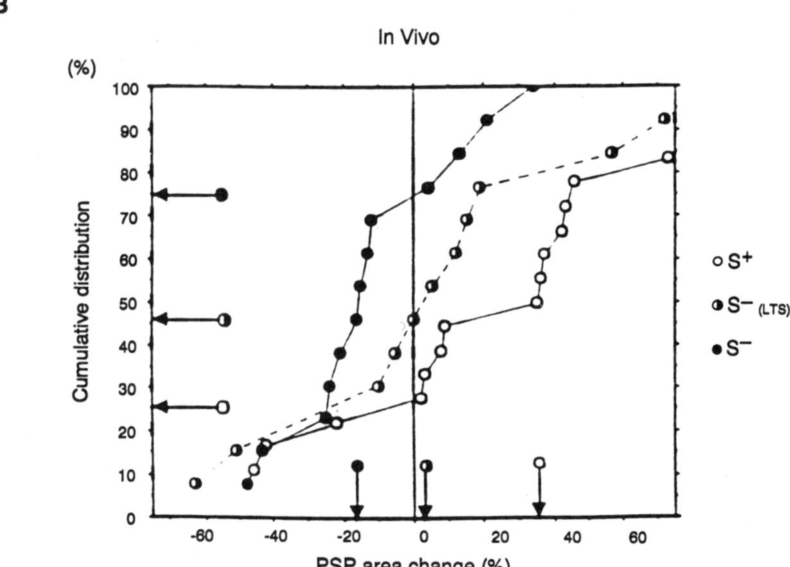

Figure 14.6 Compared homosynaptic effects of the pairing procedure in vivo and in vitro. Cumulative distribution of changes in paired visually evoked PSPs. Open circles correspond to changes in test-PSPs following S+ protocols, filled circles following S− protocols. Half-moon symbols correspond to cases where a mixed action was imposed during pairing (i.e., the end of the S− pulse being followed by a low threshold spike). Ordinates are expressed as a percentage of the total population of experimental cases for a given pairing condition. Arrows on the ordinate axis indicate the proportion of pairings for which a decrease in test PSP occurred. In abscissa, relative change in PSP peak amplitude (in vitro data from Frégnac et al., 1990 and Burke et al., 1991), or area (in vivo data from Baranyi et al., 1991). Arrows on this latter axis indicate median values of the changes observed in each condition.

digms employed in the in vivo preparation (Frégnac et al., 1990; Burke et al., 1991; Friedlander et al., 1993).

In our cortical slices, spontaneous activity observed in bicuculline-free ACSF appeared to be almost absent, and most recorded cells were of the "regular" type (McCormick et al., 1985). In spite of a better control in temporal correlation, the probability of inducing a significant change was similar to that found in vivo (39% of cases). The induction of these modifications was clearly temporally associative, since pseudo-pairing and fixed delay pairings where the current pulse was applied 120 msec out of phase with the stimulation of the test pathway, remained without effect. Input specificity was demonstrated using a dual stimulation protocol (white matter and intracortical stimulation sites) for which differential changes were observed in the paired and unpaired pathways. The major result was that the sign of the change in the paired PSP amplitude could be predicted by the sign of the change in the postsynaptic membrane potential during pairing (average potentiation of 18% and increase in the test PSP size in 68% of cases following S^+, average depression of 19% and decrease in the test PSP size in 84% of cases following S^-; see cumulative distribution of figure 12.6a).

The cells we studied often exhibited a "read-only" state for a range of membrane polarization, across which changes in the correlation between presynaptic activity and postsynaptic membrane potential did not affect synaptic efficacy. Such a "dead zone" in the synaptic gain function seems crucial to avoid constant alteration of synaptic weights during normal sensory processing. Visual cortical neurons may be assumed to undergo an adaptive state when one of two distinct postsynaptic plasticity thresholds is reached, above which (upward arrow in figure 14.6) potentiation occurs, and below which (downward arrow) depression occurs. This suggests that a particular pathway is capable of bidirectional changes of efficiency, depending on the polarity of the change in membrane potential (see model detailed in figure 14.1d).

PLASTICITY OF "EFFECTIVE" CONNECTIVITY

The conclusions that we have presented so far have been established from the measurements of the peak amplitude or the area of the composite PSP. They rely on the concept of "effective" connectivity (Aertsen et al., 1988), corresponding to the resulting efficiency of converging circuits which terminate on the recorded cell and are coactivated by the test input. Such measurement, although not directly related to the efficacy of transmission at identified synapses, has the additional interest for the modeler of providing an algebraic weighting function of the input, which can even change sign when disynaptic inhibition overcomes excitation. Such a convention has been used in previous models of visual cortex (Cooper, 1973; see also Crick and Asanuma, 1986) or in artificial neural nets (Hopfield, 1982). It might be particularly adapted to the study of composite PSPs, which result from the intricate balance between excitatory and inhibitory inputs at different delays, such as for instance in the electrosensory lobe of the electric fish (Bell et al., 1993).

The pairing-induced changes in "effective" synaptic efficacy that we observed may be due to changes in early or later components of excitatory or inhibitory (Llano et al., 1991; Konnerth et al., 1992) transmission onto the postsynaptic cell being recorded. Since our intracellular electrode only directly depolarizes (S^+) or hyperpolarizes (S^-) the cell that is impaled during the pairing protocol, major changes, when they occur, are thought to be localized at the level of synapses (fed either by monosynaptic or polysynaptic pathways) that impinge on the recorded cell. Interneurons are assumed to experience the same conditions of activation throughout the control, pairing and postpairing periods (but see Miles and Wong, 1987).

However, our electrophysiological tests of the covariance hypothesis suffer technical limitations. The current clamp-recording mode seemed easier to achieve than voltage-clamp recording in vivo (but see Baranyi et al., 1991). Consequently, the membrane potential was manipulated at the soma rather than at the active synaptic site. While the resultant somatic depolarization from our current injection pulses (average of +80–120 mV at the soma) during S^+ pairing is known to occur under physiological conditions where the cell is activated to spike, it is likely that strong somatic hyperpolarizations imposed during S^- protocols are less physiologically relevant for the cell. It should be observed that postsynaptic pulse amplitudes we used were similar to or less than those employed in the hippocampus to induce long-term potentiation, and of a shorter duration (Wigström et al., 1986; Gustafsson et al., 1987; Malenka et al., 1988; Jaffe and Johnston, 1990; Malinow and Tsien, 1990; Zalutsky and Nicoll, 1990; Huang et al., 1992). Current values applied to induce synaptic depression were also in a range similar to those used previously in DC mode to block LTP (Malinow and Miller, 1986; Zalutsky and Nicoll, 1990; Malenka, 1991).

Although the intrasomatic current values that we and others applied during pairing can cause large net changes in somatic membrane potential, the voltage drop at the relevant postsynaptic locus (i.e., corresponding to the active afferent pathway) is likely to be considerably less. The cable properties of the cortical neuron attenuate the signal imposed at the soma as it moves out to the sites of synaptic input along the dendritic shaft or spines of the neurons. Further attenuation of the signal into the dendritic spine head may also occur resulting in signals of only several millivolts at the postsynaptic site (Rall and Segev, 1985, 1987). A similar but symmetrical reasoning can be applied to the case of EPSPs and IPSPs evoked by natural visual stimulation (Ferster, 1986; Baranyi et al., 1991; Debanne et al., 1991; Douglas et al., 1991; Bringuier et al., 1992), when recorded at the soma, that is, after they have already been attenuated and temporally distorted by dendritic integration. Thus, the actual amplitude of synaptically generated postsynaptic voltage changes is probably considerably greater than the values of depolarization and hyperpolarization elicited at the synaptic site with our somatic current injections.

How does activation in the network induced by our pairing protocols compare with that imposed in previous studies of LTP/LTD in visual cortex? First, only a sequence of conjunctions of individual presynaptic pulses with postsynaptic polarization was needed in our protocols to induce a change. Second, pharmacological blockade of synaptic inhibition was not necessary in vitro to reveal this type of plasticity. Third, the effects we observed lasted for more than 10 min, but showed a decrement over time in most cases.

The intensity of the stimulation used in our in vitro experiments to condition one pathway appears to be lower than that used by Artola and coworkers, since test response amplitudes ranged between 2 and 8 mV compared to 10 and 15 mV in most of their studies (Artola et al., 1990; Artola and Singer, 1987, 1990). If the induction mode differed (low-frequency pairing vs. tetanus), the kinetics of the PSP changes were quite distinct. The use of tetanizing stimuli induced first a period of post-tetanic depression extending over several minutes whatever the change in postsynaptic membrane potential imposed during pairing (see figure 3 in Artola et al., 1990), and the test PSP slope remained altered for more than 20 min after the end of the conditioning. This differs from the transient changes observed in our experimental conditions.

Although low-frequency pairing protocols avoid masking of postsynaptic effects by post-tetanic modulation of afferent release, it is tempting to label the effects observed with our experimental protocols as short-term potentiation (STP) or depression. Indeed, the phenomena we observed appear most similar to the STP described for the hippocampus by Malenka and co-workers (Colino et al., 1992; Huang et al., 1992). These investigators also used strong postsynaptic depolarization (+70–90 mV from rest) for their pairings (Huang et al., 1992), but the pulses were applied for considerably longer durations (400 to 800 msec vs. 50 to 80 msec in our experiments) and in the presence of picrotoxin. Such conditions may be sufficient to convert the potentiation effects found in our experiments into a nondecrementing form such as LTP. These conditions may be satisfied in the in vivo visual cortex preparation, where similar up- and down-regulations of visual responses of somewhat longer duration were observed using the same pairing procedure. Such reversible rebalancing of synaptic weights in developing *and* adult primary sensory cortex may be an important property in its own right even more salient than nondecremental plasticity such as LTP, thus imparting the capacity for dynamic modulation of cortical connections. Enduring mechanisms where synaptic efficacy continuously remains at an altered level may be more relevant for structures presumed to play a primary role in the formation of memories such as the hippocampal formation.

POSTSYNAPTIC THRESHOLDS FOR SYNAPTIC PLASTICITY

Our data support the hypothesis that changes in membrane potential below or above certain threshold values can lead to differential effects on postsynaptic integration. Intracellular depolarization above spiking threshold paired with afferent activity is found to induce a significant potentiation of the evoked compound PSP, even in regular-spiking neurons, and in spite of the preserved inhibitory influence of the cortical network. Our conclusion agrees with similar observations made by others in visual (Gilbert et al., 1990; but see Artola and Singer, 1987) and frontal cortical neurons (Sutor and Hablitz, 1989). The observation that pairing afferent activity with postsynaptic hyperpolarization reduces PSP amplitude is more controversial, contradicting the hypothesis that inhibition reduces the capacity of a cell to undergo adaptive changes (Baranyi and Szente, 1987; Malinow and Miller, 1986).

As discussed earlier, the proposal that postsynaptic thresholds of homosynaptic depression and potentiation are reached when departing in opposite directions from the average past working state of the synapse (V_f and dotted line in figure 14.1) agrees closely with the covariance rule (Sejnowski, 1977a,b; Bienenstock et al., 1982; compare figure 14.1, panels b and d). The theoretical possibilities it opens for the dynamics of cortical network connectivity (Willshaw and Dayan, 1990) should in fact be distinguished from that offered by the synaptic plasticity scheme favored by Artola and co-workers (Artola et al., 1990, shown in figure 14.1c). According to the so-called ABS rule, depression and potentiation thresholds should be crossed in succession when depolarizing the postsynaptic cell further away from its initial resting potential. This latter rule bears interesting error-correcting properties, since it decreases the efficiency of active synapses associated with intermediate depolarization, that is, which could correspond to false-positive responses in a simple supervised recognition task (Hancock et al., 1991). However, the biological plausibility of the model implies that the level of postsynaptic activation reached in the control test situation (i.e., equivalent to the in vivo condition when the cortical neuron transmits reliably visual information) would be less than that needed for such weight corrections to occur. The depression threshold (downward arrow in figure 14.1c) should, however, correspond to a level of postsynaptic activation lower than that required to induce synaptic potentiation (upward arrow). This latter threshold has been linked with that required for NMDA receptor activation (Bear et al., 1987; Bear et al., 1990) or for low threshold spike activity (Komatsu et al., 1991). These requirements appear difficult to concile with two physiological observations: (1) activity of supragranular layer neurons evoked by natural stimuli in vivo is known to trigger NMDA-receptor activation without inducing long-lasting changes in the visual responses (Miller et al., 1989); (2) reducing the contrast of an effective visual stimulus in vivo or the intensity of a test stimulation in vitro has not been shown to induce depression of the evoked cellular responses (visual discharge or PSP).

In hippocampus the progressive increase in the intensity of the postsynaptic current pulse used in low-frequency pairing experiments does not reveal an intermediate range of postsynaptic depolarization, which would be optimal in inducing depression (Wigström et al., 1986), but rather elicits STP changes that eventually will be converted into stable LTP for higher values of currents (Colino et al., 1992). Larger current intensities above spike inactivation level will lead to no change (Malenka et al., 1988; Perkel, et al., 1991) or ultimately to transient unspecific depression (Kullmann et al., 1992).

The use of long-duration repetitive stimulation yields a different picture, promoting depression for low frequency of activation (1–3 Hz) (Staubli and Lynch, 1990; Fujii et al., 1991; Dudek and Bear 1992; Mulkey and Malenka, 1992; Grover and Teyler, unpublished observations; but see Kullmann et al., 1992), and LTP for higher frequencies (20–50 Hz) (Dudek and Bear, 1992; Mulkey and Malenka, 1992; Xie et al., 1992). This phenomenology, which could be taken as supportive of the ABS model, does not seem to apply simply to visual cortex, at least when using white matter activation (see Kirkwood et al., 1993): prolonged period of stimulation of optic radiations at 2 Hz is most effective in inducing LTP (Komatsu et al., 1981, 1988; Perkins and Teyler, 1988; Berry et al., 1989).

DEFINING RELEVANT POSTSYNAPTIC VARIABLES

Although the diversity in the intensity and frequency of stimulation and in the parameters of control of the postsynaptic state imposed by various experimenters makes direct interstudy comparisons almost impossible, general agreement seems to have been reached in considering the rise of free intracellular calcium during pairing or tetanus as a critical variable in the induction of synaptic potentiation and depression, as well as in the duration of the change in synaptic efficacy (Lisman, 1989; Malenka, 1991; Malenka et al., 1992). However a common interpretation of blockade of LTP and LTD induction following the intracellular injection of BAPTA, an efficient calcium chelator, assumes some undefined requisite in the level(s) of postsynaptic depolarization to be reached to induce a synaptic change (Lynch et al., 1983; Malenka et al., 1988; Hirsch and Crépel, 1991; Bröcher et al., 1992; Xie et al., 1992). The key question, which has also to be addressed in interpreting our experiments, remains the relative threshold values in $(Ca^{2+})_i$ increase for inducing respectively LTP and LTD (see discussion in Bindman et al., this volume). The demonstration of synaptic depression in partially calcium chelated (figure 3 in Bröcher et al., 1992) or almost completely calcium chelated neurons (Kimura et al., 1990; Yoshimura et al., 1991) makes one wonder how the damped increase of the free calcium concentration in BAPTA-injected cells compares with the *normal rise* resulting from the test synaptic activation in untreated cells.

Recent calcium measurements during synaptic activation in CA1 pyramidal cells show that somatic depolarization caused by current injection (as is the case in our S^+ protocol) promotes a maximal increase of calcium in proximal dendrites, which is amplified by spike activity (Regehr et al., 1989; Jaffe et al., 1992). In contrast, hyperpolarization (as in our S^- protocol), in spite of the increase in subthreshold EPSP size it may produce, reduces the synaptically evoked voltage-gated calcium influx (Miyakawa et al., 1992). Synaptic depression observed in our S^- protocol as well in a number of independent studies hyperpolarizing the postsynaptic cell either through current injection (Stanton and Sejnowski, 1989; Xie et al., 1992) or blocking cortical postsynaptic activity via $GABA_A$ agonist application (Reiter and Stryker, 1988; Tamura et al., 1993) might be the result of the reduction in this particular calcium influx *below* the level associated with normal synaptic transmission. The rather high level of hyperpolarization needed at the soma in our experiments might be explained by the distal synaptic location of this non spike related calcium component (Miyakawa et al., 1992).

CONCLUSION

Although it is not known to what degree comparable activity-dependent situations may occur under "unsupervised" physiological conditions, similar "supervised" cellular learning techniques have been applied successfully

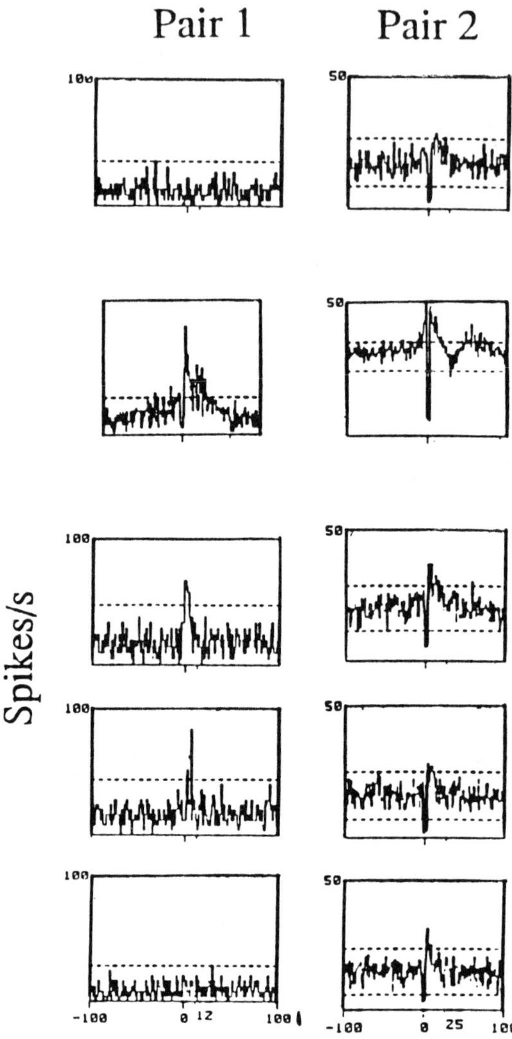

Figure 14.7 Effects of forced levels of coactivity on the effective connectivity between pairs of cells in the attentive behaving monkey. Pairs of cells have been recorded simultaneously in the auditory cortex of the behaving monkey (Ahissar, 1991; Ahissar et al., 1992). The ordinates of the cross-correlograms indicate the probability of firing of the postsynaptic neuron, conditional to the fact that the presynaptic one fired at time zero. Indirect measure of the "effective" connectivity between the two neurons is given by the area of the short latency peak. During pairing, contiguity of firing between the two neurons was controlled by stimulating the postsynaptic neuron using an auditory stimulus.

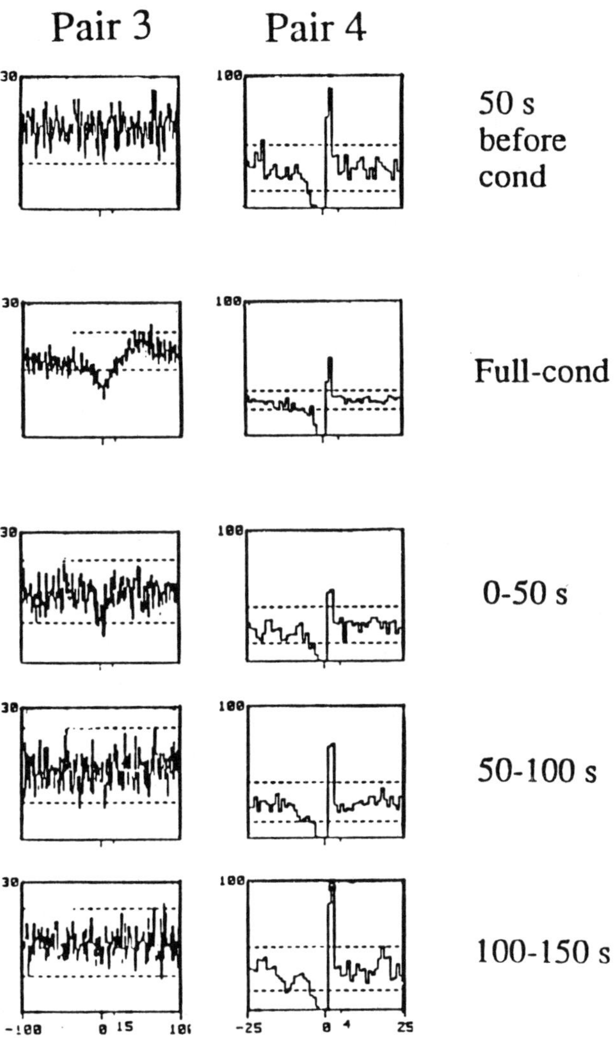

This stimulus was applied immediately after the presynaptic neuron had fired. An imposed increase in the contiguity (as in our S+ situation) caused strengthening of the functional connection (pairs 1 and 2, left panel), whereas a decrease in the contiguity (as in our S− protocols) caused a weakening of the connection's effective efficacy (pairs 3 and 4, right panel). These short-lasting modifications were mostly induced when the monkey had to attend to the train of auditory stimuli used in the cellular pairing procedure in order to get a reward. (Courtesy of Ehud Ahissar)

in vivo in the anesthetized or behaving animal (Greuel et al., 1988; Calhusac and Rolls, 1989; Sillito and Jones, 1992; Cruikshank and Weinberger, 1993). In a recent ingenious cross-correlation study, Ahissar and collaborators demonstrated that the functional coupling between two cortical sensory neurons could be up- and down-regulated during attentive behavior by a supervised control between presynaptic and postsynaptic activity (Ahissar et al., 1992). Not only did they found that the Hebbian-like rule appears six times more efficient in the attentive than in the nonattentive state, but they also observed significant short-lasting reduction in effective coupling following periods of negative covariance (figure 14.7).

These observations and our own in vivo and in vitro results are compatible with the hypothesis that a fixed level of presynaptic activity can induce either up- or down-regulation of effective connectivity depending on the level of postsynaptic membrane polarization imposed repetitively at the time of synaptic activation. Our data demonstrate that periods of maintained negative correlation of presynaptic and postsynaptic activity can induce a selective weakening of the efficacy of the active synapses, as has been reported in CA1 pyramidal cells and granule cells of the dentate gyrus of the hippocampus (Chattarji et al., 1989; Sejnowski et al., 1989; Stanton and Sejnowski, 1989; Xie et al., 1992). So, most importantly, our results suggest that the covariance hypothesis (Sejnowski, 1977b; Bienenstock et al., 1982) may have biological validity in the neocortex. The modification to be introduced in the model is to consider as the relevant postsynaptic variable the local level of membrane potential or the rate of change in free intracellular calcium level evoked at the active synaptic site rather than the postsynaptic firing frequency as initially assumed by modelers.

ACKNOWLEDGMENTS

The in vivo work was supported by grants from HFSP and CEE (Brain ST2J-0416C) to Yves Frégnac. Visits of Y. F. to the Neurobiology Research Center (UAB, Birmingham, USA) were supported by HFSP. Visits of Attila Baranyi in France were funded by CNRS, MRT, and FRM. Dominique Debanne was supported by MRT and Fondation Singer-Polignac during his PhD work. We thank Michael Friedlander, Nicole Ropert, and Vincent Bringuier for helpful comments, Kirsty Grant for help with the English, and Michèle Gautier for editing the bibliography and figures.

REFERENCES

Aertsen, A. M. H. J., Gerstein, G. L., Habib, M., and Palm, G. (1988) Dynamics of neuronal firing correlation: Modulation of "effective" connectivity. *J. Neurophysiol.* 61:900–917.

Ahissar, E. E. (1991) Examination of models for learning in behaving monkey's cortex. PhD thesis, Hebrew University, Jerusalem, Israel.

Ahissar, E., Vaadia, E., Ahissar, M., Bergman, H., Arieli, A., and Abeles, M. (1992) Dependence of cortical plasticity on correlated activity of single neurons and on behavioral context. *Science* 257:1412–1415.

Albus, J. (1971) A theory of cerebellar function. *Math. Biosci.* 10:25–61.

Artola, A., and Singer, W. (1987) Long-term potentiation and NMDA receptors in rat visual cortex. *Nature* 330:649–652.

Artola, A., and Singer, W. (1990) The involvement of *N*-methyl-D-aspartate receptors in induction and maintenance of long-term potentiation in rat visual cortex. *Eur. J. Neurosci.* 2:254–269.

Artola, A., Bröcher, S., and Singer, W. (1990) Different voltage-dependent thresholds for inducing long-term depression and long-term potentiation in slices of the rat visual cortex. *Nature* 347:69–72.

Baranyi, A., and Feher, O. (1981) Synaptic facilitation requires paired activation of convergent pathways in the neocortex. *Nature* 290:413–415.

Baranyi, A., and Szente, M. B. (1987) Long-lasting potentiation of synaptic transmission requires postsynaptic modifications in the neocortex. *Brain Res.* 423:378–384.

Baranyi, A., Debanne, D., Shulz, D., and Frégnac, Y. (1991a) Postsynaptic membrane potential regulates potentiation and depression of visually evoked synaptic potentials in kitten cortical neurons recorded in vivo. *Soc. Neurosci. Abstr.* 17:1470.

Baranyi, A., Woody, C. D., and Szente, M. B. (1991b) Properties of associative long-lasting potentiation induced by cellular conditioning in the motor cortex of conscious cats. *Neuroscience* 42:321–334.

Bear, M., Kleinschmidt, A., Gu, Q., and Singer, W. (1990) Disruption of experience-dependent synaptic modifications in striate cortex by infusion of an NMDA receptor antagonist. *J. Neurosci.* 10:909–925.

Bear, M. F., Cooper, L. N., and Ebner F. F. (1987) A physiological basis for a theory of synapse modification. *Science* 237:42–48.

Bear, M. F., Press, W. A., and Connors, B. W. (1992) Long-term potentiation in slices of kitten visual cortex and the effects of NMDA receptor blockade. *J. Neurophysiol.* 67:841–851.

Bell, C. C., Caputi, A., Grant, K., and Serrier, J. (1993) Storage of a sensory pattern by anti-hebbian synaptic plasticity in an electric fish. *Proc. Natl. Acad. Sci. USA* 90:4650–4654.

Berry, R. L., Teyler, T. J., and Taizhen, H. (1989) Induction of LTP in rat primary visual cortex: Tetanus parameters. *Brain Res.* 481:221–227.

Bienenstock, E., Cooper, L. N., and Munro, P. (1982) Theory for the development of neuron selectivity: Orientation specificity and binocular interaction in visual cortex. *J. Neurosci.* 2:32–48.

Blakemore, C., and Cooper, G. F. (1970) Development of the brain depends on the visual environment. *Nature* 228:419–478.

Bode-Greuel, K. M., Singer, W., and Aldenhoff, J. B. (1987) A current source density analysis of field potentials evoked in slices of visual cortex. *Exp. Brain Res.* 69:213–219.

Bonhoeffer, T., Kossel, A., Bolz, J., and Aertsen, A. (1990) Modified hebbian rule for synaptic enhancement in the hippocampus and the visual cortex. In *The Brain,* Cold Spring Harbor Press (eds.), 15, pp. 137–145.

Bowling, D. B., and Caverhill, J. I. (1989) ON/OFF organization in the cat lateral geniculate nucleus: Subliminae versus columns. *J. Comp. Neurol.* 283:161–168.

Bringuier, V., Frégnac, Y., Debanne, D., Shulz, D., and Baranyi, A. (1992) Synaptic origin of rhythmic visually evoked activity in kitten area 17 neurons. *NeuroReport* 3:1065–1068.

Bröcher, S., Artola, A., and Singer, W. (1992) Intracellular injection of Ca^{2+} chelators blocks induction of long-term depression in rat visual cortex. *Proc. Natl. Acad. Sci. USA* 89:123–127.

Brown, T. H., Ganong, A. H., Kairiss, E. W., and Keenan, C. L. (1990) Hebbian synapses: Biophysical mechanisms and algorithms. *Annu. Rev. Neurosci.* 13:475–511.

Buisseret, P., and Imbert, M. (1976) Visual cortical cells: Their developmental properties in normal and dark reared kittens. *J. Physiol. (Lond.)* 255:511–525.

Bullier, J., and Henry, G. H. (1979a) Laminar distribution of first-order neurons and afferent terminals in cat striate cortex. *J. Neurophysiol.* 42:1271–1281.

Bullier, J., and Henry, G. H. (1979b) Ordinal position of neurons in cat striate cortex. *J. Neurophysiol.* 42:1251–1263.

Burke, J. P., Frégnac, Y., and Friedlander, M. J. (1991) Spatial input specificity of synaptic potentiation and depression in the visual cortex. *Soc. Neurosci. Abstr.* 17:114.

Calhusac, P. M. B., and Rolls, E. T. (1989) Modifications of neuronal responses to natural inputs paired with activation of the postsynaptic neurone by iontophoretic L-glutamate in the macaque hippocampus. *J. Physiol. (Lond.)* 420:45P.

Callaway, E. M., and Katz, L. C. (1990) Emergence and refinement of clustered horizontal connections in cat striate cortex. *J. Neurosci.* 10:1134–1153.

Carew, T. J., Hawkins, R. D., Abrams, T. W., and Kandel, E. R. (1984) A test of Hebb's postulate at identified synapses which mediate classical conditioning in *Aplysia. J. Neurosci.* 4:1217–1224.

Chattarji, S., Stanton, P. K., and Sejnowski, T. J. (1989) Commissural synapses, but not mossy fiber synapses, in hippocampal field CA3 exhibit associative long-term potentiation and depression. *Brain Res.* 495:145–150.

Colino, A., Huang, Y. Y., and Malenka, R. C. (1992) Characterization of the integration time for the stabilization of long-term potentiation in area CA1 of the hippocampus. *J. Neurosci.* 12:180–187.

Cooper, L. N. (1973) A possible organization of animal memory and learning. In *Proceedings of the Nobel Symposium on collective properties of physical system,* B. L. Lindquist and S. Lindquist (eds.). London, pp. 252–264.

Crick, F. H. C., and Asanuma, C. (1986) Certain aspects of the anatomy and physiology of the cerebral cortex. In *Parallel Distributed Processing,* J. L. M. Clelland and D. E. Rumelhart (eds.). Cambridge, Mass.: MIT Press, pp. 333–371.

Cruikshank, S. J., and Weinberger, N. M. (1993) Induction of auditory cortical receptive field plasticity: Hebb rules? *Soc. Neurosc. Abstr.* 19:164.

Cynader, M. S. (1979) Competitive interactions in postnatal development of kitten's visual system. In *Developmental Neurobiology of Vision,* R. D. Freeman (ed.). New York: Plenum Press, pp. 109–120.

Dan, Y., and Poo, M. (1992) Hebbian depression of isolated neuromuscular synapses in vitro. *Science* 256:1570–1573.

Debanne, D., Bringuier, V., Shulz, D., Baranyi, A., and Frégnac, Y. (1991) Intracellular study of visual responses properties and oscillatory behavior of kitten area 17 neurons at the peak of critical period. *Soc. Neurosci. Abstr.* 17:1470.

Douglas, R. J., Martin, K. A. C., and Whitteridge, D. (1991) An intracellular analysis of the visual responses of neurones in cat visual cortex. *J. Physiol. (Lond.)* 440:659–696.

Dudek, S. M., and Bear, M. F. (1992) Homosynaptic long-term depression in area CA1 of hippocampus and effects of *N*-methyl-D-aspartate receptor blockade. *Proc. Natl. Acad. Sci. USA* 89:4363–4367.

Ferster, D. (1986) Orientation selectivity of synaptic potentials in neurons of cat primary visual cortex. *J. Neurosci.* 6:1284–1301.

Frégnac, Y. (1979) Development of orientation selectivity in the primary visual cortex of normally and dark reared kittens. I. Kinetics. *Biol. Cybern.* 34:187–193.

Frégnac, Y. (1987) Cellular mechanisms of epigenesis in cat visual visual cortex. In *Imprinting and Cortical Plasticity*, J. P. Rauschecker and P. Marler (eds.). New York: Wiley, pp. 221–266.

Frégnac, Y., and Debanne, D. (1993) Potentiation and depression in visual cortical neurons: a functional approach to synaptic plasticity. In *Brain Mechanisms of Perception and Memory: from Neuron to Behavior*, T. Ono et al. (ed.). Oxford University Press, pp. 553–561.

Frégnac, Y., and Imbert, M. (1984) Development of neuronal selectivity in the primary visual cortex of the cat. *Physiol. Rev.* 64:325–434.

Frégnac, Y., and Shulz, D. (1993). Models of synaptic plasticity and cellular analogs of learning in the developing and adult vertebrate visual cortex. In *Advances in Neural and Behavioral Development*, V. Casagrande and P. Shinkman (eds.). New Jersey: Ablex, Vol. 4, pp. 149–235.

Frégnac, Y., Shulz, D., Thorpe, S., and Bienenstock, E. (1988) A cellular analogue of visual cortical plasticity. *Natue* 333:367–370.

Frégnac, Y., Shulz, D., and Debanne, D. (1989) The role of neuronal co-activity in shaping visual cortical receptive field properties. *Biomed. Res.* 10:29–35.

Frégnac, Y., Smith, D., and Friedlander, M. J. (1990) Postsynaptic membrane potential regulates synaptic potentiation and depression in visual cortical neurons. *Soc. Neurosci. Abstr.* 16:798.

Frégnac, Y., Shulz, D., Thorpe, S., and Bienenstock, E. (1992) Cellular analogs of visual cortical epigenesis: I. Plasticity of orientation selectivity. *J. Neurosci.* 12:1280–1300.

Friedlander, M. J., Sayer, R. J., and Redman, S. J. (1990) Evaluation of long term potentiation of small compound and unitary EPSPs at the hippocampal CA3-CA1 synapse. *J. Neurosci.* 10:814–825.

Friedlander, M. J., Frégnac, Y., and Burke, J. P. (1993) Temporal covariance of post-synaptic membrane potential and synaptic input. Role in synaptic efficacy in visual cortex. *Prog. Brain Res.* 95:207–223.

Fujii, S., Saito, K., Miyakawa, H., Ito, K., and Kato, H. (1991) Reversal of long-term potentiation (depotentiation) induced by tetanus stimulation of the input to CA1 neurons of guinea-pig hippocampal slices. *Brain Res.* 555:112–122.

Gally, J. A., Montague, R. P., Reeke, G. N., and Edelman, G. M. (1990) The NO hypothesis: Possible effects of a short-lived, rapidly diffusible signal in the development and function of the nervous system. *Proc. Natl. Acad. Sci. USA* 87:3547–3551.

Gilbert, C. D., and Wiesel, T. N. (1983) Clustered intrinsic connections in cat visual cortex. *J. Neurosci.* 3:1116–1133.

Gilbert, C. D., Hirsch, J. A., and Wiesel, T. N. (1990) Lateral interactions in visual cortex. In *The Brain*, Cold Spring Harbor Press (eds.), Vol. 15, pp. 663–677.

Grant, K., Frégnac, Y., Hester, F., Friedlander, M. J. F., Smith, D., and Debanne, D. (1990) Morphology of visual cortical neurons during the critical period, labelled by intracellular biocytin injections. *Soc. Neurosci. Abstr.* 16:494.

Greuel, J. M., Luhmann, H. J., and Singer, W. (1988) Pharmacological induction of use-dependent receptive field modifications in the visual cortex. *Science* 242:74–77.

Gustafsson, B., Wigström, H., Abraham, W. C., and Huang, Y. Y. (1987) Long-term potentiation in the hippocampus using depolarizing current pulses as the conditioning stimulus to single volley synaptic potentials. *J. Neurosci.* 7:774–780.

Hancock, P. J. B., Smith, L. S., and Phillips, W. A. (1991) A biologically supported error-correcting learning rule. *Neural Comp.* 3:201–212.

Hebb, D. O. (1949) *The Organization of Behavior.* New York: Wiley.

Hirsch, J. C., and Crepel, F. (1991) Blockade of NMDA receptors unmasks a long-term depression in synaptic efficacy in rat prefrontal neurons in vitro. *Exp. Brain Res.* 85:621–624.

Hopfield, J. J. (1982) Neural networks and physical systems with emergent collective computational abilities. *Proc. Natl. Acad. Sci. USA* 79:2554–2558.

Huang, Y. Y., Colino, A., Selig, D. K., and Malenka, R. C. (1992) The influence of prior synaptic activity on the induction of long-term potentiation. *Science* 255:730–733.

Hubel, D. H., and Wiesel, T. N. (1970) The period of susceptibility to the physiological effects of unilateral eye closure in kittens. *J. Physiol. (Lond.).* 206:419–436.

Jaffe, D., and Johnston, D. (1990) Induction of long-term potentiation at hippocampal mossy-fiber synapses follows a Hebbian rule. *J. Neurophysiol.* 64:948–960.

Jaffe, D. B., Johnston, D., Lasser-Ross, N., Lisman, J. E., Miyakawa, H., and Ross, W. N. (1992) The spread of Na^+ spikes determines the pattern of dendritic Ca^{2+} entry into hippocampal neurons. *Nature* 357:244–246.

James, W. (1890) *Psychology: Briefer Course.* Cambridge, Mass.: Harvard University Press.

Kelso, S. R., Ganong, A. H., and Brown, T. H. (1986) Hebbian synapses in hippocampus. *Proc. Natl. Acad. Sci. USA* 83:5326–5330.

Kimura, F., Tsumoto, T., Nishigori, A., and Yoshimura, Y. (1990) Long-term depression but not potentiation is induced in Ca^{2+}-chelated visual cortex neurons. *NeuroReport* 1:65–68.

Kirkwood, A., Dudek, S. M., Gold, J. T., Aizenman, C. D., and Bear, M. F. (1993) Common forms of synaptic plasticity in hippocampus and neocortex in vitro. *Science* 260:1518–1521.

Kisvarday, Z. F., and Eysel, U. T. (1992) Cellular organization of reciprocal networks in layer III of cat visual cortex (area 17). *Neuroscience* 46:275–286.

Komatsu, Y., Toyama, K., Maeda, J., and Sakaguchi, H. (1981) Long-term potentiation investigated in a slice preparation of striate cortex of young kittens. *Neurosci. Lett.* 26:269–274.

Komatsu, Y., Fujii, K., Maeda, J., Sakaguchi, H., and Toyama, K. (1988) Long-term potentiation of synaptic transmission in kitten visual cortex. *J. Neurophysiol.* 59:124–141.

Komatsu, Y., Nakajima, S., and Toyama, K. (1991) Induction of long-term potentiation without participation of *N*-methyl-D-aspartate receptors in kitten visual cortex. *J. Neurophysiol.* 65:20–32.

Konnerth, A., Dreessen, J., and Augustine, G. J. (1992) Brief dendritic calcium signals initiate long-lasting synaptic depression in cerebellar purkinje cells. *Proc. Natl. Acad. Sci. USA* 89:7051–7055.

Kossel, A., Bonhoeffer, T., and Bolz, J. (1990) Non-hebbian synapses in rat visual cortex. *NeuroReport* 1:115–118.

Kullmann, D. M., Perkel, D. J., Manabe, T., and Nicoll, R. A. (1992) Ca^{2+} entry via postsynaptic voltage-sensitivity Ca^{2+} channels can transiently potentiate excitatory synaptic transmission in the hippocampus. *Neuron* 9:1175–1183.

Linsker, R. (1986a) From basic network principles to neural architecture: Emergence of spatial opponent cells. *Proc. Natl. Acad. Sci. USA* 83:7508–7512.

Linsker, R. (1986b) From basic network principles to neural architecture: Emergence of orientation selective cells. *Proc. Natl. Acad. Sci. USA* 83:8390–8394.

Linsker, R. (1986c) From basic network principles to neural architecture: Emergence of orientation columns. *Proc. Natl. Acad. Sci. USA* 83:8779–8783.

Lisman, J. (1989) A mechanism for the Hebb and the anti-Hebb processes underlying learning and memory. *Proc. Natl. Acad. Sci. USA* 86:9574–9578.

Llano, I., Leresche, N., and Marty, A. (1991) Calcium entry increases the sensitivity of cerebellar Purkinje cells to applied GABA and decreases inhibitory synaptic currents. *Neuron* 6:565–574.

Löwel, S., and Singer, W. (1992) Selection of intrinsic horizontal connections in the visual cortex by correlated neuronal activity. *Science* 255:209–212.

Lynch, G., Larson, J., Kelso, S., Barrionuevo, G., and Schottler, F. (1983) Intracellular injections of EGTA block induction of hippocampal long-term potentiation. *Nature* 305: 719–721.

Maffei, L., and Galli-Resta, L. (1990) Correlation in the discharges of neigboring rat retinal ganglion cells during prenatal life. *Proc. Natl. Acad. Sci. USA* 87:2861–2864.

Malenka, R. C. (1991) Postsynaptic factors control the duration of synaptic enhancement in area CA1 of the hippocampus. *Neuron* 6:53–60.

Malenka, R. C., Kauer, J. A., Zucker, R. S., and Nicoll, R. A. (1988) Postsynaptic calcium is sufficient for potentiation of hippocampal synaptic transmission. *Science* 242:81–84.

Malenka, R. C., Lancaster, B., and Zucker, R. S. (1992) Temporal limits on the rise in postsynaptic calcium required for the induction of long-term potentiation. *Neuron* 9:121–128.

Malinow, R. (1991) Transmission between pairs of hippocampal slice neurons: Quantal levels, oscillations, and LTP. *Science* 252:722–724.

Malinow, R., and Miller, J. P. (1986) Postsynaptic hyperpolarization during conditioning reversibly blocks induction of long-term potentiation. *Nature* 320:529–530.

Malinow, R., and Tsien, R. W. (1990) Presynaptic enhancement shown by whole-cell recordings of long-term potentiation in hippocampal slices. *Nature* 346:177–180.

Marr, D. C. (1970) A theory for cerebral cortex. *Proc. R. Soc. Lond. B* 176:161–234.

Mason, A., Nicoll, A., and Stratford, K. (1991) Synaptic transmission between individual pyramidal neurons of rat visual cortex in vitro. *J. Neurosci.* 11:72–84.

McCormick, D. A., Connors, B. W., Lighthall, J. W., and Prince, D. A. (1985) Comparative electrophysiology of pyramidal and sparsely spiny stellate neurons of the neocortex. *J. Neurophysiol.* 54:782–806.

Meister, M., Wong, R. O. L., Baylor, D. A., and Shatz, C. J. (1990) Synchronous bursting activity in ganglion cells of the developing mammalian retina. *Invest. Ophthalmol. Vis. Sci. (Suppl.)* 31:115.

Miles, R., and Wong, R. K. S. (1987) Latent synaptic pathways revealed after tetanic stimulation in the hippocampus. *Nature* 329:724–726.

Miller, K. D., Keller, J. B., and Stryker, M. P. (1989) Models of ocular dominance column formation: Analytical and computation results. In *Advances in Neural Information Processing Systems*, D. S. Touretzky (ed.). San Mateo, Calif.: Morgan Kaufman vol. I, pp. 375–383.

Milleret, C., Gary-Bobo, E. and Buisseret, P. (1988) Comparative development of cell properties in cortical area 18 of normal and dark-reared kittens. *Exp. Brain Res.* 71:8–20.

Miyakawa, H., Ross, W. N., Jaffe, D., Callaway, J. C., Lasser-Ross, N., Lisman, J. E., and Johnston, D. (1992) Synaptically activated increases in Ca^{2+} concentration in hippocampal CA1 pyramidal cells are primarily due to voltage-gated Ca^{2+} channels. *Neuron* 9:1163–1173.

Montague, P. R., Gally, J. A., and Edelman, G. M. (1991) Spatial signalling in the development and function of neural connections cerebral cortex. *Cerebral Cortex* 1:199–220.

Mulkey, R. M., and Malenka, R. C. (1992) Mechanisms underlying induction of homosynaptic long-term depression in area CA1 of the hippocampus. *Neuron* 9:967–975.

Müller, C. M. (1992) A role for glial cells in activity-dependent central nervous plasticity? Review and hypothesis. *Int. Rev. Neurobiol.* 34:215–281.

Perkel, D. J., Manabe, T., and Nicoll, R. A. (1991) Role of membrane potential and calcium in the induction of long-term potentiation (LTP). *Soc. Neurosci. Abstr.* 17:2.

Perkin, A. T., IV, and Teyler, T. J. (1988) A critical period for long-term potentiation in the developing rat visual cortex. *Brain Res.* 439:222–229.

Rall, W., and Segev, I. (1985) Space clamp problems when voltage clamping branched neurons with intracellular microelectrodes. In *Voltage and Patch Clamping with Microelectrodes*, J. T. G. Smith, H. Lecar, S. J. Redman and P. Gage (eds.). Bethesda: American Physiological Society, pp. 191–215.

Rall, W., and Segev, I. (1987) Functional possibilities for synapses on dendrites and dentritic spines. In *Synaptic Function*, G. M. Edelman, W. E. Gall and W. M. Cowan (eds.). New York: Wiley, pp. 605–636.

Ramoa, A. S., Shadlen, M., and Freeman, R. D. (1987) Dark-reared cats: Unresponsive cells become visually responsive with microiontophoresis of an excitatory amino acid. *Exp. Brain Res.* 65:658–665.

Rauschecker, J. P., and Singer, W. (1981) The effects of early visual experience on the cat's visual cortex and their possible explanation by Hebb synapses. *J. Physiol. (Lond.)* 310:215–239.

Regehr, W. G., Connor, J. A., and Tank, D. W. (1989) Optical imaging of calcium accumulation in hippocampal pyramidal cells during synaptic activation. *Nature* 341:533–536.

Reiter, H. O., and Stryker, M. P. (1988) Neural plasticity without postsynaptic action potentials: Less-active inputs become dominant when kitten visual cortical cells are pharmacologically inhibited. *Proc. Natl. Acad. Sci. USA* 85:3623–3627.

Rochester, N., Holland, J. H., Haibt, L. H., and Duda, W. L. (1956) Tests on a cell assembly theory of the action of the brain, using a large digital computer. *IRE Trans. Inform. Theory*. IT 2:80–93.

Sayer, R. J., Friedlander, M. J., and Redman, S. J. (1990) The time course and amplitude of EPSPs evoked at synapses between pairs of CA3/CA1 neurons in the hippocampal slice. *J. Neurosci.* 10:826–836.

Schiller, P. H. (1982) Central connection of the retinal ON and OFF pathways. *Nature* 297:580–583.

Schmidt, J. T., and Edwards, D. L. (1983) Activity sharpens the map during the regeneration of the retino-tectal projection in goldfish. *Brain Res.* 269:29–39.

Schmidt, J. T., and Eisele, L. E. (1985) Stroboscopic illumination and dark rearing block the sharpening of retinotectal map in goldfish. *Neuroscience* 14:535–546.

Sejnowski, T. J. (1977a) Statistical constraints on synaptic plasticity. *J. Theoret. Biol.* 69: 387–389.

Sejnowski, T. J. (1977b) Storing covariance with non-linearly interacting neurons. *J. Math. Biol.* 4:303–321.

Sejnowski, T. J., Chattarji, S., and Stanton, P. K. (1989) Induction of synaptic plasticity by Hebbian covariance in the hippocampus. In *The Computing Neuron*, R. Durbin, C. Miall, and G. Mitchison (eds.). Reading, Mass.: Addison-Wesley, pp. 105–124.

Shatz, C. J. (1990) Impulse activity and patterning of connections during CNS development. *Neuron* 5:745–756.

Shatz, C. (1992) The developing brain. *Sci. Am.* 267:34–41.

Shinkman, P. G., and Bruce, C. J. (1977) Binocular differences in cortical receptive fields of kittens after rotationnally disparate binocular experience. *Science* 197:285–287.

Shulz, D., and Frégnac, Y. (1992) Cellular analogs of visual cortical epigenesis: II. Plasticity of binocular integration. *J. Neurosci.* 12:1301–1318.

Shulz, D., Debanne, D., and Frégnac, Y. (1993). Cortical convergence of ON- and OFF-pathways, and functional adaptation of receptive field organization in cat area 17. *Progress in Brain Research* 95, pp. 191–205.

Sillito, A. M., and Jones, H. E. (1992) Potentiation and depression of the strength of the transfer of the retinal input in the dorsal lateral geniculate nucleus (dLNG). *Soc. Neurosci. Abstr.* 18:142.

Singer, W., and Tretter, F. (1976) Unusually large receptive fields in cats with restricted visual experience. *Exp. Brain Res.* 26:171–184.

Stanton, P. K., and Sejnowski, T. J. (1989) Associative long-term depression in the hippocampus induced by Hebbian covariance. *Nature* 339:215–218.

Staubli, U., and Lynch, G. (1990) Stable depression of potentiated synaptic responses in the hippocampus with 1–5 Hz stimulation. *Brain Res.* 513:113–118.

Stent, G. (1973) A physiological mechanism for Hebb's postulate of learning. *Proc. Natl. Acad. Sci. USA* 70:997–1001.

Stern, P., Edwards, F. A., and Sakman, B. (1992) Fast and slow components of unitary EPSCs on stellate cells elicited by focal stimulation in slices of rat visual cortex. *J. Physiol. (Lond.)* 449:247–278.

Stryker, M. P., and Harris, W. A. (1986) Binocular impulse blockade prevents the formation of ocular dominance columns in cat visual cortex. *J. Neurosci.* 6:2117–2133.

Stryker, M. P., and Strickland, S. L. (1984) Physiological segregation of ocular dominance columns depends on the pattern of afferent electrical activity. *Invest. Ophthalmol. Vis. Sci. (Suppl.)* 25:278.

Sutor, B., and Hablitz, J. J. (1989) EPSP's in rat neocortical neurons in vitro. I. Electrophysiological evidence for two distinct EPSP's. *J. Neurophysiol.* 61:607–620.

Tamura, H., Tsumoto, T., and Hata, Y. (1992) Activity-dependent potentiation and depression of visual cortical neurons to optic nerve stimulation in kittens. *J. Neurophysiol.* 68: 1603–1612.

Tsumoto, T., and Freeman, R. D. (1987) Dark-reared cats: Responsivity of cortical cells influenced pharmacologically by an inhibitory antagonist. *Exp. Brain Res.* 65:666–672.

Tsumoto, T., and Suda, K. (1979) Cross-depression: An electrophysiological manifestation of binocular competition in the developing visual cortex. *Brain Res.* 168:190–194.

Tsumoto, T., and Suda, K. (1982) Laminar differences in development of afferent innervation to striate cortex neurons in kittens. *Exp. Brain Res.* 45:433–466.

Von der Malsburg, C. (1973) Self-organization of orientation sensitive cells in the striate cortex. *Kybernetik* 4:85–100.

Von der Malsburg, C., and Bienenstock, E. (1986) Statistical coding and short-term synaptic plasticity: A scheme for knowledge representation in the brain. In *Disordered Systems and Biological Organization*, E. Bienenstock, F. Fogelman-Soulié and G. Weisbuch. (eds.). Berlin: Springer-Verlag, pp. 247–272.

Wässle, H., Peichl, L., and Boycott, B. B. (1981a) Morphology and topography of ON- and OFF-alpha cells in the cat retina. *Proc. R. Soc. Lond. B* 212:157–175.

Wässle, H., Boycott, B. B., and Illing, R. B. (1981b) Morphology and mosaic of ON- and OFF-beta cells in the cat retina and some functional considerations. *Proc. R. Soc. Lond. B* 212:177–195.

Wiesel, T. N., and Hubel, D. H. (1963) Single-cell responses in striate cortex of kittens deprived of vision in one eye. *J. Neurophysiol.* 26:1003–1017.

Wigström, H., Gustafsson, B., Huang, Y. Y., and Abraham, W. C. (1986) Hippocampal long-lasting potentiation is induced by pairing single afferent volleys with intracellularly injected depolarizing current pulses. *Acta Physiol. Scand.* 26:317–319.

Williams, J. H., Errington, M. L., Lynch, M. A., and Bliss, T. V. P. (1989) Arachidonic acid induces a long-term activity-dependent enhancement of synaptic transmission in the hippocampus. *Nature* 341:739–742.

Willshaw, D., and Dayan, P. (1990) Optimal plasticity from matrix memories: What goes up must come down. *Neural Comput.* 2:85–93.

Willshaw, D. G., and Von der Malsburg, C. (1976) How patterned neural connections can be set up by self-organization. *Proc. R. Soc. Lond. B* 194:431–445.

Wurtz, R. H., Castellucci, V. F., and Nusrala, J. M. (1967) Synaptic plasticity: The effect of the action potential in the postsynaptic neuron. *Exp. Neurol.* 18:350–368.

Xie, X., Berger, T. W., and Barrionuevo, G. (1992) Isolated NMDA receptor-mediated synaptic responses express both LTP and LTD. *J. Neurophysiol.* 67:1009–1013.

Yoshimura, Y., Tsumoto, T., and Nishigori, A. (1991) Input-specific induction of long term depression in Ca^{2+}-chelated visual cortex neurons. *NeuroReport* 2:393–396.

Zalutsky, R. A., and Nicoll, R. A. (1990) Comparison of two forms of long-term potentiation in single hippocampal neurons. *Science* 248:1619–1624.

IV Relationships between Long-Term Potentiation and Learning and Memory

15 Long-Term Potentiation in Piriform Cortex: Characterization and Functional Speculations

Lewis B. Haberly, Kevin L. Ketchum, and Evan D. Kanter

In recent years there has been intensive study of long-term potentiation (LTP) at cellular and molecular levels, but relatively little work attempting to relate this process to neuronal circuit or behavioral level questions. One system that appears particularly well-suited for relating LTP to higher level processes is the piriform (primary olfactory) cortex. As detailed in previous reviews (Haberly, 1985, 1990a), the piriform cortex has an orderly three-layered architecture that, like the hippocampus, lends itself well to the study of cellular and local circuit processes. Factors that should facilitate the analysis of its role in discrimination and learning include: its single-synapse separation from sensory receptors, the spatial and temporal ordering of integrative processes in its pyramidal cells (Rodriguez and Haberly, 1989; Ketchum and Haberly, 1993a), a close relationship to the hippocampal formation (Eichenbaum et al., 1991; Lynch and Granger, 1991), inputs from monoaminergic and cholinergic systems (Haberly and Price, 1978), and a temporal pattern of activation by natural stimulation that resembles the theta-burst paradigm commonly used to evoke LTP. Finally, certain "neural network" models with capabilities for discrimination of complex spatial and temporal patterns provide a theoretical framework to guide the analysis of piriform cortex as a consequence of the parallels in their architecture to its circuitry (Haberly, 1985; Haberly and Bower, 1989).

Studies of activity-related plasticity in the piriform cortex are at a relatively early stage, but the presence of NMDA-dependent homosynaptic and associative LTP has been documented for afferent and intrinsic associational fiber systems (Jung et al., 1990; Kanter and Haberly, 1990, 1993). The impairment of olfactory learning after antagonism of NMDA receptors (Staubli et al., 1989) and the dependence of persistent LTP on the behavioral relevance of stimuli (Roman et al., 1987) suggest that this LTP has a role in learning.

In this chapter, studies of LTP in piriform cortex are summarized and possible relationships discussed between this process and the analysis and storage of olfactory information.

OVERVIEW OF ANATOMY

Input-Output Relationships

Piriform cortex is primary olfactory cortex by virtue of the direct projection it receives from the olfactory bulb (figure 15.1A). This projection is from mitral and tufted cells that receive direct input from olfactory receptor cells. Piriform cortex gives rise to extensive projections to the entorhinal cortex, insular and orbitofrontal areas of neocortex, and other olfactory cortical areas including the cortical amygdala (Luskin and Price, 1983a; Price, 1985; Takagi, 1986; Price et al., 1991). There are lighter projections to the lateral hypothalamus and mediodorsal nucleus of the thalamus.

Cytoarchitecture

Piriform cortex is usually described in terms of three layers (figure 15.1B, 15.2). Layer I is a superficial plexiform layer that contains apical dendrites of pyramidal cells, tangentially distributed fiber systems, and a small number of GABAergic neurons. Layer II contains a high density of cell bodies, most of which are those of so-called superficial pyramidal cells. Apical dendrites of these cells extend through layer I to the cortical surface. Layer III contains a much lower density of neuronal somata than layer II and extensive neuropil. The superficial part of layer III is dominated by pyramidal cells, termed deep pyramidal cells, whose apical dendrites arborize together with those of superficial pyramidal cells in layer I. Multipolar cells, many of which are GABAergic, dominate the deep part of layer III.

A striking feature of piriform cortex is the contrast in organization of fiber systems in the vertical (perpendicular to the surface) and horizontal dimensions. In the vertical dimension, fiber systems as well as different neuronal types and their processes are highly ordered (Price, 1973; Haberly and Price, 1978; Luskin and Price, 1983b; Haberly et al., 1987). Within layer I, afferent fibers from the olfactory bulb are restricted to a sublamina termed layer Ia where they synapse on distal apical dendrites of pyramidal cells. The deep part of layer I, termed layer Ib, contains a dense matrix of associational axons that originate from pyramidal cells throughout the piriform cortex as well as other olfactory cortical areas. Synapses of these axons are predominantly on proximal apical dendritic segments of

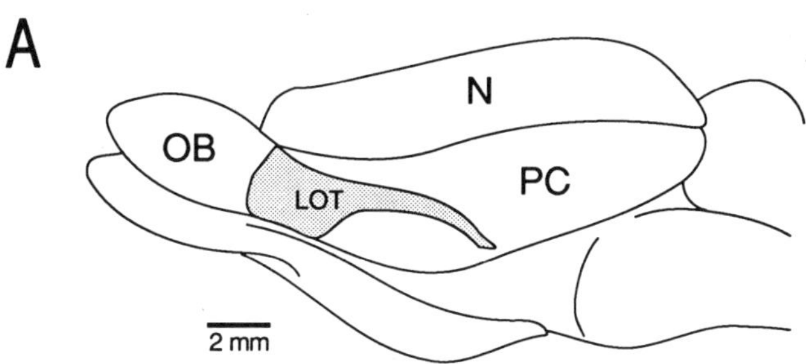

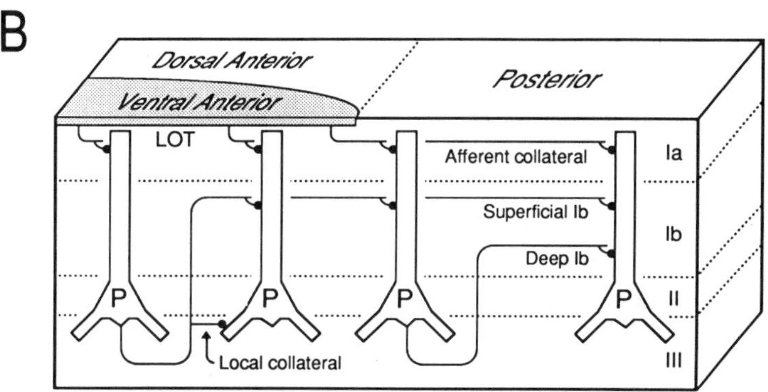

Figure 15.1 (*A*) Lateral oblique view of rat brain showing piriform cortex. Piriform cortex (PC) receives afferent input from the olfactory bulb (OB) via the lateral olfactory tract (LOT). N, neocortex. (*B*) Schematic representation of piriform cortex showing the differential laminar termination of excitatory fiber systems on pyramidal cells. Anterior and posterior parts are separated by a plane (dotted line) at the caudal end of the LOT (stippled area). Fine collateral branches from afferent fibers in the LOT synapse on the distalmost segments of pyramidal cell apical dendrites in layer Ia. Long associational axons from pyramidal cells in the ventral anterior piriform cortex synapse predominantly in the superficial part of layer Ib throughout the piriform cortex. Long associational axons from pyramidal cells in the dorsal part of anterior and the posterior piriform cortex synapse predominantly in the mid-to-deep part of layer Ib. (Reprinted from Ketchum and Haberly, 1993a)

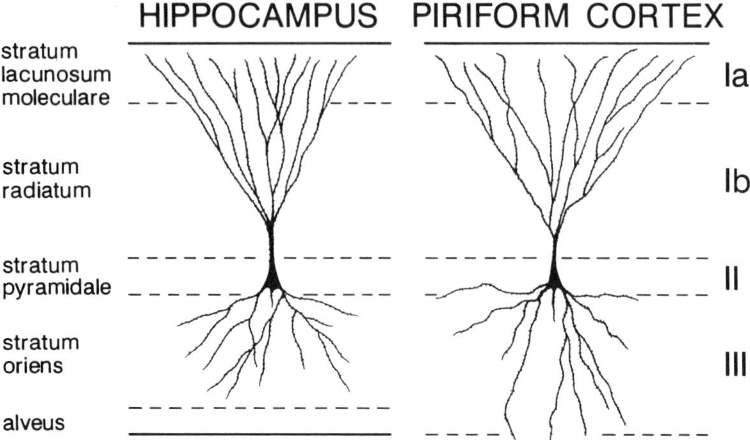

Figure 15.2 Comparison of hippocampus and piriform cortex. Layer Ia of piriform cortex and stratum lacunosum-moleculare of hippocampus receive afferents of extrinsic origin; layer Ib and stratum radiatum receive intrinsic associational and commissural fibers; layer II and stratum pyramidale are densely packed laminae of pyramidal cell somata; layer III and stratum oriens contain pyramidal cell basal dendrites, but layer III of piriform cortex contains pyramidal cells in its superficial part while stratum oriens does not. (Reprinted from Haberly, 1990)

the same pyramidal cells that receive afferent input on their distal apical segments.

In the horizontal dimension, afferent and associational fiber systems in piriform cortex are highly distributed spatially, with no orderly point-to-point topographical ordering as observed in other primary sensory areas (Haberly and Price, 1977, 1978; Scott et al., 1980). Each point in the cortex receives afferent input from the entire extent of the olfactory bulb and from pyramidal cells throughout the piriform cortex. It is this highly distributed spatial organization with convergence of input and distributed positive feedback systems onto the same neuronal elements that has inspired models with a likeness to artificial neural networks (Haberly, 1985; Granger et al., 1988; Wilson and Bower, 1988; Haberly and Bower, 1989; Ambros-Ingerson et al., 1990; Hasselmo et al., 1991; Ketchum and Haberly, 1991).

OVERVIEW OF PHYSIOLOGY

Following shock activation of afferent fibers, a complex series of postsynaptic currents is evoked in pyramidal cells of piriform cortex (figure 15.3) (Haberly and Shepherd, 1973; Rodriguez and Haberly, 1989; Ketchum and Haberly, 1993a). Those of relevance to the present analysis include a monosynaptic excitatory postsynaptic current (EPSC) in distal apical den-

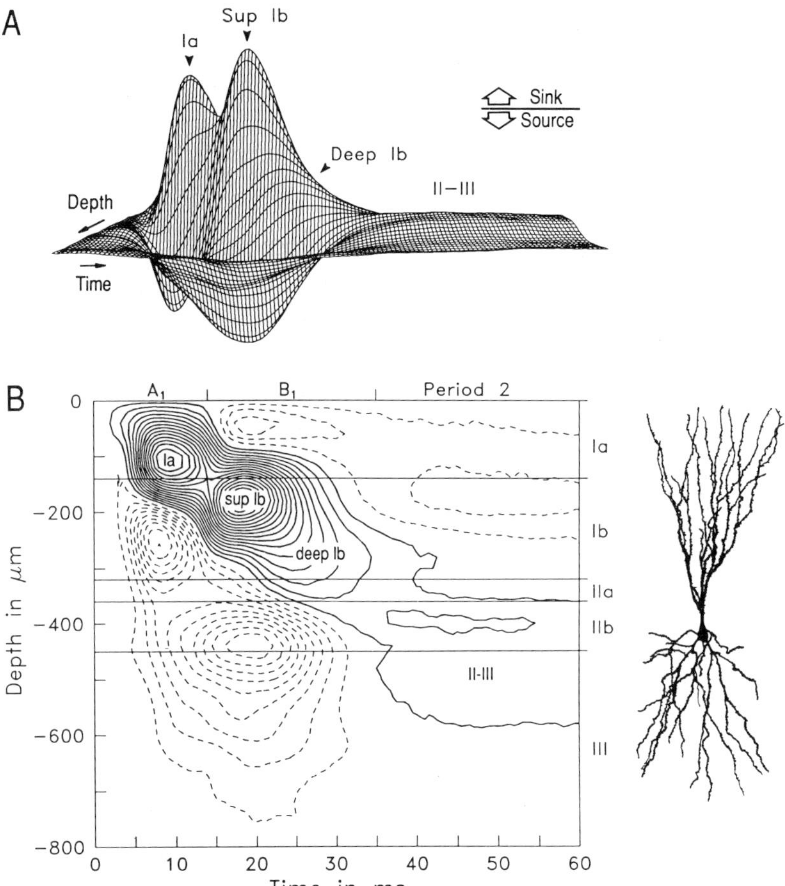

Figure 15.3 Net membrane currents evoked by shock stimulation of afferent fibers as revealed by current source-density analysis. Inward currents (sinks) are represented in the upward direction in the surface plot (*A*) and as solid lines in the contour plot (*B*). Time periods of the evoked potential are indicated at the top of the contour plot (A_1 and B_1 are peaks of the dichrotic, surface negative phase; period 2 is the surface positive phase). The pyramidal cell at the right indicates correspondence between neuronal elements and the cortical lamination (Roman numerals in *A* and *B*). The layer Ia sink (Ia) is the monosynaptic EPSC in distal apical dendrites; the sinks in the superficial part (sup Ib) and mid to deep part of layer Ib (deep Ib) are disynaptic EPSCs mediated by long associational fibers in proximal apical dendrites. The sink in layers II–III and associated outward current in layer I are associated with inhibitory processes. (Reprinted from Ketchum and Haberly, 1993a)

dritic segments, a large disynaptic EPSC in more proximal apical segments mediated by intrinsic associational fibers, and fast, Cl^--mediated and slow, K^+-mediated inhibitory postsynaptic currents (IPSCs).

Excitatory Processes

Afferent input reaches the piriform cortex by way of myelinated axons of the lateral olfactory tract (LOT) that courses over the ventral part of the anterior half of piriform cortex (termed anterior piriform cortex). Fine collaterals from these axons provide afferent input to the remainder of the anterior piriform cortex and to the posterior piriform cortex. The entire extent of the anterior piriform cortex is activated with relatively synchrony as a consequence of the overlying LOT and the short lengths of its collateral axons that provide afferent input (Ketchum and Haberly, 1993a,b). By contrast, the posterior half is sequentially activated at a relatively slow rate by long, unmyelinated collaterals from the LOT.

The disynaptic EPSC consists of at least three components (Rodriguez and Haberly, 1989: Ketchum and Haberly, 1993a), the strongest of which is mediated by a heavy, widespread system of associational axons that originates exclusively from pyramidal cells confined to the ventral part of the anterior piriform cortex (Haberly and Feig, in preparation). This component terminates in a thin subzone at the superficial margin of layer Ib that slightly overlaps the afferent fiber termination zone in layer Ia. The mean conduction velocity of these associational axons (0.6 m/sec) is comparable to that of the afferent fiber collaterals (0.75 m/sec) that provide input to the posterior piriform cortex (Ketchum and Haberly, 1993b). As a result, the mono- and disynaptic EPSCs evoked by discrete bursts of afferent activity propagate across the cortex as two successive waves that maintain a relatively constant 8 msec peak-to-peak latency (Ketchum and Haberly, 1993a).

Inhibitory Processes

GABAergic inhibition in piriform cortex is mediated by interneurons in feedforward and feedback relationships with respect to pyramidal cells (Satou et al., 1982, 1983; Tseng and Haberly, 1988). Both feedforward and feedback pathways mediate both fast ($GABA_A$-Cl^-) and slow ($GABA_B$-K^+) inhibitory postsynaptic potentials (IPSPs) (Satou et al., 1982, 1983; Tseng and Haberly, 1988; Hoffman and Haberly, 1989a). In pyramidal cells, the slow IPSP is generated predominantly in the dendritic tree and the fast IPSP, predominantly in the vicinity of cell bodies (Tseng and Haberly, 1988), although as described below, there is also a fast IPSP component in the dendritic tree that strongly regulates the NMDA component of the excitatory postsynaptic potential (EPSP).

Olfactory Coding

Odor-induced activity in the olfactory bulb and cortex is patterned in a complex way in both spatial and temporal dimensions. In the rat, odors trigger a rhythmic sniffing at a frequency on the order of 5 Hz that is synchronized with hippocampal theta activity during odor discrimination performance (figure 15.4A) (Macrides et al., 1982; Eichenbaum et al., 1991). Each inspiratory phase evokes a fast, approximately 50-Hz oscillation (figure 15.4B) that is tightly correlated in the bulb and cortex (Boudreau, 1964; Bressler and Freeman, 1980).

Studies with the 2-deoxyglucose method and various physiological recording methods at receptor, bulbar, and cortical levels of the olfactory system (e.g., Adrian, 1953; Sharp et al., 1975; Jourdan, 1982; Mackay-Sim et al., 1982; Cattarelli et al., 1988) suggest that the spatial distribution

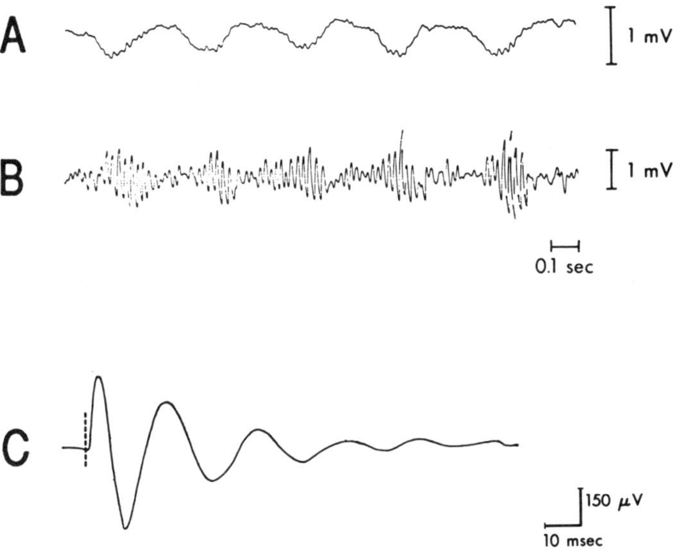

Figure 15.4 Slow and fast oscillatory rhythms in piriform cortex. (*A*) Field potential recorded from piriform cortex of the rat during odor stimulation. The signal has been filtered to accentuate low frequencies. The slow (ca. 5 Hz) oscillation is driven by rhythmic sniffing. (*B*) Fast oscillation (ca. 50 Hz) recorded simultaneously with the record in *A* with a bipolar electrode positioned across the pyramidal cell layer. Note the large increase in amplitude during each inspiratory phase of the sniff cycle. (*C*) Fast oscillation elicited in piriform cortex by weak shock stimulation of afferent fibers. The dashed line indicates position of the stimulus. The average frequency and other characteristics of shock evoked oscillations match those of the fast oscillation evoked by odors. (Modified from Woolley and Timiras 1965)

of activity evoked by different odorants is different, but these patterns are exceedingly complex and highly distributed. The degree to which "hot spots" or particularly active neurons participate in discrimination is an area of ongoing debate (Shepherd, 1991). Single-unit responses to different odorants can also display distinctly different temporal patterns over sniff cycles (Macrides and Chorover, 1972; Macrides, 1977). Temporal patterns (Harrison and Scott, 1986), and probably spatial patterns as well, can change with the strength of stimulation, thus further complicating the central discrimination task.

LTP IN PIRIFORM CORTEX

Homosynaptic LTP

In a study of post-tetanic potentiation in a slice preparation of guinea pig piriform cortex, Richards (1972) noted that the population EPSP failed to decline to baseline level for an "indefinite" period after application of long duration (> 20 sec), high-frequency trains to afferent fibers. Recent studies in rat piriform cortex slices using more physiologic stimuli (Jung et al., 1990; Kanter and Haberly, 1990) have confirmed that this potentiation in the afferent fiber system is LTP, and that it also occurs in intrinsic association fiber systems (figure 15.5). In both studies, theta-burst stimuli (4 shocks at 100 Hz, repeated 10 times at 5 Hz) potentiated population EPSPs for the duration of experiments (up to 4 hr). The induction of LTP in both afferent and association fiber systems was blocked by APV, indicating a dependence on the activation of NMDA receptors. Expression of the LTP was documented for the fast, non-NMDA component of the EPSP. It has not been determined if the small NMDA component present at resting membrane potential (Hoffman and Haberly, 1989b; Collins, 1991) is also potentiated. LTP was synapse specific; that is, theta-burst stimuli confined to afferent fibers did not potentiate responses to association fiber stimulation and vice versa. A large, rapidly decaying post-tetanic potentiation (PTP) was seen in piriform cortex only from 100-Hz, 1-sec stimulation and not from theta-burst stimulation (Jung et al., 1990; Kanter, unpublished).

Although the primary findings were the same for studies from the two laboratories, there were some differences that leave certain aspects of LTP in piriform cortex unresolved, the most significant of which relate to the ability of the afferent fiber system to support LTP. In order to consistently induce afferent fiber LTP in normal bathing medium, Jung and coworkers found it necessary to concomitantly deliver theta-burst stimulation to association fibers, after which LTP required 20–30 min to fully develop.

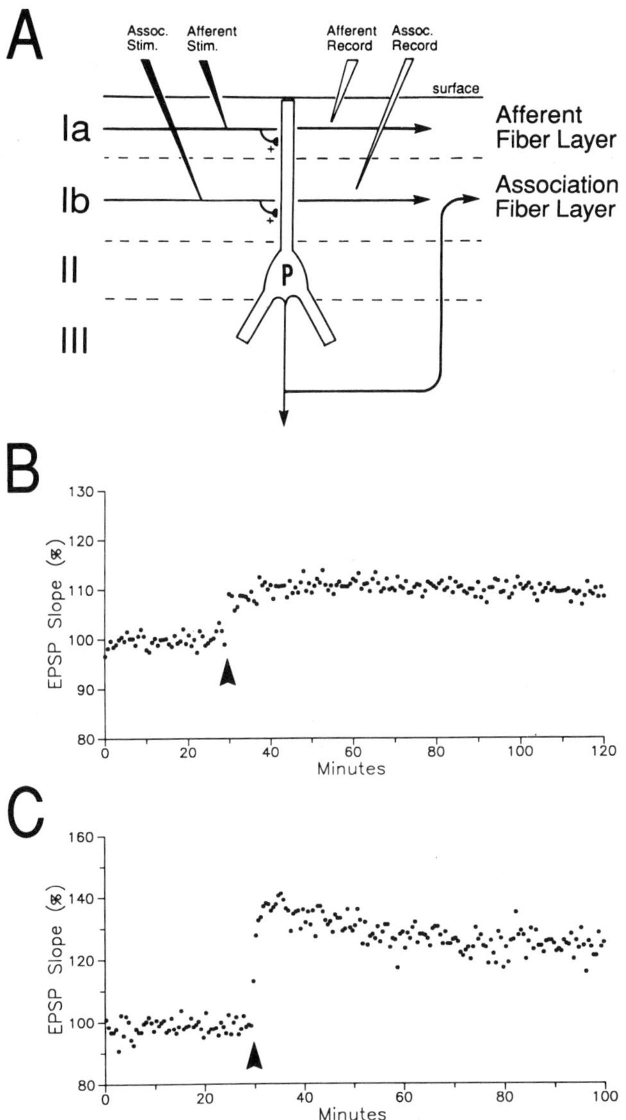

Figure 15.5 Homosynaptic LTP in piriform cortex. (*A*) Cortical lamination and placement of stimulating and recording electrodes. P, superficial pyramidal cell. Afferent fibers were stimulated in the lateral olfactory tract; association fibers were stimulated in the deep part of layer Ib to minimize current spread to layer Ia. (*B*) Slope of the rising phase of the population (extracellular) EPSP evoked by test shocks to afferent fibers in the LOT as a function of time. At the arrow a potentiating theta-burst stimulus (10 repetitions at 5 Hz of 4 shock, 100 Hz bursts) was applied through the same electrode. (*C*) Same as *B*, but test and theta-burst stimuli applied to association fibers in layer Ib. (Adapted from Kanter and Haberly, 1990)

Afferent fiber LTP was more readily induced after reducing the concentration of Mg^{2+} to 50 μM. Kanter and Haberly also reported that LTP in the afferent system was induced less consistently than in the associational system, but it was observed with stimulation of afferent fibers alone in standard bathing medium with an overall success rate of 23/28 slices. No delay in development of potentiation was seen. Instead, afferent fiber LTP approached maximal within seconds, more rapidly than association fiber LTP, which was observed in both studies to develop over a period of approximately 1 min.

By virtue of its synapse-specific nature and the NMDA-dependent induction of an increase in the non-NMDA component of the EPSP, LTP in piriform cortex resembles that characterized in the CA1 field and dentate gyrus of the hippocampal formation (for review see Madison et al., 1991). The only apparent difference is in the magnitude of potentiation. In the piriform cortex of adult rats a potentiation of 10–15% was typically obtained for afferent fibers after a single theta-burst train, with a saturation level of 15–25% after multiple trains. A potentiation of 20–40% was typically obtained for association fibers after a single train, with saturation at 50–75%. In the hippocampal formation potentiation of 100–200% is commonly reported; however, the standard stimulus used is a 100-Hz train of 1-sec duration. Studies utilizing the presumably more physiological theta-burst stimulus (e.g., Larson et al., 1986; Staubli et al., 1990) report a lesser degree of potentiation, which may be comparable in magnitude to that seen in the association fiber system in piriform cortex.

An NMDA-dependent potentiation of the non-NMDA component of the EPSP comparable in magnitude to that in piriform cortex has been observed in the visual cortex, although only after blockage of $GABA_A$ inhibition (Artola and Singer, 1987, 1990). There is also a readily apparent potentiation of the NMDA component in many cells in the visual cortex (Artola and Singer, 1990). In the entorhinal cortex, NMDA-dependent LTP has been evoked by stimulation of fibers in the molecular layer that include association fibers from piriform cortex (Alonso et al., 1990). However, in contrast to the LTP observed in piriform cortex, expression was exclusively in the NMDA component.

Associative LTP

Prompted by anatomical demonstrations of the spatial juxtaposition of afferent and association fibers, and predictions from parallels with neural network models, a recent study has shown that "associative" LTP can be induced in piriform cortex (figure 15.6) (Kanter and Haberly, 1993). Application of a strong theta-burst stimulus to one pathway consistently potentiated synapses of the other pathway that were concomitantly activated by weak, otherwise nonpotentiating shocks. This associative LTP was similar to homosynaptic LTP in NMDA-dependence of induction and non-NMDA expression, as well as characteristic magnitudes and time courses for afferent and association fiber pathways, suggesting that the same mechanism is activated by single pathway and associative paradigms.

An intriguing feature of this associative form of LTP is that, unlike homosynaptic LTP, which is consistently observed under standard recording conditions, blockage of the fast, Cl^--mediated IPSP with the $GABA_A$ antagonist bicuculline was required for its induction. In normal bathing medium persistent potentiation was never induced by the paired strong-weak paradigm, while it was consistently observed (27/28 slices) after disinhibition by focal delivery of bicuculline near the recording site. As in the hippocampus, blockage of the fast IPSP had no effect on the magnitude of homosynaptic LTP induced by strong potentiating trains, although it may increase the probability of LTP induction by weaker stimuli. Blockage of the slow IPSP with $GABA_B$ antagonists had no facilitating effect on associative LTP induction in this study.

In the hippocampus, disinhibition may also be required for certain associative interactions (e.g., Gustafsson and Wigström, 1986). White and coworkers (1990) have demonstrated in vivo that induction of associative LTP in the temporodentate pathway from entorhinal cortex to dentate gyrus depends on the degree of spatial overlap of synapses. Kanter and Haberly (1993) suggest the hypothesis that segregation of afferent and association fiber inputs onto different segments of apical dendrites serves to allow regulation of associative interactions by centrifugal inputs that modulate the activity of GABAergic processes.

Analysis of responses to potentiating theta-burst trains (Kanter and Haberly, 1993) revealed that in the disinhibited condition where associative LTP can be induced, a slow potential was evoked that was blocked by APV (figure 15.7). This suggests that facilitation of the NMDA component of the EPSP is at least partly responsible for the enabling of associative LTP by disinhibition. Recent studies (Kapur et al., 1992, and in preparation) employing locally applied bicuculline provide evidence for a previously unidentified, fast-feedforward IPSP in the apical dendritic region which restrains facilitation of the NMDA component in burst responses. By contrast, the fast-feedback IPSP in the vicinity of cell bodies (Tseng and Haberly, 1988) appears to have little effect on this facilitation process. It can therefore be proposed that the fast-feedforward IPSP regulates the induction of associative LTP.

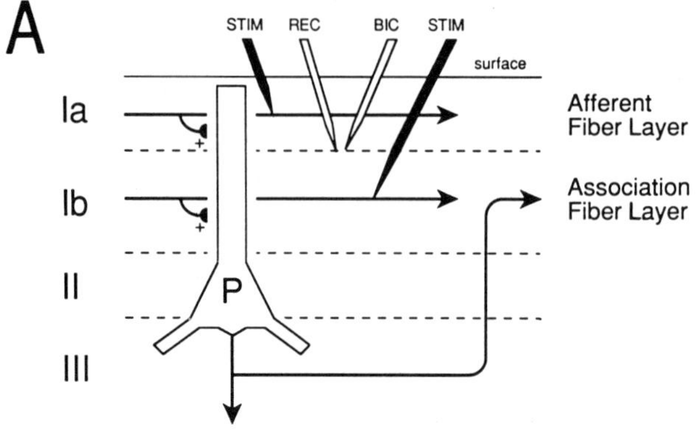

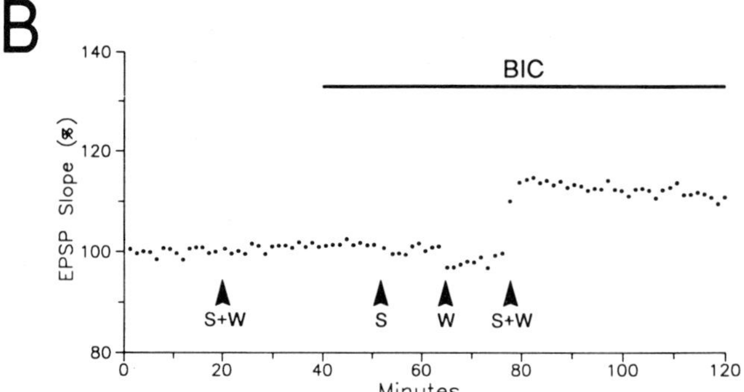

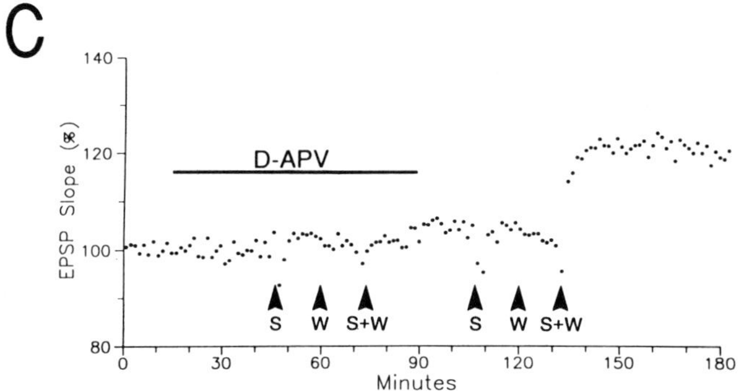

Figure 15.6 Associative LTP in piriform cortex. (*A*) Placement of stimulating and recording electrodes (as described in figure 15.5) and a bicuculline-containing pipette (BIC) used to locally block the fast IPSP. (*B*) Slope of the rising phase of the EPSP evoked by weak test shocks to afferent fibers as a function of time. S, strong (potentiating), theta-burst stimulus to association fibers; W, 10 weak shocks to afferent fibers at 5 Hz; S + W, paired stimulation. The fast IPSP in the vicinity of the recording electrode was blocked by iontophoresis of bicuculline from the pipette illustrated in A during the interval indicated by the horizontal line. Neither theta-burst stimulation to association fibers alone (S) nor 5 Hz weak shocks to afferent fibers alone (W) produced potentiation of afferent responses. LTP was produced only by paired stimulation and only in the presence of bicuculline. (*C*) Same paradigm as in *B*, but D-APV (15 μM) was added to the bath during the period

In Vivo Studies of LTP

Studies of LTP in piriform cortex in vivo have revealed some intriguing parallels and differences from those in vitro. In a urethane anesthetized rat preparation, theta-burst stimuli induced a potentiation of the response evoked by afferent fiber stimulation (Kapur and Haberly, 1991). As in the slice studies, induction was dependent on the activation of NMDA receptors and expression was in the fast, non-NMDA component. The magnitude of potentiation was also comparable to that observed in vitro. However, in contrast to findings in slices, the potentiation typically (n = 9/12) declined to near baseline level within a 1- to 2-hr period. Local disinhibition that did not evoke epileptiform activity did not increase the duration of potentiation. Curiously, following disinhibition of larger areas that elicited epileptiform activity, persistent LTP was consistently evoked. This was true when the potentiating theta-burst train was applied after periods of up to 5 hr following washout of the surface-applied $GABA_A$ antagonist when epileptiform activity had ceased. This suggests that it was the occurrence of epileptiform activity and not a direct effect of the disinhibition that was responsible for the increased duration.

In studies in which behavioral relevance of theta-burst shock stimulation was established by reinforcement in a learning paradigm, persistent potentiation of the population EPSP evoked by afferent fiber stimulation was obtained (Roman et al., 1987). When an identical set of bursts was applied without reinforcement, no LTP was observed after delays of 30 min and 24 hr. Together with the findings for the anesthetized preparation, these results suggest that, in vivo, shock stimuli alone are not sufficient to evoke persistent LTP.

In other studies in unanesthetized animals it has been shown that 30 repetitions of 100-msec duration, 100-Hz shock trains to the deep layers of the olfactory bulb at 5- to 10-sec intervals induce a large, persistent potentiation of a long-latency field potential component in piriform cortex (Stripling et al., 1988, 1991; Patneau and Stripling, 1992). It has been postulated that this potentiation is in an inhibitory process based in part on an increase in paired pulse inhibition of the disynaptic field EPSP (Patneau and Stripling, 1992). It is unclear how this process, referred to by

of the horizontal line. Bicuculline was applied locally throughout the experiment. APV blocked the associative LTP indicating a dependence on NMDA receptors. Similar experiments (not illustrated) revealed that associative LTP could also be induced in the association fiber system by pairing weak shocks to layer Ib with theta-burst stimulation to afferent fibers. This effect also required disinhibition and was blocked by APV. (Modified from Kanter and Haberly, 1993)

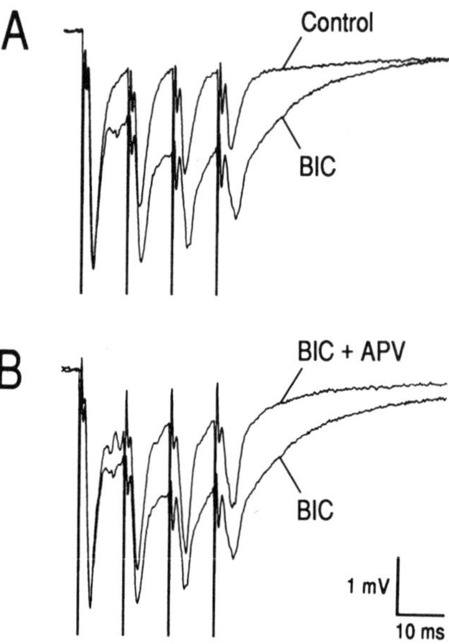

Figure 15.7 Theta-burst stimuli evoke a slow, NMDA-dependent response component after local blockage of the fast IPSP with bicuculline. (*A*) Response to 4 shock, 100 Hz burst to association fibers in layer Ib under control conditions and after focal application of bicuculline (BIC) as described in figure 15.6. (*B*) The slow response component evoked in the presence of bicuculline was blocked by bath application of 15 μM D-APV. (Modified from Kanter and Haberly, 1993)

the authors as "selective LTP," relates to LTP of monosynaptic EPSPs and further discussion is beyond the scope of this chapter.

Modulation of LTP

If LTP serves as the basis for memory storage, then mechanisms must be present to prevent the saturation of synaptic efficacies by behaviorally irrelevant repetitive activity. For example, bursting activity during seizures might otherwise block subsequent learning. For homosynaptic LTP, the study of Roman et al. (1987) as described above suggests that one form of regulation by behavioral significance could be in the persistence of LTP.

In the case of associative LTP, the requirement for disinhibition in the slice preparation suggests the possibility of control via regulation of inhibitory pathways. Cholinergic and monoaminergic fiber systems that are thought to play a role in learning and memory have been shown to modulate inhibitory processes in the hippocampus by a number of mechanisms (Sheldon and Aghajanian, 1990; Andreasen and Lambert, 1991; Droze et al., 1991; Ropert and Guy, 1991; Pitler and Alger, 1992). Blockage of the fast, dendritic IPSP by an increase or decrease in activity in one or more of these pathways could therefore enable associative LTP. Since associative and homosynaptic forms of LTP appear to have a common mechanism of induction and expression, it can be speculated that associative LTP would also be subject to the same control in persistence as homosynaptic LTP.

THE ROLE OF LTP IN DISCRIMINATION AND LEARNING

Dispersive Propagation and Convergence

In view of the need for spatial and temporal convergence of synaptic inputs for the induction of NMDA-dependent LTP, an understanding of its functional role will require the visualization of patterns of activity in afferent and associational fiber systems during natural stimulation. The spatial and temporal patterning of activity in these fiber systems has been considered in a recent modeling study of piriform cortex (Ketchum and Haberly, 1993b). In this study it was shown that the distribution in axonal conduction velocities in both systems has the effect of dispersing activity as it propagates over the cortex. As a consequence of this dispersive propagation and the spatially distributed projection patterns of both fiber systems, it can be concluded that there is extensive spatial and temporal convergence of activity within the piriform cortex: at each point, activity arriving at a given time has originated over widespread areas of the olfactory cortex and bulb over a range of times.

Oscillatory Ordering

One might expect that as a consequence of the slow time course of olfactory stimuli, this convergence would be an asychronous, continuous process. However, a study of postsynaptic currents by current source-density analysis (Ketchum and Haberly, 1993c) suggests that oscillatory rhythms imposed on the information stream order integrative processes. This study examined the 50-Hz oscillation elicited by single weak shocks (see figure 15.4C) which resembles that evoked by odors in frequency, amplitude, and phase progression across the piriform cortex (Boudreau and Freeman, 1963; Woolley and Timiras, 1965; Freeman, 1975, 1978). It was found that each cycle of the oscillation was generated by a stereotyped sequence of postsynaptic currents in different parts of pyramidal cells (figure 15.8). Each cycle started with an EPSC in distal apical dendrites mediated by afferent fibers, followed by an EPSC at fixed latency in proximal apical dendrites mediated by intrinsic association fibers. These EPSCs were fol-

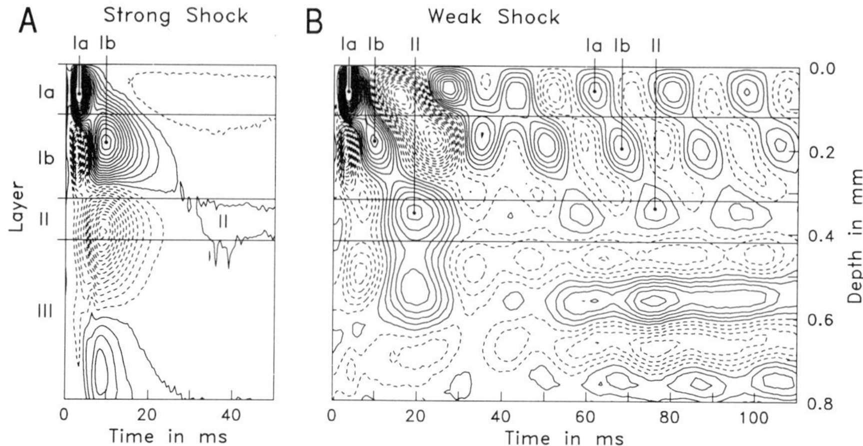

Figure 15.8 Current source-density analysis of oscillatory field potential evoked by weak shocks to afferent fibers in piriform cortex. Inward membrane currents are represented as solid lines; outward currents as dashed lines. (*A*) Nonoscillatory response to strong shock as in figure 15.3. Contour interval, 25 mV/mm^2. (*B*) Oscillatory response to single weak shock at same site as in *A*. Contour interval 2.5 mV/mm^2. Note the stereotyped sequence of sinks in layers Ia, Ib, and II within each oscillatory cycle. Compare with figure 15.3 for relationship of membrane currents to pyramidal cell morphology. (Reprinted from Ketchum and Haberly, 1993c)

lowed by a tentatively identified, fast Cl^--mediated IPSC in the vicinity of cell bodies. From an analysis of temporal relationships it was concluded that the oscillation is generated by circuitry in the olfactory bulb and the resulting periodic activation of afferent fibers drives the oscillation in piriform cortex.

The natural patterning of activity over time in afferent and association fibers may be similar to the theta-burst paradigm used to induce LTP in experimental studies. The fast, circuit-induced oscillation, together with the theta-frequency sniff cycle, result in an odor-induced activity pattern consisting of an average of 7 fast (50 Hz) oscillatory cycles (Woolley and Timiras, 1965) repeated at 150- to 200-msec intervals. While all studies of LTP have employed a 100-Hz repetition rate, it has been confirmed that 50 Hz is also effective (Kanter, unpublished).

It can also be postulated that the fast oscillation produces a repetitive 50 Hz pairing of afferent and association fiber activity based on the finding of successive EPSCs mediated by these systems in the CSD analysis of fast oscillations (Ketchum and Haberly, 1993c). From in vitro studies it is known that a single weak stimulus in the middle of theta-burst trains can evoke associative LTP; therefore, the more extensive pairing with the postulated natural stimulation pattern should be at least as effective.

Spatial Analysis

If there is an alternation of IPSCs with EPSCs during the fast oscillation as suggested by the results of CSD analysis, then the opportunity for LTP-inducing, homosynaptic convergence and across-pathway association would appear to be confined to comparatively narrow, recurring time windows. As previously discussed (Ketchum and Haberly, 1991), the experimental data from CSD analysis (Ketchum and Haberly, 1993a,c) and results of quantitative analysis (Ketchum and Haberly, 1993b) indicate that for the afferent fiber system and the major association fiber system (anteroventral origin), the relative latencies and extents of temporal dispersion are such that activity would be concentrated within the period of each oscillatory cycle. Each fast oscillatory cycle can therefore be thought of as a "snapshot" of the complex, temporally evolving spatial patterns of activity evoked by odor stimulation. Homosynaptic and associative interactions could allow an analysis of spatial patterns within each cycle using a "content-addressable memory" approach as considered in recent models (Granger et al., 1988; Wilson and Bower, 1988; Ambros-Ingerson et al., 1990; Hasselmo et al., 1991).

Temporal Analysis

Activity from the afferent and anteroventral association fiber systems remains concentrated within single oscillatory cycles only because the mean conduction velocities of both systems are comparable. As a result, posteriorally directed waves of activity in these systems remain temporally contiguous. While the anteroventral system is the heaviest association fiber system component, there is evidence for a more slowly conducting component (0.3 m/sec vs. 0.75 m/sec) that originates in the dorsal part of anterior piriform cortex (Haberly, 1978; Ketchum and Haberly, 1991, 1993b). As a consequence of the slow propagation velocity in this system, a posteriorally directed volley of activity in its fibers associated with a given fast oscillatory cycle could be overtaken by a volley in afferent fibers associated with the following cycle (figure 15.9, left). This opens up the possibility for between-cycle association (figure 15.9, right) that could allow the analysis of temporal patterns in the olfactory code. Experimental evidence in support of this hypothesis has been presented (Ketchum and Haberly, 1991).

Insight into how such between-cycle association could contribute to discrimination can be gained from an artificial network model developed by Kleinfeld (1986) (figure 15.10). This model, which is reminiscent of the circuitry of piriform cortex, has the capacity for the recognition of se-

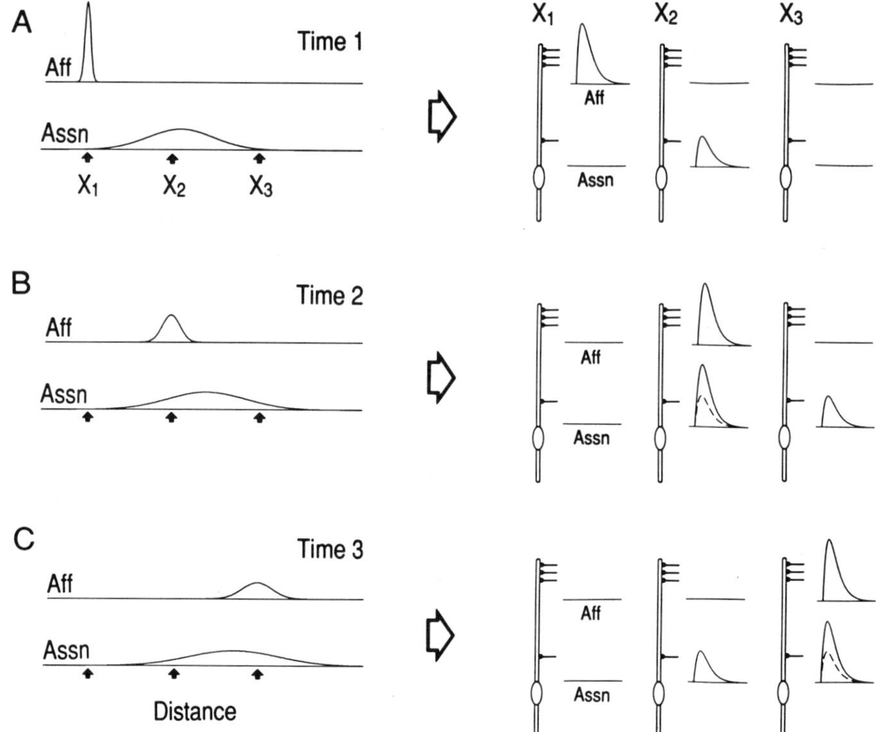

Figure 15.9 An hypothesis for associative interactions between successive cycles of the fast (50 Hz) oscillation in piriform cortex that could allow the analysis of temporal patterns in the olfactory code. The traces in the left column represent the distribution of activity in the afferent fiber system (Aff) and the slowly conducting association fiber system (Assn) (origin in anterodorsal piriform cortex) as a function of rostral-to-caudal distance within the posterior piriform cortex. At time 1 (*A*), activity in association fibers during one oscillatory cycle is centered at distance X_2, and activity in afferent fibers from the *subsequent* oscillatory cycle is centered at X_1. As time progresses, the afferent volley catches up to (*B*) and passes (*C*) the more slowly conducting association volley. As illustrated in the right half of the figure, the temporal contiguity could evoke a potentiation of synaptic strengths. Such a between-cycle associative process could underlie the discrimination of temporally evolving spatial patterns in the olfactory code. (Reprinted from Ketchum and Haberly, 1991)

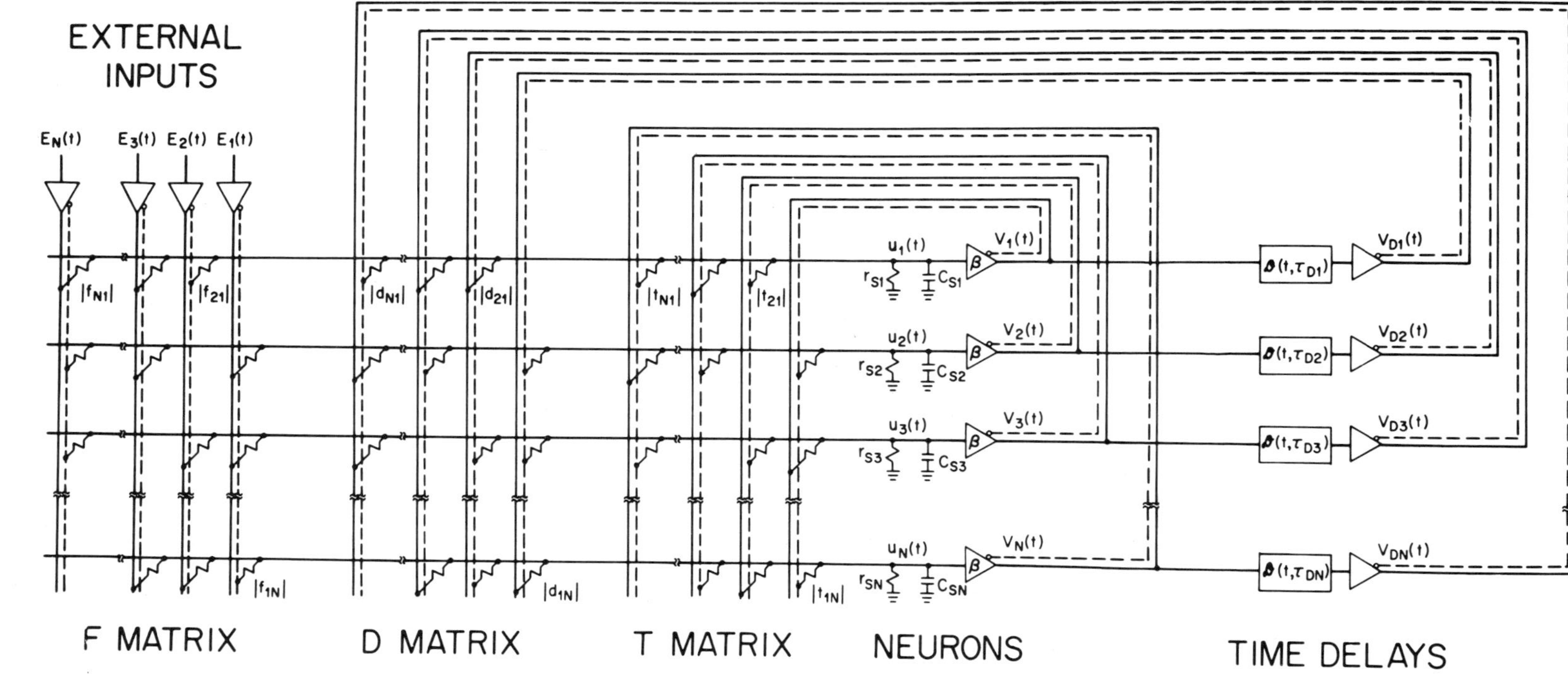

Figure 15.10 A neural network model resembling piriform cortex with capabilities for the generation or recognition of series of spatial patterns. The horizontally depicted processing elements can be likened to pyramidal cells in piriform cortex. These elements receive external inputs, equivalent to the afferent fiber system, that are modified by the strengths of the connections defined by the F matrix. Individual spatial patterns are stored in the T matrix, which modifies input from an immediate feedback system that can be likened to the fast association fiber system. The association between the current spatial pattern and the next pattern in a sequence is stored in the D matrix, which modifies input from a system of delayed feedback that can be likened to the slow association fiber system. (Reprinted from Kleinfeld, 1986)

quences of complex spatial patterns (see review by Ketchum and Haberly, 1991). Input to the network consists of a series of spatially distributed patterns that can be likened to the changing spatial patterns within successive cycles of the fast olfactory oscillation. Individual spatial patterns are stored in the strengths of feedback connections between elements (T matrix in figure 15.10). This feedback system can be likened to the fast association fiber system in piriform cortex. A second, delayed feedback system, whose junctional strengths (D matrix in figure 15.10) are determined by relationships between successive spatial patterns in the input sequence, imparts the ability for temporal analysis. This delayed system, which can be likened to the slow association fiber system in piriform cortex, drives transitions between spatial patterns, thus allowing the recognition of temporal sequences.

SUMMARY AND CONCLUSIONS

The piriform cortex appears to be ideally suited for assessing functional roles of LTP. Both afferent and associational fiber systems display NMDA-dependent LTP, which can form the basis of "Hebbian" synaptic plasticity on which many neural network models with pattern recognition abilities are based. Models of this form are under development for the analysis of spatial and temporal components of the olfactory code by piriform cortex. Recent findings concerning the regulation of LTP, and the spatial and temporal ordering of afferent and associational inputs to pyramidal cells, should facilitate this development.

ACKNOWLEDGMENT

This work was supported by grant NS19865 from NINDS.

REFERENCES

Adrian, E. D. (1953) Sensory messages and sensation. The response of the olfactory organ to different smells. *Acta Physiol. Scand.* 29:5–14.

Alonso, A., Curtis, de M., and Llinas, R. (1990) Postsynaptic Hebbian and non-Hebbian long-term potentiation of synaptic efficicy in the entorhinal cortex in slices and in the isolated adult guinea pig brain. *Proc. Natl. Acad. Sci. USA* 87:9280–9284.

Ambros-Ingerson, J., Granger, R., and Lynch, G. (1990) Simulation of paleocortex performs hierarchical clustering. *Science* 247:1344–1348.

Andreasen, M., and Lambert, J. D. C. (1991) Noradrenaline receptors participate in the regulation of GABAergic inhibition in area CA1 of the rat hippocampus. *J. Physiol.* 439: 649–669.

Artola, A., and Singer, W. (1987) Long-term potentiation and NMDA receptors in rat visual cortex. *Nature* 330:649–652.

Artola, A., and Singer, W. (1990) The involvement of *N*-methyl-D-aspartate receptors in induction and maintenance of long-term potentiation in rat visual cortex. *Eur. J. Neurosci.* 2:254–269.

Boudreau, J. C. (1964) Computer analysis of electrical activity in the olfactory system of the cat. *Nature* 201:155–158.

Boudreau, J. C., and Freeman, W. J. (1963) Spectral analysis of electrical activity in the prepyriform cortex of the cat. *Exp. Neurol.* 8:423–439.

Bressler, S. L., and Freeman, W. J. (1980) Frequency analysis of olfactory system EEG in cat, rabbit, and rat. *Electroencephalogr. Clin. Neurophysiol.* 50:19–24.

Cattarelli, M., Astic, L., and Kauer, J. S. (1988) Metabolic mapping of 2-deoxyglucose uptake in the rat piriform cortex using computerized image processing. *Brain Res.* 442: 180–184.

Collins, G. G. S. (1991) Pharmacological evidence that NMDA receptors contribute to mono- and di-synaptic potentials in slices of mouse olfactory cortex. *Neuropharmacology* 30:547–555.

Droze, V. A., Cohen, G. A., and Madison, D. V. (1991) Synaptic localization of adrenergic disinhibition in the rat hippocampus. *Neuron* 6:889–900.

Eichenbaum, H., Otto, T. A., Wible, C. G., and Piper, J. M. (1991) Building a model of the hippocampus in olfaction and memory. In *Olfaction: A Model System for Computational Neuroscience*, J. L. Davis and H. Eichenbaum (eds.), Cambridge, Mass.: MIT Press, pp. 167–210.

Freeman, W. J. (1975) *Mass Action in the Nervous System*. New York: Academic Press.

Freeman, W. J. (1978) Spatial properties of an EEG event in the olfactory bulb and cortex. *Electroencephalogr. Clin. Neurophysiol.* 44:586–605.

Granger, R., Ambros-Ingerson, J., and Lynch, G. (1988) Derivation of encoding characteristics of layer II cerebral cortex. *J. Cog. Neurosci.* 1:61–87.

Gustafsson, B., and Wigström, H. (1986) Hippocampal long-lasting potentiation produced by pairing single volleys and brief conditioning tetani evoked in separate afferents. *J. Neurosci.* 6:1575–1582.

Haberly, L. B. (1978) Application of collision testing to investigate properties of multiple association axons originating from single cells in the piriform cortex of the rat. *Soc. Neurosci. Abstr.* 4:75.

Haberly, L. B. (1985) Neuronal circuitry in olfactory cortex: Anatomy and functional implications. *Chem. Senses* 10:219–238.

Haberly, L. B. (1990a) Olfactory cortex. In *Synaptic Organization of the Brain* (3rd edition), G. M. Shepherd (ed.). London: Oxford, pp. 317–345.

Haberly, L. B. (1990b) Comparative aspects of olfactory cortex. In *Cerebral Cortex, Vol 8B*, E. G. Jones and A. Peters (eds.). New York: Plenum Press, pp. 137–166.

Haberly, L. B. and Bower, J. M. (1989) Olfactory cortex: Model circuit for study of associative memory? *Trends Neurosci.* 12:258–264.

Haberly, L. B., and Price, J. L. (1977) The axonal projection patterns of the mitral and tufted cells of the olfactory bulb in the rat. *Brain Res.* 129:152–157.

Haberly, L. B., and Price J. L. (1978) Association and commissural fiber systems of the olfactory cortex of the rat. I. Systems originating in the piriform cortex and adjacent areas. *J. Comp. Neurol.* 178:711–740.

Haberly, L. B. and Shepherd, G. M. (1973) Current density analysis of opossum prepyriform cortex. *J. Neurophysiol.* 36:789–802.

Haberly, L. B., Hansen, D. J., Feig, S. L., and Presto, S. (1987) Distribution and ultrastructure of neurons in opossum displaying immunoreactivity to GABA and GAD and high affinity tritiated GABA uptake. *J. Comp. Neurol.* 266:269–290.

Harrison, T. A., and Scott, J. W. (1986) Olfactory bulb responses to odor stimulation: Analysis of response pattern and intensity relationships. *J. Neurophysiol.* 56:1571–1589.

Hasselmo, M. E., Wilson, M. A., Anderson, B. P., and Bower, J. M. (1991) Associative memory function in piriform olfactory. cortex: Computational modeling and neuropharmacology. *Cold Spring Harbor Symp. Quant. Biol.* 55:599–610.

Hoffman, W. H., and Haberly, L. B. (1989a) Do changes in synaptic excitation or inhibition underlie generation of long-lasting bursting-induced late EPSPs in piriform cortex? *Soc. Neurosci. Abstr.* 15:700.

Hoffman, W. H., and Haberly, L. B. (1989b) Bursting induces persistent all-or-none EPSPs by an NMDA-dependent process in piriform cortex. *J. Neurosci.* 9:206–215.

Jourdan, F. (1982) Spatial dimension in olfactory coding: A representation of the 2- deoxyglucose patterns of glomerular labeling in the olfactory bulb. *Brain Res.* 240:341–344.

Jung, M. W., Larson, J., and Lynch, G. (1990) Long-term potentiation of monosynaptic EPSPs in rat piriform cortex in vitro. *Synapse* 6:279–283.

Kanter, E. D., and Haberly, L. B. (1990) NMDA-dependent induction of long-term potentiation in afferent and association fiber systems of piriform cortex in vitro. *Brain Res.* 525:175–179.

Kanter, E. D., and Haberly, L. B. (1993) Associative long-term potentiation in piriform cortex slices requires GABA-A blockade. *J. Neurosci.* 13:2477–2482.

Kapur, A., and Haberly, L. B. (1991) Duration of LTP in piriform cortex in vivo is increased after epileptiform bursting. *Soc. Neurosci. Abstr.* 17:386.

Kapur, A., Kanter, E. D., and Haberly, L. B. (1992) Nonlinear summation of NMDA responses to burst stimulation in piriform cortex slices requires $GABA_A$ blockade. *Soc. Neurosci. Abstr.* 18:1497.

Ketchum, K. L., and Haberly, L. B. (1991) Fast oscillations and dispersive propagation in olfactory cortex: A functional hypothesis. In *Olfaction: A Model System for Computational Neuroscience,* J. L. Davis and H. Eichenbaum (eds.). Cambridge, Mass.: MIT Press, pp. 69–100.

Ketchum, K. L., and Haberly, L. B. (1993a) Membrane currents evoked by afferent fiber stimulation in rat piriform cortex. I. Current source-density analysis. *J. Neurophysiol.* 69: 248–260.

Ketchum, K. L., and Haberly, L. B. (1993b) Membrane currents evoked by afferent fiber stimulation in rat piriform cortex. II. Analysis with a system model. *J. Neurophysiol.* 69: 261–281.

Ketchum, K. L., and Haberly, L. B. (1993c) Synaptic events that generate fast oscillations in piriform cortex. *J. Neurosci.* 13:3980–3985.

Kleinfeld, D. (1986) Sequential state generation by model neural networks. *Proc. Natl. Acad. Sci. USA* 83:9469–9473.

Larson, L., Wong, D., and Lynch, G. (1986) Patterned stimulation at the theta frequency is optimal for the induction of hippocampal long-term potentiation. *Brain Res.* 368:347–350.

Luskin, M. B., and Price, J. L. (1983a) The topographic organization of associational fibers of the olfactory system in the rat, including centrifugal fibers to the olfactory bulb. *J. Comp. Neurol.* 216:264–291.

Luskin, M. B., and Price, J. L. (1983b) The laminar distribution of intracortical fibers originating in the olfactory cortex of the rat. *J. Comp. Neurol.* 216:292–302.

Lynch, G., and Granger, R. (1991) Serial steps in memory processing: Possible clues from studies of plasticity in the olfactory-hippocampal circuit. In *Olfaction: A Model System for Computational Neuroscience,* J. L. Davis and H. Eichenbaum (eds.). Cambridge, Mass.: MIT Press, pp. 141–165.

Mackay-Sim, A., Shaman, P., and Moulton, D. G. (1982) Topographic coding of olfactory quality: Odorant-specific patterns of epithelial responsivity in the salamander. *J. Neurophysiol.* 48:584–596.

Macrides, F. (1977) Dynamic aspects of central olfactory processing. In *Chemical Signals in Vertebrates,* D. Muller-Schwarze and M. M. Mozell (eds.). New York: Plenum Press.

Macrides, F., and Chorover, S. L. (1972) Olfactory bulb units: Activity correlated with inhalation cycles and odor quality. *Science* 175:84–87.

Macrides, F., Eichenbaum, H. B., and Forbes, W. B. (1982) Temporal relationship between sniffing and the limbic theta rhythm during odor discrimination reveral learning. *J. Neurosci.* 2:1705–1717.

Madison, D. V., Malenka, R. C., and Nicoll, R. A. (1991) Mechanisms underlying long-term potentiation of synaptic transmission. *Annu. Rev. Neurosci.* 14:379–397.

Patneau, D. K., and Stripling, J. S. (1992) Functional correlates of selective long-term potentiation in the olfactory cortex and olfactory bulb. *Brain Res.* 585:219–228.

Pitler, T. A., and Alger, B. E. (1992). Cholinergic excitation of GABAergic interneurons in the rat hippocampal slice. *J. Physiol.* 450:127–142.

Price, J. L. (1973) An autoradiographic study of complementary laminar patterns of termination of afferent fibers to the olfactory cortex. *J. Comp. Neurol.* 150:87–108.

Price, J. L. (1985) Beyond the primary olfactory cortex: Olfactory-related areas in the neocortex, thalamus and hypothalamus. *Chem. Senses* 10:239–258.

Price, J. L., Carmichael, S. T., Carnes, K. M., Clugnet, M.-C., Kuroda, M., and Ray, J. P. (1991) Olfactory input to the prefrontal cortex. In *Olfaction: Model System for Computational Neuroscience,* J. L. Davis and H. Eichenbaum (eds.). Cambridge, Mass.: MIT Press, pp. 101–120.

Richards, C. D. (1972) Potentiation and depression of synaptic transmission in the olfactory cortex of the guinea-pig. *J. Physiol.* 222:209–231.

Rodriguez, R., and Haberly, L. B. (1989) Analysis of synaptic events in the opossum piriform cortex with improved current source density techniques. *J. Neurophysiol.* 61:702–718.

Roman, F., Staubli, U., and Lynch, G. (1987) Evidence for synaptic potentiation in a cortical network during learning. *Brain Res.* 418:221–226.

Ropert, N., and Guy, N. (1991) Serotinin facilitates GABAergic transmission in the CA1 region of rat hippocampus in vitro. *J. Physiol.* 441:121–136.

Satou, M., Mori, K., Tazawa, Y., and Takagi, S. F. (1982) Two types of postsynaptic inhibition in pyriform cortex of the rabbit: Fast and slow inhibitory postsynaptic potentials. *J. Neurophysiol.* 48:1142–1156.

Satou, M., Mori, K., Tazawa, Y., and Takagi, S. F. (1983) Interneurons mediating fast postsynaptic inhibition in pyriform cortex of the rabbit. *J. Neurophysiol.* 50:89–101.

Scott, J. W., McBride, R. L., and Schneider, S. P. (1980) The organization of projections from the olfactory bulb to the piriform cortex and olfactory tubercle in the rat. *J. Comp. Neurol.* 194:519–534.

Sharp, F. R., Kauer, J. S., and Shepherd, G. M. (1975) Local sites of activity-related glucose metablism in rat olfactory bulb during olfactory stimulation. *Brain Res.* 98:596–600.

Sheldon, P. W., and Aghajanian, G. K. (1990) Serotonin 5-HT induces IPSPs in pyramidal layer cells of rat piriform cortex: Evidence for the involvement of a 5-HT2-activated interneuron. *Brain Res.* 506:62–69.

Shepherd, G. M. (1991) Computational structure of the olfactory system. In *Olfaction: A Model System for Computational Neuroscience,* J. L. Davis and H. Eichenbaum (eds). Cambridge, Mass.: MIT Press, pp. 3–41.

Staubli, U., Thibault, O., DiLorenzo, M., and Lynch, G. (1989) Antagonism of NMDA receptors impairs acquisition but not retention of olfactory memory. *Behav. Neurosci.* 103: 54–60.

Staubli, U., Vanderklish, P., and Lynch, G. (1990) An inhibitor of integrin receptors blocks long-term potentiation. *Behav. Neural Biol.* 53:1–5.

Stripling, J. S., Patneau, D. K., and Gramlich, C. A. (1988) Selective long-term potentiation in the pyriform cortex. *Brain Res.* 441:281–291.

Stripling, J. S., Patneau, D. K., and Gramlich, C. A. (1991) Characterization and anatomical distribution of selective long-term potentiation in the olfactory forebrain. *Brain Res.* 542: 107–122.

Takagi, S. F. (1986) Studies on the olfactory nervous system in the old world monkey. *Prog. Neurobiol.* 27:195–250.

Tseng, G.-F. and Haberly, L. B. (1988) Characterization of synaptically mediated fast and slow inhibitory processes in piriform cortex in an in vitro slice preparation. *J. Neurophysiol.* 59:1352–1376.

White, G., Levy, W. B., and Steward, O. (1990) Spatial overlap between populations of synapses determines the extent of their associative interaction during the induction of long-term potentiation and depression. *J. Neurosci.* 64:1186–1198.

Wilson, M. A. and Bower, J. M. (1988) A computer simulation of olfactory cortex with functional implications for storage and retrieval of olfactory information. In *Neural Information Processing Systems,* D. Z. Anderson (ed.). New York: American Institute of Physics, pp. 114–126.

Woolley, D. E., and Timiras, P. S. (1965) Prepyriform electrical activity in the rat during high altitude exposure. *Electroencephalogr. Clin. Neurophysiol.* 18:680–690.

16 Long-Term Potentiation and Long-Term Depression in Piriform Cortex In Vivo

François S. Roman, Franck A. Chaillan, and Bernard Soumireu-Mourat

Memory storage is believed to be the result of a set of synaptic modifications in the nervous system (see review in Squire, 1987). The question of whether specific engrams are localized or distributed—either in specific structures (e.g., cerebellum, cortical area) or in large neural circuitries including different nervous structures, notably limbic structures such as the hippocampus (Thompson, 1991)—is a continuing debate. In any case, it is generally agreed that for most learning episodes to be stored in long-term memory requires modifications of synapses in the cerebral cortex (Hebb, 1949; Greenough, 1984).

The complexity of the neocortex and the intricacy of the connections from sensory organs to the hippocampal archicortex exclude these two types of cerebral cortex from use in observing putative modifications of monosynaptic responses of a restricted population of cortical neurons while learning is occurring. On the contrary, the piriform cortex (PC) is a part of the olfactory paleocortex in which such problems can be solved. Only two synapses are involved from the olfactory sensory epithelium to the PC (Shepherd, 1983), and the projections of the olfactory bulb through the lateral olfactory tract (LOT) generate a dense but restricted terminal field into the plexiform layer of the PC, making connections on apical dendrites of pyramidal cells located in deeper layers II and III (Haberly and Shepherd, 1973; Haberly, 1985). This situation creates an ideal opportunity for extracellular recordings and stimulations by single pulse sent to the LOT to elicit monosynaptic excitatory postsynaptic potentials (EPSPs) of the PC pyramidal cells followed by disynaptic EPSPs due to the resulting activation of the PC association fiber system. In addition, the LOT and PC association fibers produce feedforward and feedback inhibition of pyramidal cells via activation of inhibitory interneurons (Haberly, 1985; Tseng and Haberly, 1988; Stripling et al., 1991). On the other hand, the processing of natural olfactory information involves spatially distributed patterns of activity in the olfactory bulb (Mozell, 1964; MacKay-Sim et al., 1982; Jourdan, 1982), and the anatomical organization of the olfactory cortex contributes to the spreading of olfactory information (Ketchum and Haberly, 1991; Shepherd, 1991) within the first few synapses. Consequently, working with natural odors does not allow the delivery of punctual stimuli required to elicit field potentials in piriform cortex. To create such a punctual stimulus, we chose to replace odors by electrical patterned stimulations of the LOT (Roman et al., 1987) in order to reproduce several relevant properties of natural olfactory stimulations, and we developed an olfactory behavioral task allowing us to observe learning and memory performance.

THE OLFACTORY DISCRIMINATION TASK

Natural Odors

In this associative learning task, animals learn to associate a correct odor with a reward (water), the animals being water-deprived for 48 hr before the first training session. Rats are trained to discriminate two odors, one designated as positive and the other as negative, using a successive Go–No Go paradigm (figure 16.1). Response to the positive odor is rewarded by 0.1 ml of water; response to the negative odor results in a 10-sec

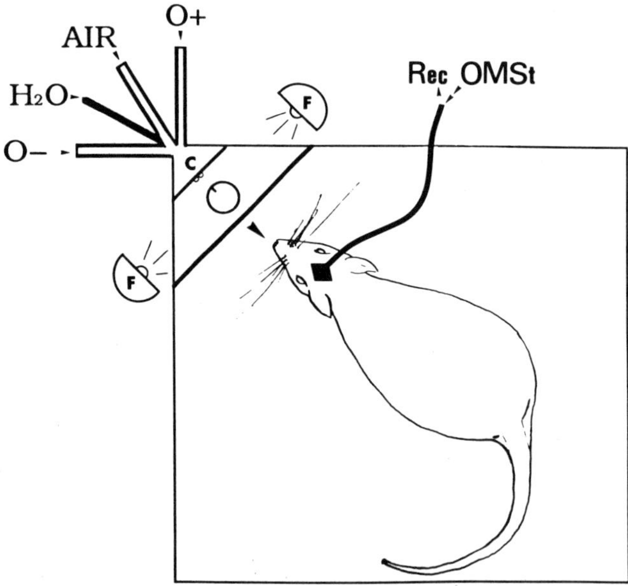

Figure 16.1 A rectangular box made from wire mesh (30 × 30 × 55) serves as olfactory training apparatus, with a photosensitive cell for detecting the animal mounted in front of a combined water spout/odor outlet in one corner. O+, positive natural odor; O−, negative natural odor; H_2O, water; AIR, ejected air; F, flash light; C, electrical photosensitive cell; OMSt, olfacto mimetic stimulation; Rec, record.

nonaversive light presentation. Each odor is presented for 10 sec in a random fashion with an intertrial interval of 30 sec. In this experiment, animals are submitted to the training for five daily sessions with each daily session consisting of 60 trials ± 5. Correct responses are both "Go" for the positive odor and "No Go" for the negative odor. Incorrect responses include both "Go" for the negative odor and "No Go" for the positive odor. Performance of the animals is defined as the percentage of correct responses per session.

Electrical Patterned Stimulations versus Natural Odor

Rats are implanted unilaterally with two stimulating electrodes in the LOT and one recording electrode in layer I of PC. Ten days later, they are connected to stimulation and recording equipment during three daily sessions of 20 min to habituate them to move easily in the training apparatus and to suppress eventual secondary exploratory effects (Barnes et al., 1991) during the following training sessions. In addition, at the end of the last habituation session, twenty single biphasic pulses are delivered every 15 sec to the stimulating electrode in order to elicit a negative-going field EPSP in PC with an intensity producing a 1-mV amplitude (about 25% of the maximum). Rats do not respond behaviorally to single pulse stimulations, but patterned stimulation consisting of 36-msec bursts of 4 pulses (at 100 Hz) delivered with an interburst interval of 160 msec elicits a robust sniffing reaction followed by an approach to the water delivery site. Positive stimulations versus negative odor are used in a discrimination paradigm similar to that used with the two natural odors.

The results obtained in these two experiments are compared in figure 16.2. As for the discrimination with two natural odors, stimulated animals were able to discriminate the positive electrical patterned stimulation versus the negative natural odor, and only a slight, not statistically significant difference was observed in the time course of learning between the two sets of experiments, specially in sessions 2 and 3. The electrical patterned stimulations (short bursts of high frequency [100–200 Hz] with 5 bursts per second) were similar to the sampling frequency used by the olfactory system of the rat (Macrides et al., 1982; Eeckman and Freeman, 1990) and, although unilaterally distributed, appeared to be processed by the olfactory system as a natural odor. Consequently, the stimulated rats performed the discrimination task as well as the control rats with natural odors. Following these observations, we defined these electrical patterned stimulations distributed into the LOT as olfactomimetic stimulations (OMSt). In addition, prior to and after every training session with the OMSt, we used single biphasic pulses to routinely determine eventual modifications of

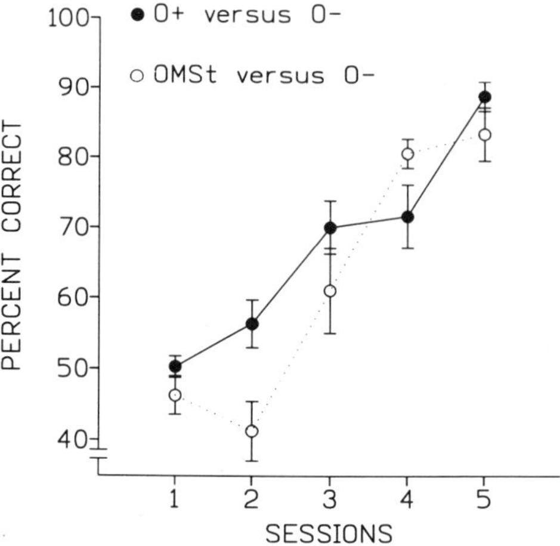

Figure 16.2 Statistical comparison of the percentage of correct responses between the group conditioned to discriminate two natural odors (positive odor: O+ versus negative odor: O−) or the electrical olfacto mimetic stimulation versus negative odor (OMSt versus O−) showed no significant difference across training sessions.

synaptic efficiency of specific cortical neurons directly resulting from the learning of the positive OMSt meaning.

OLFACTORY DISCRIMINATION AND LONG-TERM POTENTIATION IN PIRIFORM CORTEX: THE LINK

Three experiments were conducted using the OMSt. The first consisted in training the rats to discriminate a positive OMSt versus a negative natural odor from the first training session. The second consisted in training rats with successive natural odor pairs before replacing a new natural positive odor by the OSMt. Finally, in the last experiment, implanted rats were submitted to the same amount of OMSt, negative odor, flash light and water, without any association, in other words, a pseudo-conditioning.

In experiment 1, according to the procedure previously described, baseline values of EPSPs in PC were recorded 24 hr prior to the first training session by delivering single pulses to the stimulating electrode that was going to provide the OMSt during the behavioral task, and also to another stimulating electrode positioned at ±1 mm of distance into the LOT—that is, a control electrode. Prior to the start of the first training session, respective EPSPs elicited by both electrodes were again recorded to verify the stability of baseline values. Animals were trained to discriminate the OMSt versus the negative odor. The decision to associate the OMSt with the positive reward (water) was motivated by preliminary results showing that this association was more easily accomplished than the association

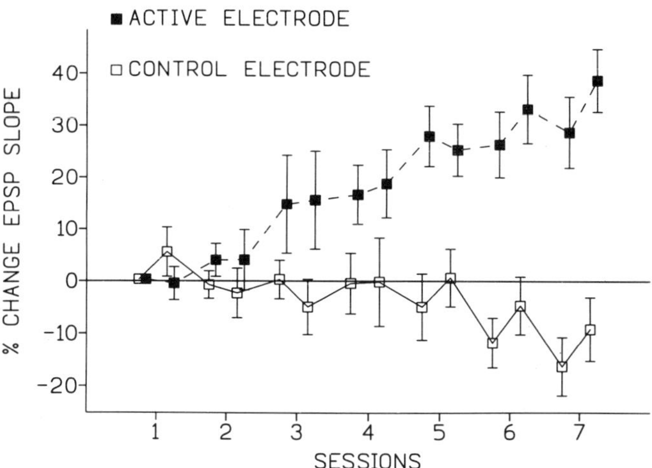

Figure 16.3 Change in evoked potentials induced by the active electrode in which during the discrimination session, the OMSt were delivered and by the control electrode. The slopes of 20 evoked responses (EPSPs) to single pulses were collected for each of the two stimulating electrodes in the LOT prior to session 1 of behavioral testing. The average value of these EPSP slopes was used as a baseline. At the end of this first training session, 20 single pulse responses to each of the two electrodes were again collected. These responses were averaged and expressed as a percentage of the baseline values (mean $\pm$ SEM). Seven rats were tested. This procedure was used prior to, during, and after each subsequent session. A statistically significant change in EPSP slope evoked by the active electrode was found prior to session 4 ($P < .05$ comparing the responses collected prior to session 4 with the baseline collected prior to session 1 with the same electrode; Mann-Whitney U-test, 2-tailed). This increase was maintained after the end of the session ($P < .05$). An increase of this potentiation was observed prior to session 5 ($P < .001$) and at the end of sessions 6 and 7 ($P < .001$). No statistically significant change was observed in the EPSPs evoked by the control electrode, except the responses collected prior to training sessions 6 and 7 in comparison with the baseline collected prior to the first session with the same electrode ($P < .05$).

with the negative reward (flash light). Finally, at the end of the training, both EPSPs were again recorded and compared with their baseline values. This procedure was maintained during seven sessions, and the results of these experiments are shown in figure 16.3.

In experiment 2, implanted rats were intensively trained to discriminate seven successive odor pairs (one each day) to acquire primatelike "learning sets" until only a few trials (between 5 and 10 trials) were required to learn a new odor pair. Then a new positive natural odor was replaced by the OMSt to be discriminated with another new negative natural odor. The learning occurred without obvious problems, although 20 additional trials were necessary to reach a level of 80% of correct responses within the session. The respective EPSPs, as in experiment 1, were recorded at the end of this first training session and compared to their respective baseline

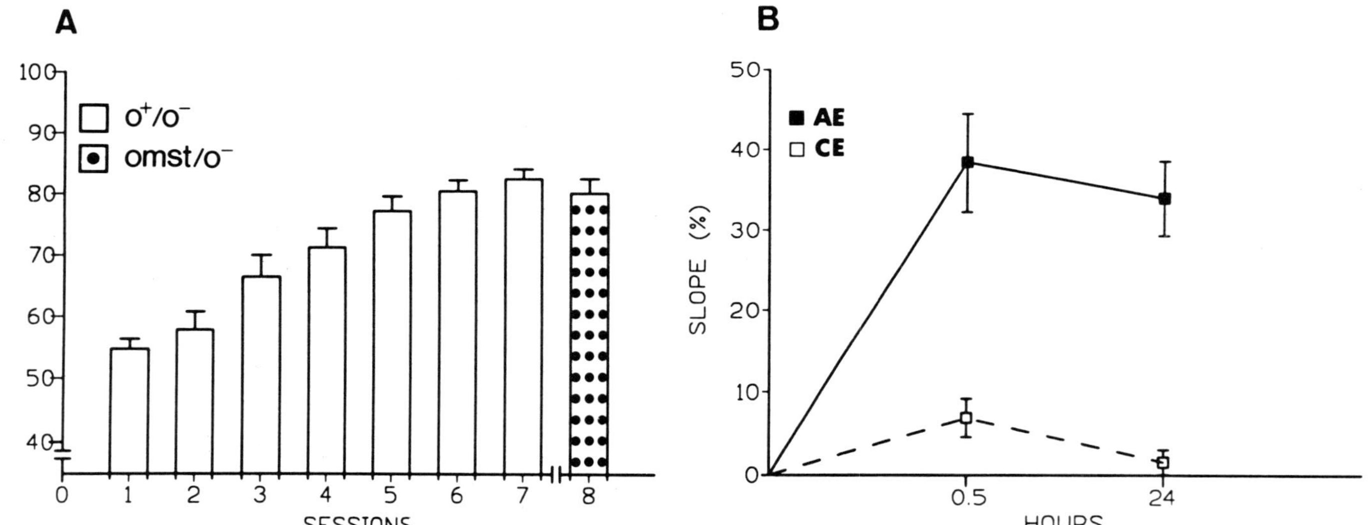

Figure 16.4 (*A*) Percentage of correct responses performed by implanted rats (n = 12) required to discriminate between a new odor pair for 7 sessions (O+/O−). In session 8, the rats had to discriminate the OMSt delivered on one electrode (active electrode) versus a new negative odor (OMSt/O−). A level of about 80% ±3 (mean ± SEM) of percent correct was observed since the 6th session. In session 8, the rats performed as well as in session 6. (*B*) Change in evoked potentials induced by the active electrode (AE), which in session 8 delivered the OSMt, and by the control electrode (CE). The slopes of 20 evoked responses (EPSPs) to single pulses were collected for each of the two stimulating electrodes in the LOT prior to session 8. The average value of these EPSP slopes was used as a baseline. At the end of session 8, 20 single pulse responses to each of the two electrodes were again collected. These responses were averaged and expressed as a percentage of the baseline values (mean ± SEM) and statistically different for the active electrode ($P < .01$ Mann-Whitney U-test, 2-tailed). This increase was maintained 24 hr later. A slight, not significant increase was observed in the EPSPs evoked to the control electrode, at the end of session 8; 24 hr later the slight increase disappeared.

values. Twenty-four hours later, the evoked responses were also recorded (figure 16.4).

In the pseudo-conditioning experiment, animals received the same number of training sessions (i.e., seven) with the same number and duration of cue presentations (i.e., OMSt and negative odor) and the same amounts of reward (i.e., water and flash light), taking into account the respective amounts distributed in experiment 1, session after session. However, the main difference was that they experienced explicitly unpaired presentations of patterned electrical stimulation with water and negative odor with flash light.

In both experiments 1 and 2, the successful learning of the association between the OSMt and the water reward was followed by a mean increase in slope of evoked potential of about 35 ± 5% (means ± SEM) and this increase was still present 24 hr later. In addition, a long-term potentiation-like phenomenon in PC was observed only with the electrode to which the OMSt were delivered. Although the occurrence of the LTP could have been produced by the delivery of patterned stimulation of the LOT alone, several reports from different laboratories indicate that the monosynaptic LOT–piriform cortex system does not sustain LTP (Racine et al., 1983; Stripling et al., 1988). The data from experiment 3 further confirmed that additive behavioral components were necessary to elicit LTP in piriform cortex in vivo. Moreover, in the pseudo-conditioning experiment, when the implanted rats were stimulated, with the OMSt unpaired with the water reward, a progressive homosynaptic decrease in EPSPs or long-term depression (LTD) occurred (figure 16.5).

LTP IN PIRIFORM CORTEX: DIFFERENCES WITH LTP IN OTHER BRAIN STRUCTURES

It thus appears that additional conditions associated with learning are required to elicit LTP of monosynaptic EPSPs in the rat PC in vivo. The increase in efficiency of specific synapses in a restricted cortical field is not simply due to the stimulation itself, as demonstrated by the data obtained in the experiment with pseudo-conditioned rats. Moreover, in the pseudo-conditioning group (experiment 3), the imposed patterned stimulation was followed by a LTD phenomenon, and in experiments 1 and 3, a progressive although weaker depression was recorded with the control electrode, revealing a heterosynaptic depression in the surrounding territory. Although a positive correlation was found between behavioral performance and LTP magnitude, it would be difficult to conclude that the behavioral performance is directly related to the slope or amplitude of the EPSP. In fact, it seems more accurate to propose that LTP indicates the formation of

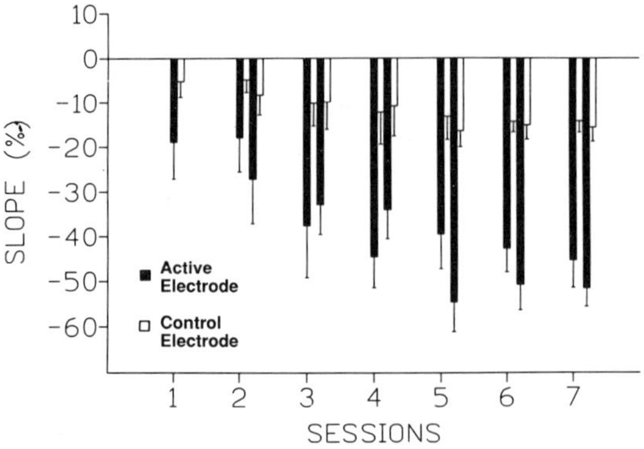

Figure 16.5 Evoked responses in piriform cortex were elicited by single pulse stimulation through two electrodes in the lateral olfactory tract in six pseudo-conditioned rats. OMSt were delivered through only one electrode (active electrode) during the 7 successive sessions. Twenty single evoked potentials were collected and average using the same paradigm as experiment 1. A progressive depression of the evoked potentials was observed since the end of session 1. A statistically significant difference with the active electrode appeared after the end of session 2 ($P < .01$; Mann Whitney U-test, 2-tailed). A statistically significative depression was observed with the control electrode since the end of session 5 ($P < .05$).

a long-term association between the cue (in this case the OMSt) and its reward (the water).

Another specific property of LTP in piriform cortex is its developmental time course. In experiment 1, the development of LTP was progressive and correlated with the improvement of behavioral performance, while in experiment 2, when the rats were previously trained to discriminate natural odors, the occurrence of LTP was immediate. Thus, in the first circumstance, LTP develops slowly, while in the second, the increase is immediate, as observed when similar patterned stimulations are delivered to hippocampal pathways in naive rats (i.e., without learning circumstances) and induce an immediate and extremely stable LTP.

POSSIBLE MECHANISMS UNDERLYING LTP AND LTD IN PIRIFORM CORTEX

One of the most obvious questions to be answered is why LTP cannot be induced in piriform cortex by brief high-frequency stimulation of excitatory pathways, whereas this pattern of stimulation produces LTP in other brain structures with neural circuitries similar to those found in piriform cortex. An explanation based on the phylogenetic origin of the PC is not likely since LTP has been observed not only in paleocortex, as in hippocampus, but also in phylogenetically recent neocortical areas. Nevertheless, the olfactory system is likely to be the oldest sensory system in mammals, and its primitive and specific anatomical organization might explain this specific feature. Sensory information requires several synapses to achieve access to neocortex as well as to archicortex, whereas olfactory information traverses only two synapses to reach paleocortex and especially piriform cortex. Thus, most sensory information undergoes more filtering or selection before reaching its respective cortical territory than does olfactory information. The pathways followed by the sensory modalities different from olfaction are usually longer, creating both a potential loss of electrical efficiency and numerous possibilities of modulation, by direct or indirect influences from other brain structures, at every synapse. Although the processing of olfactory information by the olfactory bulb (Astic and Saucier, 1986; Saucier and Astic, 1986) can be electrically modulated through cortical and subcortical influences (Holley and MacLeod, 1977; Gervais et al., 1988), the possibilities of modulation are clearly less important in the olfactory system compared to the other sensory modalities. Consequently, pyramidal neurons should be excessively solicited by permament stimuli originating in the natural environment. Under these conditions, it is possible that pyramidal cells exhibit periods of massive depolarization followed by long periods of depression, thus losing their capacity to integrate olfactory information. An adaptive solution to improve the efficiency of this system and to preserve the target cells could be to generate a powerful modulation of the input signal by some other structures, modulation acting either presynaptically on the axonal endings of the LOT or postsynaptically on the pyramidal cells in piriform cortex.

Three types of in vitro or in vivo experiments in rats support this hypothesis and specify such a mechanism. In piriform cortex slices (i.e., when piriform cortex is disconnected from other brain structures), short- and long-term potentiation of the population synaptic responses can be easily observed using patterned stimulation similar to those used in behaving rats (Hasselmo and Bower, 1990; Jung et al., 1990; Kanter and Haberly, 1990). In vivo, no LTP was elicited by electrical stimulation of the LOT (Racine et al., 1983, 1991), even with strong intensity, and only a long-term enhancement of the field potential due to an increase of the polysynaptic inhibitory postsynaptic potential (IPSP) was observed but not in monosynaptic responses (Stripling et al., 1988, 1991). Moreover, when rats were stimulated for long periods with the OMSt in the pseudoconditioning experiment, a substantial homosynaptic LTD was observed. Finally, it is only when some learning conditions were added to the stimulation paradigm that the OMSt elicited LTP in piriform cortex. Furthermore, this form of LTP occured either progressively or rapidly in correlation with the learning paradigm (or procedural components) of the task. Thus, patterned electrical stimulation in piriform cortex can elicit either LTP or LTD, and recent reports suggest that the two processes are convertible, depending on the postsynaptic membrane potential or the prevailing concentration of intracellular Ca^{2+} (Artola et al., 1990; Frégnac et al., 1990; Kimura et al., 1990; Yoshimura et al., 1991; Tsumoto, 1992).

An exclusive automodulation of postsynaptic membrane of the pyramidal cells by feedforward or feedback could not explain why the same parameters of patterned electrical stimulation induce either LTP or LTD. Thus another possibility, as we suggested previously, is that the modulation is under the influence of other structures activated during the learning task.

Among the structures directly connected to the olfactory paleocortex, the diagonal band of Broca (DBB) is of particular interest. First, this structure receives direct afferents from CA3 pyramidal cells of hippocampus (Gaykema et al., 1991); second, the horizontal part of the DBB sends direct afferents to piriform cortex (Luskin and Price, 1982). In addition, this pathway is GABAergic (Brashear et al., 1986), and GABA has been shown both in hippocampus and neocortex to play an important role in the mechanisms responsible for LTP induction by regulating the threshold for activation of NMDA receptors (Thompson, 1986; Artola and Singer, 1987). In this regard, it is important to note that the piriform cortex as well as hippocampus is a structure rich in NMDA receptors (Collins and Howlett, 1988).

A hypothesis consistent with the literature on learning and memory (Scoville and Milner, 1957; see review in Squire, 1987) and the present results proposes that the hippocampus, once procedural learning is acquired, can activate the horizontal DBB, which in turn could unlock the cortical olfactory matrix, allowing the occurrence of a LTP-like phenomenon. Under this condition, the piriform cortex would be "ON" in order to associate an odor with its valence by modifying the weight of the activated synapses. This would increase the efficiency of this odor to drive a specific behavior in relation to the particular learning condition, as a result of the activation of some other network. Without this active influence from limbic structures, the olfactory cortex would be locked or "OFF," and usual stimulation by natural odor would be neutral, while excessive stimulation either by natural odor or OMSt would be followed by an LTD phenomenon, thus preventing the electrical propagation of olfactory information and consequently the occurrence of a specific behavior.

Behavioral experiments with rats with bilateral lesions in the horizontal DBB have demonstrated a severe deficit on the olfactory task (figure 16.6), and physiological experiments are in progress to observe the modulation, if not the occurrence, of LTP and LTD in piriform cortex by activation of limbic structures.

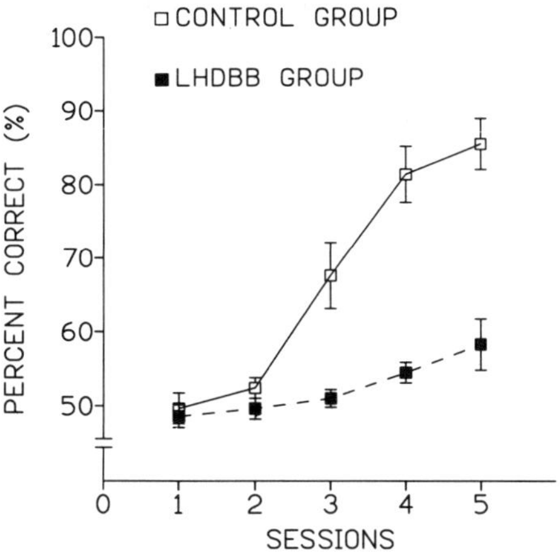

Figure 16.6 Percentage of correct responses performed during 5 sessions by a group of rats (n = 10) with bilateral lesions of the horizontal diagonal band of Broca and by a group of sham-operated control animals (n = 10). Lesioned animals were never able to learn the discrimination problem, in contrast to the control animals, with a statistical difference across the five sessions as indicated by a multivariate analysis of variance, MANOVA [$F(4, 72) = 17.3$, $P < .001$].

CONCLUDING REMARKS

We have used an olfactory discrimination task with the OMSt in order to understand the processes by which the occurrence of LTP in a cortical network (piriform cortex) might give rise to long-term memory. This approach has clearly indicated that long-term storage of olfactory information involves more than single synapses or neurons and in fact depends on complex circuitries and rules of plasticity. Two features of LTP in PC can be distinguished in vivo. First, the occurrence of LTP in piriform cortex is dependent on the learning and memory of the the cue-reward association. Second, the circuitries involved in the learning and memory of these associations required not only the olfactory system itself, but also the cooperativity of the limbic circuitries. Two comments can be made about these features. First, using a simulation of the olfactory bulb–olfactory cortex system, Lynch and Granger (1989) observed that learning of large number of odors takes place as a result of potentiation of active synapses on sufficiently depolarized cells via a rule based on hippocampal LTP; they indeed verify experimentally that a significant increment in synaptic strength follows even a single 50-msec burst of activity (Larson and Lynch, 1989), the time interval used during each operation cycle in their model.

Thus, associative learning and LTP occurrence require only a few training trials per cue to be observed (Slotnick and Katz, 1974; Roman et al., 1987, 1989; Staubli et al., 1987) in vivo (when the animals showed a learning-set formation) as well as in simulation studies. Second, most of the associative learning in rats appears to be correlated with changes in synapse properties in hippocampus (Laroche et al., 1991) which could be the decisive switch allowing LTP in some associative cortical territory and consequently long-term memory storage. This olfactory model should be extremely useful to perform a spatiotemporal analysis of the physiological events (i.e., LTP) resulting from behavioral performances in the limbic circuitry and PC in order to determine whether these events are occurring sequentially or in parallel in the olfactory and limbic networks.

ACKNOWLEDGMENTS

This work was supported by CNRS and by grants from the DRET and INSERM. The authors are grateful for the expert secretarial assistance of Michelle Bauget.

REFERENCES

Artola, A., and Singer, W. (1987) Long-term potentiation and NMDA receptors in rat visual cortex. *Nature* 330:649–652.

Artola, A., Brochier, S., and Singer, W. (1990) Different voltage-dependent thresholds for inducing long-term depression and long-term potentiation in slices of rat visual cortex. *Nature* 347:69–72.

Astic, L., and Saucier, D. (1986) Anatomical mapping of the neuroepithelial projection to the olfactory bulb in the rat. *Brain Res. Bull.* 16:445–454.

Barnes, C. A., McNaughton, B. L., and Erickson, C. A. (1991) Behaviorally induced synaptic enhancement: Memory or modulation? *Long-term Potentiation: A Debate of Current Issues,* M. Baudry and J. L. Davis (eds.). Cambridge, Mass.: MIT Press, pp. 247–257.

Brashear, H. R., Zaborsky, L., and Heimer, L. (1986) Distribution of GABAergic and cholinergic neurons in the rat diagonal band. *Neuroscience* 17:439–451.

Collins, G. G. S., and Howlett, S. J. (1988) The pharmacology of excitatory transmission in the rat olfactory cortex slice. *Neuropharmacology* 27:697–705.

Eeckman, F. H., and Freeman, W. J. (1990) Correlations between unit firing and EEG in the rat olfactory system. *Brain Res.* 528:238–244.

Frégnac, Y., Smith, D., and Friedlander, M. J. (1990) Postsynaptic membrane potential regulates synaptic potentiation and depression in visual cortical neurons. *Soc. Neurosci. Abstr.* 16:798.

Gaykema, R. P. A., Van Der Kuil, J., Hersh, L. B., and Luiten, P. G. M. (1991) Patterns of direct projections from the hippocampus to the medial septum-diagonal band complex: Anterograde tracing with *Phaseolus vulgaris* leucoagglutinin combined with immunohistochemistry of choline acetyltransferase. *Neuroscience* 43:349–360.

Gervais, R., Holley, A., and Keverne, B. (1988) The importance of central noradrenergic influences on the olfactory bulb in the processing of learned olfactory cues. *Chem. Senses* 13:3–12.

Greenough, W. (1984) Structural correlates of information storage in mammalian brain: A review and hypothesis. *Trends Neurosci.* 7:229–233.

Haberly, L. B. (1985) Neuronal circuitry in olfactory cortex: Anatomy and functional implications. *Chem. Senses* 10:219–238.

Haberly, L. B., and Shepherd, G. M. (1973) Current density analysis of summed evoked potentials in opossum prepyriform cortex. *Neurophysiol.* 36:789–802.

Hasselmo, M. E., and Bower, J. M. (1990) Afferent and association fiber differences in short-term potentiation in piriform (olfactory) cortex of the rat. *J. Physiol.* 64:179–190.

Hebb, D. O. (1949) *The Organization of Behavior.* New York: Wiley.

Holley, A., and MacLeod, P. (1977) Transduction et codage de l'information olfactive chez les vertébrés. *J. Physiol. (Paris)* 73:725–848.

Jourdan, F. (1982) Spatial dimension in olfactory coding: A representation of the 2-deoxyglucose patterns of glomerular labelling in the olfactory bulb. *Brain Res.* 240:341–344.

Jung, M. W., Larson, J., and Lynch, G. (1990) Long-term potentiation of monosynaptic EPSPs in rat piriform cortex in vitro. *Synapse* 6:279–283.

Kanter, E. D., and Haberly, L. B. (1990) NMDA-dependent induction of long-term potentiation in afferent and association fiber systems in piriform cortex in vitro. *Brain Res.* 525:175–179.

Ketchum, K. L., and Haberly, L. B. (1991) Fast oscillations and dispersive propagation in olfactory cortex and other cortical areas: A functional hypothesis. In *Olfaction: A Model System for Computational Neuroscience,* H. Eichenbaum and J. L. Davis (eds.). Cambridge, Mass.: MIT Press, pp. 69–101.

Kimura, F., Tsumoto, T., Nishigori, A., and Yoshimura, Y. (1990) Long-term depression but not potentiation is induced in Ca^{2+}-chelated visual cortex neurons. *NeuroReport* 1:65–68.

Laroche, S., Doyère, V., and Rédini Del Négro, C. (1991) What role for long-term potentiation in learning and the maintenance of memories? In *Long-term Potentiation: A Debate of Current Issues,* M. Baudry and J. L. Davis (eds.). Cambridge, Mass.: MIT Press, pp. 301–316.

Larson, J., and Lynch, G. (1989) Theta pattern stimulation and the induction of LTP: The sequence in which synapses are stimulated determines the degree to which they potentiate. *Brain Res.* 489:49–58.

Luskin, M. B., and Price, J. L. (1982) The topographic organization of associational fibers of the olfactory system in the rat, including centrifugal fibers to the olfactory bulb. *J. Comp. Neurol.* 216:264–291.

Lynch, G., and Granger, R. (1989) Simulation and analysis of a cortical network. *Psychol. Learning Motivation* 23:205–241.

MacKay-Sim, A., Shaman, P., and Moulton, D. G. (1982) Topographic coding of olfactory quality: Odorant-specific patterns of eptithelial responsivity in the salamander. *J. Neurophysiol.* 48:584–596.

Macrides, F., Eichenbaum, H., and Forbes, W. (1982) Temporal relationship between sniffing and the limbic theta rhythm during odor discrimination reversal learning. *J. Neurosci.* 2:1705–1717.

Mozell, M. M. (1964) Olfactory discrimination: Electrophysiological spatio-temporal basis. *Science* 143:1336–1337.

Racine, R. J., Milgram, W. N., and Hafner, S. (1983) Long-term potentiation phenomena in the rat limbic forebrain. *Brain Res.* 260:217–231.

Racine, R. J., Moore, K. A., and Evans, C. (1991) Kindling-induced potentiation in the piriform cortex. *Brain Res.* 556:218–225.

Roman, F., Staubli, U., and Lynch, G. (1987) Evidence for synaptic potentiation in a cortical network during learning. *Brain Res.* 418:221–226.

Roman, F., Han, D., and Baudry, M. (1989) Effects of two ACTH analogs on successive odor discrimination learning in rats. *Peptides* 10:303–307.

Saucier, D., and Astic, D. (1986) Analysis of the topographical organization of olfactory epithelium projections in the rat. *Brain Res. Bull.* 16:455–462.

Scoville, W. B., and Milner, B. (1957) Loss of recent memory after bilateral hippocampal lesions. *J. of Neurol. Neurosurg. Psychiat.* 20:11–21.

Shepherd, G. M. (1983) *Neurobiology*. Oxford: Oxford University Press.

Shepherd, G. M. (1991) Computational structure of olfactory system. In *Olfaction: A Model System for Computational Neuroscience*, H. Eichenbaum and J. L. Davis (eds.). Cambridge, Mass.: MIT Press, pp. 3–41.

Slotnick, B. M., and Katz, H. M. (1974) Olfactory learning-set formation in rats. *Science* 185:796–798.

Staubli, U., Fraser, D., Faraday, R., and Lynch, G. (1987) Olfactory and the "data" memory system in rats. *Behav. Neurosci.* 101:757–765.

Squire, L. (1987) *Memory and Brain*. Oxford: Oxford University Press.

Stripling, J. S., Patneau, D. K., and Gramlich, C. A. (1988). Selective long-term potentiation in the piriform cortex. *Brain Res.* 441:281–291.

Stripling, J. S., Patneau, D. K., and Gramlich, C. A. (1991) Characterization and anatomical distribution of selective long-term potentiation in the olfactory forebrain. *Brain Res.* 542:107–122.

Thompson, A. M. (1986) A magnesium-sensitive polysynaptic potential in rat cerebral cortex resembles neuronal responses to *N*-methyl-D-aspartate. *J. Physiol. (Lond.)* 370:531–549.

Thompson, R. F. (1991) Are memory traces localized or distributed? *Neuropsychologia* 29:571–582.

Tseng, G. F., and Haberly, L. B. (1988) Characterization of synaptically mediated fast and slow inhibitory processes in piriform cortex in an in vitro slice preparation. *J. Neurophysiol.* 59:1352–1376.

Tsumoto, T. (1992) Long-term potentiation and long-term depression in the neocortex. *Prog. Neurobiol.* 39:209–228.

Yoshimura, Y., Tsumoto, T., and Nishigori, A. (1991) Input-specific induction of long-term depression in Ca^{2+}-chelated visual cortex neurons. *NeuroReport* 2:393–396.

17 The Hippocampus, Long-Term Potentiation, and Memory: Enhancing the Connection

Tim Otto and Howard Eichenbaum

Among the many prominent breakthroughs in memory research was the discovery that the brain supports at least two memory systems, only one of which depends on hippocampal function. A second, equally important finding involved the discovery of a use-dependent long-term potentiation (LTP) of synaptic efficacy within the hippocampus that shares many properties with those of behaviorally defined memory. Until very recently, however, there has been little effort toward developing a detailed, cohesive synthesis of the data on LTP and the distinctions between hippocampal-dependent and hippocampal-independent memory. Toward that end, we will attempt to enhance the connection between LTP and memory processes by describing a tentative model of the way in which synaptic plasticity within hippocampal-cortical circuits might underlie hippocampal-dependent and hippocampal-independent forms of memory.

There is accumulating evidence that many features of hippocampal LTP bear at least a superficial similarity to those of behaviorally defined memory; however, the precise role played by LTP in memory function is still largely unknown. Among the qualities LTP shares with memory are rapid induction (Larson and Lynch, 1986; Diamond et al., 1988), strengthening following repetition (Bliss and Lømo, 1973; Barnes, 1979), and a gradual decay in the absence of repetition (Barnes, 1979; Racine et al., 1983). In addition, like memory, LTP demonstrates *specificity* for the inputs activated during induction (Andersen et al., 1977; Lynch et al., 1977), and can demonstrate *associativity* in that near-simultaneous activation of separate, convergent inputs produces more robust potentiation than does stimulation to either path individually (McNaughton et al., 1978). Furthermore, patterns of hippocampal neuronal activity during learning mirror the patterns of activation optimal for inducing LTP in vitro (Larson and Lynch, 1986; Diamond et al., 1988; Otto et al., 1991b), and in some situations both LTP and learning share a common dependence on activation of the NMDA receptor (Morris, 1989; Robinson et al., 1989; Gilbert and Mack, 1990).

Collectively, these data suggest that LTP and some forms of memory have common physiological and molecular profiles.

Perhaps resulting from a combination of its discovery in hippocampus and the known role of the hippocampus in some forms of memory, initial attempts to relate LTP and memory processes were characterized by an implicit yet widespread acceptance of the notion that enhancements of synaptic strength likely serve a memory function only for those forms of memory that depend critically on hippocampal circuitry. However, a careful consideration of the neuropsychological literature suggests that this view may need considerable elaboration if LTP does indeed serve as a physiological substrate for the storage of memory. Specifically, considerable data on the effects of hippocampal system damage indicate that long-term storage of some types of information can proceed in the absence of hippocampal function, suggesting that our consideration of LTP as a memory storage device must be expanded to include mechanisms by which plastic changes in extra-hippocampal brain areas support hippocampal-independent forms of learning and memory. Correspondingly, several investigators have demonstrated LTP in a number of brain areas outside the hippocampal system (Racine et al., 1983; Chapman et al., 1990; Teyler, 1989; Kanter and Haberly, 1989). Moreover, other neuropsychological data indicate that the hippocampus is not the final storage site even for those memories for which its processing is critical (see Squire et al., 1984; Eichenbaum et al., 1992). Rather, in those situations in which the hippocampal system plays a critical role in memory processing, storage within the hippocampal system likely serves as a temporary "memory buffer" while mediating the consolidation and long-term storage of those memories elsewhere. Many investigators have suggested that long-term storage of associative memories is supported by a distributed network of interconnected neuronal elements within cortical areas that are interconnected with the hippocampal system (Squire et al., 1984).

In the present chapter we propose a model of hippocampal-cortical interactions in memory in which associative LTP plays a pivotal role. We propose that hippocampal-independent memories involving representations of individual stimuli are mediated by potentiated responses within the cortical areas that process perceptual information; some simple forms of learned performance are supported entirely by these representations. However, other, more elaborate forms of performance that involve or require an organization of multiple memories depend on hippocampal processing of these individual representations. We propose that hippocampal system participation in the organization and storage of relationships among these individual representations involves two phases: during the process of consolidation the organization of these cortical representations is stored by potentiated pathways within the hippocampal system itself; then, following consolidation, these organizations are stored within cortex. A graphic, simplified conception of this model is presented in figure 17.1. At a minimum such a model would involve three successive stages: (1) A set of parallel perceptual pathways that are facilitated by use and would independently support habitual, hippocampal-independent behavioral responses to individual stimuli. (2) Projections from these separate nodal representations reach the hippocampal system; this stage involves both nodal representations and weak associative connections between them that are strengthened and temporarily stored as a result of hippocampal processing. Associative retrieval during this stage depends on hippocampal mediation of the relations between these representations. (3) Projections are directed back to the cortex from the hippocampal system and converge with incoming sensory input during repetitions of experience with the same stimuli; this coactivation results in the storage of these associations within cortex. After considerable repetition these associations can ultimately be retrieved without hippocampal involvement.

The following sections review the neuropsychological and electrophysiological evidence supporting this model, and focus primarily on our own data with respect to olfactory learning and memory in the rat. We have chosen to explore the neural substrates of olfactory memory in part because, as outlined in the next section, the anatomy of the olfactory system and its projections to the hippocampal system closely parallels that of the idealized memory circuit illustrated in figure 17.1. Subsequent sections review the neuropsychological evidence suggesting that hippocampal-dependent olfactory memory differs from hippocampal-independent olfactory memory in that the former represents the encoding of associations or relations among multiple stimuli; corresponding electrophysiological data are presented demonstrating that this relational encoding is reflected in the activity of hippocampal CA1 neurons during olfactory learning. Additional behavioral and electrophysiological data are discussed, supporting the notion that, prior to consolidation, storage of hippocampal-dependent memory likely occurs within the parahippocampal cortical areas which, together with the hippocampal formation, make up the hippocampal system (see below). Furthermore, in support of the notion that extra-hippocampal enhancements of synaptic efficacy might underlie hippocampal-independent olfactory learning, we present data from our recent studies on plasticity within the olfactory system demonstrating that cortical LTP follows training in a task in which performance is not dependent on the hippocampal system. Finally, in support of our proposal that

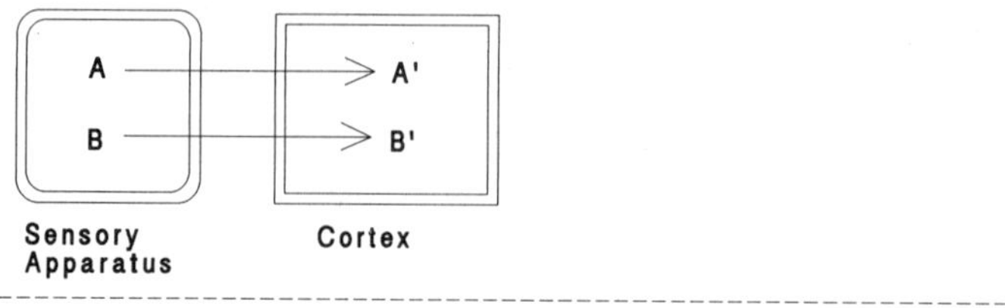

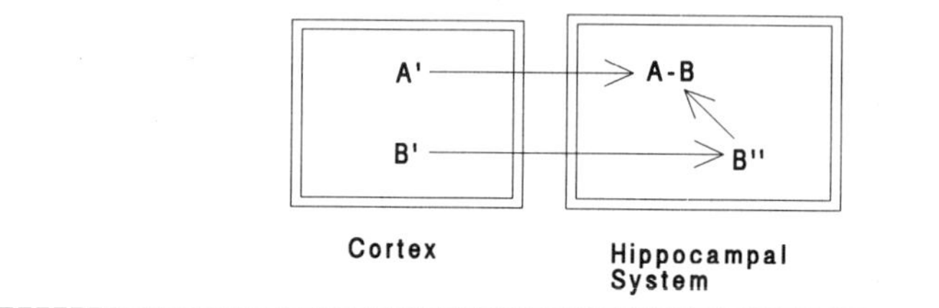

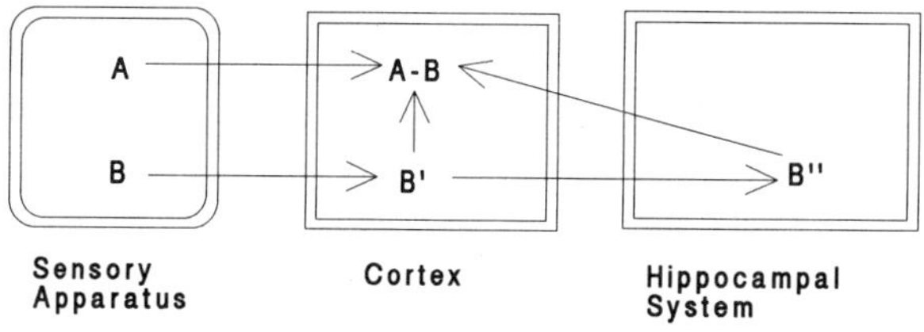

Figure 17.1 An idealized memory circuit containing the minimum connections required for encoding hippocampal-independent and hippocampal-dependent memories. For all panels, A and B refer to individual stimuli, A′, B′, and B″ refer to stored representations of those stimuli, and A-B refers to a stored association between stimuli A and B. (*Top*) Permanent, nonassociative representations of the individual stimuli can be stored in cortex without hippocampal influence. (*Middle*) The relations among multiple stimuli can be temporarily stored within the hippocampal system as a function of a hippocampally mediated strengthening of weak associative connections. (*Bottom*) The permanent storage of hippocampal-dependent associative memories results from a repetitive activation of converging sensory and hippocampal inputs to cortex.

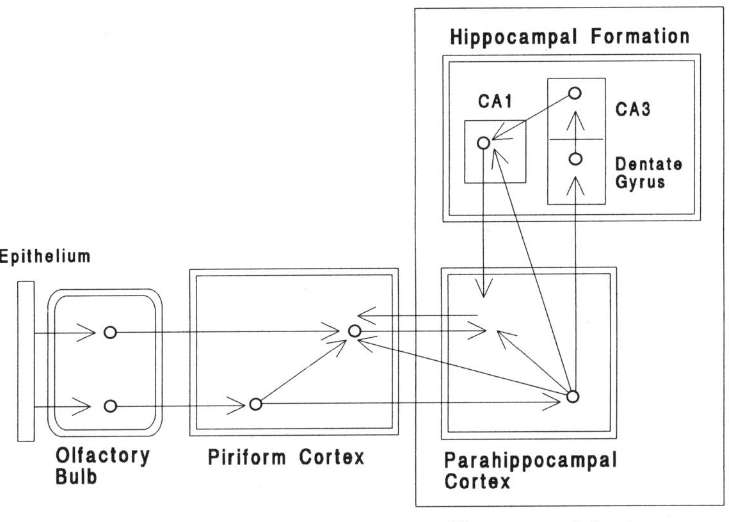

Figure 17.2 A simplified diagram of the routes of projection within the olfactory-hippocampal circuit for two hypothetical odor stimuli, with emphasis on the parallel between these connections and those of the idealized memory circuit presented in figure 17.1. Some reciprocal projections have been omitted for clarity. In accord with recent neuropsychological and anatomical data (reviewed in Eichenbaum and Otto, in press), we differentiate the *hippocampal formation*, composed of the dentate gyrus, CA3, CA1, and subiculum, from the *hippocampal system*, which in addition to the hippocampal formation includes the adjacent parahippocampal cortical areas (entorhinal and perirhinal cortex). Note that projections from the piriform cortex reach the hippocampus in as few as two synapses, and that the parahippocampal cortex is a site of convergence for primary sensory information and the output of its processing by the hippocampus. See text for additional details.

hippocampal activity can influence cortical storage, we present data indicating that the firing patterns of hippocampal output (CA1) cells parallel the stimulation parameters optimal for inducing LTP and are likely to support LTP in hippocampal targets, and that the projection from hippocampus to parahippocampal cortex is indeed capable of supporting LTP.

HYPOTHESES REGARDING THE SITES AND DURATIONS OF OLFACTORY MEMORY

As mentioned briefly above, the intrinsic anatomy of the olfactory system and its projections to the hippocampal system (figure 17.2) parallel that of the idealized minimum circuit required to account for hippocampal-independent and hippocampal-dependent memory. Moreover, this congruence permits the empirical examination of the hypotheses regarding the anatomical loci and physiological substrates of hippocampal-dependent and -independent memory alluded to above.

Several excellent reviews have detailed the intrinsic anatomy of olfactory structures and their routes of interconnection with the hippocampal system (Haberly and Price, 1978; Kosel et al., 1981; Luskin and Price, 1983; Haberly and Bower, 1989). The olfactory system is unique in that it is intimately and reciprocally connected with the hippocampal system; other sensory systems project to the hippocampal system via multiple stages of thalamic and cortical processing. The main olfactory bulb projects monosynaptically via the lateral olfactory tract (LOT) to areas along the entire extent of primary olfactory (piriform) cortex (Luskin and Price, 1982). Fibers originating from cells in both the superficial and deep layers of piriform cortex project both caudally and rostrally, forming a widely distributed intracortical associational system (Haberly and Price, 1978; Luskin and Price, 1983).

Olfactory projections to the hippocampal system provide exceptionally rapid hippocampal processing of olfactory information. The olfactory bulb monosynaptically innervates the rostral portions of the lateral entorhinal and the perirhinal cortices (Kosel et al., 1981). The caudally projecting associational fibers of the piriform cortex also project heavily to parahippocampal cortical areas including both the lateral entorhinal cortex (Haberly and Price, 1978) and the perirhinal cortex (Luskin and Price, 1983). Axonal projections arising from these parahippocampal areas constitute the primary cortical input to the hippocampal formation, projecting mainly to dentate gyrus but additionally to areas CA3 and CA1 (Witter et al., 1989).

Outputs from the hippocampal formation reach piriform cortex in as few as two synapses. Hippocampal output area CA1 sends both monosynaptic and multisynaptic (via the subiculum) projections back to cells in the superficial (Finch et al., 1986) and deep (VanGroen and Wyss, 1990; Witter et al., 1989) layers of lateral entorhinal cortex and to the perirhinal cortex (Swanson and Cowan, 1977; Witter et al., 1989; VanGroen and Wyss, 1990). These parahippocampal cortical areas project in turn back to the piriform cortex (Luskin and Price, 1983), thereby completing the loop of hippocampal system input back to cortex.

A focal point in this model is the parahippocampal cortical region, which is monosynaptically and reciprocally connected with both the piriform cortex and the hippocampal formation, and can therefore be viewed as an important site of convergence for primary sensory information and its processing by the hippocampal formation. This convergence is consistent with the view that these cortical areas are anatomically positioned to serve as storage sites for hippocampal-dependent associative olfactory memories; however, the neuropsychological data indicate that the parahippocampal region likely plays only a *temporary* storage function. On the other hand, the piriform cortex is appropriately positioned and organized for the *permanent* storage of olfactory associations. The afferent and efferent connections of the piriform cortex indicate that it is an important site of convergence of incoming sensory information and the results of sensory processing by the entire hippocampal system. This convergence, together with the elaborate intracortical associational system, render piriform cortex a likely site of long-term storage of hippocampal-dependent associative olfactory memories; indeed, the anatomical features of piriform cortex have led several investigators to conclude that it serves as a useful model for studying cortical association systems in general (Switzer et al., 1985), and cortical information storage in particular (Lynch, 1986; Haberly and Bower, 1989).

The features of the olfactory-hippocampal pathways described above provide an anatomical domain in which specific hypotheses generated by our model can be tested empirically. First, rats with hippocampal system damage should demonstrate intact acquisition of independent odor-reward associations, and the potentiation mechanisms subserving these representations should be induced without hippocampal involvement. Second, temporary storage of the associations between multiple independent cues should depend on the convergence within parahippocampal cortex of primary sensory information and its processing by the hippocampal system, and this storage should be manifest in potentiated responses within this region. Third, the projections from parahippocampal cortex back to the piriform cortex should be critical to the permanent consolidation of these associative memories.

We have explored the first two of these three related hypotheses regarding olfactory memory processing in detail. These neuropsychological and elecrophysiological studies investigating various aspects of each of these hypotheses have taken advantage of the anatomical features of the olfactory-hippocampal circuit. First, by placing selective, discrete lesions along the olfactory-hippocampal circuit, the olfactory memory processes that are dependent on the hippocampal system can be characterized and differentiated from those that can be supported by earlier stages of olfactory processing. Second, because of the relatively direct olfactory projections to the hippocampus, learning-related engagements of hippocampal cellular activity can be precisely time-locked to behavioral and paradigmatic events, thereby allowing a detailed analysis of the conditions determining hippocampal neural activity and the specific firing patterns characterizing this engagement. Third, the relatively dense, reciprocal, and convergent projections between discrete areas all along the olfactory-hippocampal circuit permit an examination of both learning- and stimulation-induced changes in monosynaptic excitatory postsynaptic potentials (EPSPs) within the piriform cortex, the parahippocampal cortex, and the hippocampus, thereby permitting an analysis of the role of synaptic enhancement in olfactory memory.

DEFINING FEATURES OF HIPPOCAMPAL-DEPENDENT OLFACTORY MEMORY

Lesions to structures within the hippocampal system can either facilitate, impair, or have no effect on learning and memory, indicating that the behavioral effects of hippocampal system damage depend critically on the nature of the memory processing demands inherent to the specific task in which they are examined. While our experiments have focused primarily on odor-guided learning and memory in rats, a review of the relevant literature indicates that the fundamental properties of hippocampal-dependent memory are the same across stimulus modalities and across species (see Eichenbaum et al., 1992). As will be discussed in detail below, neuropsychological data indicate that damage to the hippocampal system leaves intact, and in some cases facilitates, the encoding of single odor stimuli in tasks which deemphasize the encoding of relations among multiple stimuli and permit the expression of these memories in circumstances that reproduce the original learning event. By contrast, converging neuropsychological and electrophysiological data indicate that the hippocampal system participates critically in an encoding of the associations or relations among multiple, independent odor cues. Furthermore, there is accumulating evidence that the parahippocampal cortical areas serve at least a temporary memory storage function. Thus consistent with the tentative model describing the separate and interactive roles of cortex and the hippocampal system in memory, lasting representations of the simple association of individual odor stimuli and their reinforcement history can be established in the absence of normal hippocampal system function. Formation and maintenance of memories for the associations or relations among multiple odor stimuli, on the other hand, depends critically on the hippocampal system. The studies supporting this view typically involve an examination of the effects of discrete lesions of one or more of the several loci along the olfactory-hippocampal circuit on performance in odor-guided tasks that differ in the extent to which performance depends on an encoding of the relations among independent stimuli. Parallel studies also examine the functional correlates of hippocampal (CA1) unit activity during performance of the same tasks. These neuropsychological and electrophysiological data are reviewed below.

Neuropsychological Studies

One means of testing the hypothesis that the hippocampal system participates in an encoding of the relevant relations among multiple odor cues is to assess the effects of discrete lesions within the hippocampal system on variates of odor discrimination paradigms in which identical odor stimuli are used, but that differentially *encourage* or *discourage* the comparison between cues and thus favor or hinder, respectively, a representation based on relevant relations among them. A number of recent reviews describe these data in detail (Eichenbaum et al., 1992; Otto and Eichenbaum, 1992a), and they will therefore be summarized here only briefly. When explicit comparison between odor stimuli is encouraged by presenting multiple stimuli simultaneously and in close juxtaposition and requiring the subject to choose between them, damage to either of the major hippocampal afferent or efferent fiber systems (produced by lesions of the fornix or the parahippocampal region) impairs acquisition of new information (Staubli et al., 1984; Eichenbaum et al., 1988). Conversely, when comparisons between the odor cues are discouraged by presenting stimuli individually and separately across successive trials and not requiring a choice between them, the same damage produces no observable deficit and, in some cases, facilitates learning (Eichenbaum et al., 1988; Otto et al., 1991b); thus, successive-odor discriminations are retained by both normal rats and rats with hippocampal system damage for months (Otto et al., 1991a). Therefore, when accurate discrimination performance likely involves encoding the relations between cues, performance is dramatically impaired, but when performance depends simply on the assignment of reinforcement value to individual odors, hippocampal system damage fails to impair either acquisition or retention.

The hypothesis that the hippocampus participates in a representation of the associations or relations between odor stimuli is further supported by data demonstrating qualitative distinctions in the odor-memory representations of normal rats and rats with hippocampal system damage. For example, even when rats with fornix lesions succeeded in learning simultaneous olfactory discriminations, their stimulus sampling strategies indicated that they failed to compare and encode the relevant relations among odor cues: unlike normal subjects, rats with hippocampal system damage executed accurate discriminative responses without investigating each odor individually, suggesting their performance was based on separate representations for each distinct stimulus compound (i.e., each left-right juxtaposition of an odor pair) presented (Eichenbaum et al., 1989). Consistent with this interpretation, subsequent probe tests revealed that rats with fornix lesions could not recognize a set of familiar odor elements when they were presented in novel pairings. It appeared that rats with hippocampal system damage acquired an individual association for each odor-pair compound, employing a representational strategy much like that used more appropriately in successive-odor discrimination.

A recent experiment investigating the effects of parahippocampal cortical damage on odor-guided paired-associate learning in the rat (Bunsey and Eichenbaum, 1992) further supports the notion that the hippocampal system is critical to an encoding of the relations among multiple cues. Normal rats successfully learned to discriminate several paired odor-associates from a number of mispairings of the same odors. Rats with aspiration of the parahippocampal cortical areas, however, were profoundly impaired in learning the paired associates, but were unimpaired relative to controls in learning the significance of other stimulus pairs for which an association between odors was not required (figure 17.3). Thus, together with the data on simultaneous discrimination, these findings indicate that lesions of discrete components within the hippocampal system result in an impairment

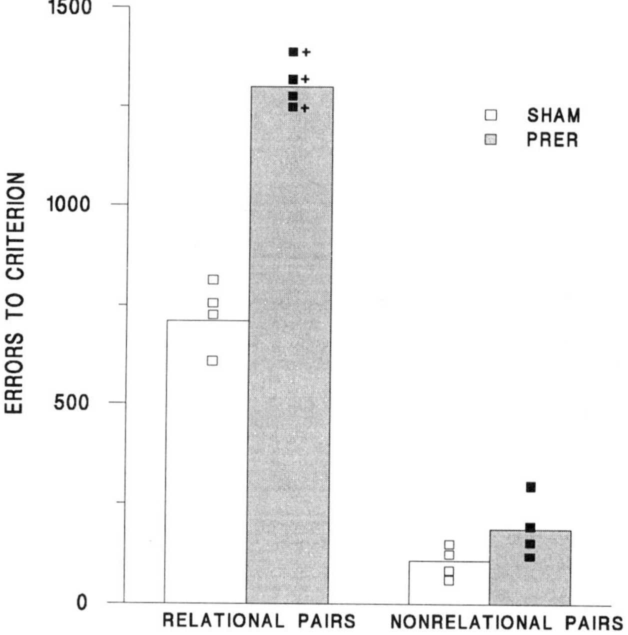

Figure 17.3 The effect of lesions of the parahippocampal cortex (PRER, perirhinal plus entorhinal cortex lesions; SHAM, operated controls) on aquisition odor-guided paired associate learning. Errors to reach a performance criterion of 80% correct within a block of 40 trials are plotted for each of two different trial types: "relational" trials in which was required, and "nonrelational" trials in which learning the significance of other stimulus pairs for which an association between odors was not required. (Data are from Bunsey and Eichenbaum, 1992)

of memories based on an encoding of the relations among multiple odor cues.

The observations described above suggest that the hippocampal formation and the associated parahippocampal cortical areas together form a functional memory system that participates selectively in the encoding and storage of memory representations based on the relations among cues, and that this system is not necessary to the acquisition of performance adaptations to individual stimuli with fixed reinforcement value. One recent study, however, revealed a behavioral dissociation between the subcomponents of this system. We (Otto and Eichenbaum, 1992b) have recently developed an odor-guided continuous delayed non-match to sample task (cDNM) that requires subjects to remember across a memory delay the odor presented on the immediately preceding trial, and respond for reinforcement only if the odor presented on the current trial is *different* (i.e., a *non-match*). Thus correct performance in this task demands a precise perceptual matching of current information to a stored representation of information presented previously, but once that matching operation is complete, the stored representation is useless with respect to predicting reinforcement on subsequent trials. Following aspiration of the parahippocampal cortical areas, rats performed as well as normal subjects when the memory delay was very brief, thus demonstrating an intact immediate memory capacity. However, these animals were markedly impaired relative to controls when longer memory delays were introduced. In contrast, after electrolytic lesions of the fornix, which significantly disrupt hippocampal activity but leave parahippocampal-neocortical communication intact, rats were unimpaired at all delays examined. This pattern of findings parallels that observed in performance of a visually guided DNMS task in monkeys with lesions of the parahippocampal cortical areas or the fornix (Gaffan et al., 1984; Zola-Morgan et al., 1989; Murray, 1991), and suggests that, unlike the hippocampal formation, the parahippocampal cortical areas are critical to the maintenance of individual stimulus representations for at least several minutes.

Collectively these data indicate that the hippocampal system is critical to an encoding of the associations among odor cues, and that learning that does not depend on such encoding can proceed independent of hippocampal function. Furthermore, it appears that the parahippocampal cortical areas can support an extended but temporary storage function for some forms of olfactory information, even in the absence of normal hippocampal formation function.

Unit Recording Studies

With the exception of the paired associate task, we have recorded the activity of single units in hippocampal area CA1 of rats during performance of each of the tasks described above. In each case, the firing repertoire of a relatively large proportion of individual hippocampal cells supports the view that the hippocampus participates in an encoding of the relations among cues. Specifically, Eichenbaum et al. (1986; see also Wiener et al., 1989) found a class of "cue-sampling cells" which fired selectively during odor sampling, but their firing repertoires clearly indicated that these cells were not simply "sensory." Rather, the firing of these cells reflected specific relations between odor cues in both the simultaneous- and the successive-cue discrimination tasks. In the sequential discrimination task these cells discharged maximally whenever the subject sampled a specific sequence of odor stimuli. In the simultaneous discrimination task firing of these cells was dependent on specific spatial configurations of the presented odors. More recently, Otto and Eichenbaum (1992c) examined the firing patterns of CA1 cells during performance in the cDNM task, and found that a large proportion of cue-sampling cells fired differentially on match versus nonmatch trials; such cells were more prevalent in sessions when performance was highly accurate. These results are consistent with the view that hippocampal cellular activity reflects the processing of comparisons between multiple cues. This is true even in tasks for which hippocampal function is not required, such as in successive discrimination and cDNM, suggesting that the hippocampus processes the comparison of discriminative stimuli even in tasks in which it does not play a critical role.

Consistent with the neuropsychological data described previously, data on the activity of single units during learning indicate that the parahippocampal cortical areas, but likely not the hippocampal formation, serve an extended but temporary storage function. Although the extent to which these parahippocampal cortical memory correlates accompany olfactory learning remains to be examined, several investigators using other species or other tasks report relatively robust "passive" memory correlates in cortex in the form of differential activity during the sampling of novel versus familiar cues. Brown and colleagues (1987; Riches et al., 1991) examined the responses of parahippocampal neurons in monkeys to repeated presentations of visual stimuli. They found that cells in parahippocampal cortical areas responded maximally to novel stimuli and exhibited relatively lower rates of firing when stimuli were repeated. Moreover, these "passive" memory correlates were observed even in instances in which there were intervening presentations of different stimuli, suggesting that these representations can withstand substantial interference. Finally, the results of a recent study by Sakai and Miyashita (1992) indicate that in monkeys performing a visually guided paired-associate task, some cells in and near the parahippocampal region fire to both elements of a paired associate. These data suggest that over the course of learning, parahippocampal cells can develop learning-dependent coding for the individual items that together make up the "pair," and are consistent with the notion that the parahippocampal areas store, at least temporarily, hippocampal-dependent memory representations.

"Passive" memory correlates may also be observed in piriform cortex during hippocampal-independent olfactory learning. Recording from well-trained rats actively engaged in learning novel (hippocampal-independent) successive odor discriminations, McCollum and colleagues (1991) found that a significant proportion of neurons in piriform cortex responded to one or more odors only during the first few trials of a session. This reduction in firing rate was strongly correlated with the acquisition of the discrimination, which typically occurred within three to five trials. Although these data might reflect simple sensory habituation processes and therefore must be viewed with caution in this context, they are superficially reminiscent of the "passive" memory correlates observed in parahippocampal cortex.

Unlike cells in the olfactory and parahippocampal cortical areas, cells within the hippocampal formation display neither "passive" memory correlates nor "active" correlates in the form of elevated firing during memory delays following stimulus presentation (Watanabe and Niki, 1985; Brown et al., 1987; Cahusac et al., 1989; Sakurai, 1990a,b; Riches et al., 1991). They do, however, fire differentially when stimuli are presented as sample or match/non-match cues in DNMS tasks (Brown et al., 1987; Riches et al., 1991; Otto and Eichenbaum, 1992c).

Combining the findings across species on "passive" memory representations as reflected in the activity of single neurons, some tentative hypotheses regarding the interactive processing of parahippocampal and hippocampal activity and the likely sites of hippocampal system–dependent memory storage can be advanced. First, hippocampal cells do not demonstrate stimulus-evoked firing that persists through a memory delay nor do they display differential firing to novel versus repeated cues, indicating that the hippocampus per se does not maintain representations of individual stimuli. Rather, the firing properties of hippocampal cells suggest that the hippocampus participates in the comparative processing of relations among stimuli stored elsewhere. Second, cells in parahippocampal cortical areas display "passive" correlates as reflected by differential activity for novel versus familiar cues; this representation is able to withstand substantial interference imposed by the presentation of intervening cues. Many parahippocampal neurons also fire preferentially to multiple items that together make an associated "pair" in the hippocampal-dependent paired associate learning task. Collectively these data suggest that the two general regions forming the hippocampal system, the hippocampal formation and the parahippocampal cortical areas, contribute differentially to stimulus comparison and storage, and are consistent with the notion that the hippocampal formation participates in the active comparison of memory representations stored in the parahippocampal cortical regions.

Hippocampal System Involvement in Olfactory Memory: Conclusions

Neuropsychological and electrophysiological data suggest that the hippocampus participates in an encoding of the relevant relations among stimuli, and provide clues regarding the likely sites of information processing and storage for both hippocampal-dependent and hippocampal-independent memories. Long-lasting retention of the significance of individual odor cues in successive discrimination was unaffected by damage to the hippocampal system, including removal of the parahippocampal cortical areas. These data are consistent with an interpretation that the representation of single items occurs outside the hippocampal system and without hippocampal mediation. A likely candidate for the storage of these single representations is piriform cortex. With respect to the cDNM task in which accurate performance depends only on a precise perceptual match of a current stimulus to the representation of one presented earlier, performance was selectively impaired in a delay-dependent manner by damage to the parahippocampal cortical areas; the hippocampal formation per se appears not to be involved in either short- or intermediate-term storage of olfactory information in this task. These data indicate that in the absence of parahippocampal cortical function, short-term odor-memory representations may be maintained by processing within the olfactory system itself. An interaction between piriform cortex and the parahippocampal cortical areas appears to be critical to intermediate-term storage and subsequent perceptual matching of this information to stimuli encountered later. Finally, the data from both the discrimination and paired associate tasks suggest that an encoding of the relations among cues depends critically on both the hippocampal formation and the parahippocampal cortical areas, that is, the hippocampal system. Temporary storage of these relational memories likely occurs not within the hippocampal formation but rather in the parahippocampal cortical areas, although the storage or retrieval processes themselves may depend upon the hippocampal formation (e.g., Winocur, 1990).

HIPPOCAMPAL-DEPENDENT AND -INDEPENDENT MEMORY: IS THERE A ROLE FOR POTENTIATION PHENOMENA?

The effects of discrete hippocampal lesions on performance in a variety of odor-guided tasks and the resultant conclusions regarding the potential loci of information processing and storage described above generate several testable predictions regarding the anatomical loci and functional role of potentiation phenomena in memory. Specifically (hippocampal-independent) successive odor discrimination learning might at a minimum be supported by enhancements in synaptic efficacy among connections within the areas of piriform cortex that code for a specific odor. Conversely (hippocampal-dependent) simultaneous odor discrimination and paired associated learning might additionally involve alterations in synaptic efficacy at the point of convergence of primary sensory and hippocampal input, that is, within the parahippocampal cortical areas. While the validity of each of these hypotheses remains to be fully explored, there is considerable data supporting the view that potentiation phenomena such as these might accompany learning. These data are reviewed below.

Learning-related Enhancements of Synaptic Efficacy within Piriform Cortex and Hippocampal-independent Olfactory Learning

The extent to which hippocampal-independent learning is accompanied by enhancements in synaptic efficacy within those areas of piriform cortex that code for a specific odor has been addressed in part by studies examining the effect of electrical stimulation of sensory afferents on EPSPs evoked in piriform cortex. Bursts of stimulation applied to the LOT have been reported to produce LTP in piriform cortex in vitro (Kanter and Haberly, 1989; Jung et al., 1990). In the awake, behaving rat, however, LTP does not normally occur following similar tetanizing stimulation (Stripling et al., 1988). Instead, LTP in the piriform cortex is found only if the electrical stimulation of the LOT serves as a learning cue (Roman et al., 1987). The results of these studies indicate that electrical stimulation of sensory afferents to the piriform cortex can serve as an olfactory stimulus in a successive discrimination task, and that this "electrical odor" discrimination learning is accompanied by an enhancement of synpatic strength within piriform cortex.

The use of electrical stimulation of olfactory structures in place of odor stimuli was originally reported by Mouly, Vigouroux, and Holley (1985), who demonstrated that stimulation of discrete patches of rat olfactory bulb served effectively as distinct, discriminable stimuli in an olfactory learning task. More recent studies have extended these findings by demonstrating that, when delivered in a pattern based on a combination of the response properties of single units in piriform cortex and the temporal characteristics of the sniffing cycle, LOT stimulation can be used effectively as an "electrical odor" during olfactory learning. Specifically, "theta-burst" stimulation, composed of short bursts of high-frequency pulses delivered to the LOT every 200 msec, produces robust sniffing reactions in rats, and can be used effectively as a discriminative stimulus during learning (Roman et al., 1987). Moreover, transfer tests indicate that this patterned stimulation is likely perceived by rats as an odor, and once learned, this "electrical odor" information is retained for extended periods as are odors (Otto et al., 1988). The question remains as to whether these "electrical odors" induce plasticity in piriform cortex, and if so, whether this plasticity is correlated with memory formation. Roman et al. (1987) were the first to report that stimulation of the LOT resulted in LTP in piriform cortex only when that stimulation served as a learning cue, and only then if the rat actually learned the significance of that cue. A similar experiment was performed by Otto et al. (1988); these data are described below.

Prior to electrical odor training, subjects were implanted bilaterally with stimulating electrodes in the LOT and recording electrodes in the superficial plexiform layer (1a) of piriform cortex; one stimulating electrode was used to deliver theta-burst stimulation during "electrical odor" training, which required a discrimination between LOT stimulation and clean air. Population EPSPs evoked in piriform cortex by single pulse stimulation of the ipsilateral LOT were collected for both hemispheres immediately before (baseline) and at several time points following electrical odor training.

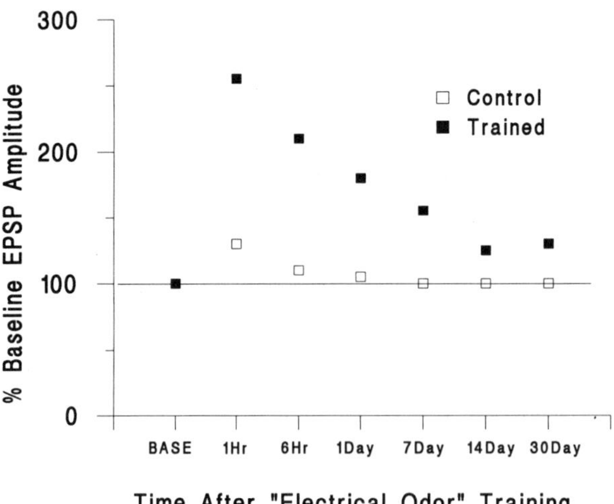

Figure 17.4 Learning-dependent LTP in piriform cortex following patterned stimulation of the LOT.

Immediately after reaching a performance criterion of 90% correct, the same number of theta-bursts as was delivered to the training electrode during "electrical odor" training were then delivered to the contralateral electrode. Thus each hemisphere received the same number and type of stimuli, but in one case the reward significance of these stimuli was learned and presumably retained, while in the other there was no obvious significance attached to the stimuli. As illustrated in figure 17.4, behaviorally irrelevant theta stimulation of the control electrode induced small yet significant synaptic potentiation in ipsilateral piriform cortex which decayed within 24 hr. In marked contrast, the potentiation observed following "electrical odor" training was more robust and persistent, strongly suggesting that synaptic modifications in piriform cortex of sufficiently long duration to account for behavioral memory accompany a form of olfactory learning for which hippocampal processing is not required.

The Activity of Hippocampal Output (CA1) Cells

The model memory circuit illustrated in figures 16.1 and 16.2 suggests a role for synaptic potentiation within the hippocampal system in hippocampal-dependent olfactory learning. Specifically, we have hypothesized that the temporary, preconsolidation storage of the associations and relations among odors is stored within the hippocampal system and should be manifest in an enhancement of synaptic efficacy within one or more hippocampal system components. We have recently begun to address this hypothesis by examining the extent to which the firing patterns of hippocampal CA1 cells during hippocampal-dependent simultaneous odor discrimination coincide with the patterns of activation optimal for inducing hippocampal LTP; as described below, the results indicate that the hippocampus discharges in patterns that might indeed support a naturally occurring analog of experimentally induced LTP.

While in hippocampus LTP can be elicited over a wide range of stimulation parameters, recent studies indicate that LTP is induced preferentially by electrical stimulation which is based on three separate but related patterns of naturally occurring cell activity. LTP in CA1 is preferentially induced by high-frequency bursts of stimulation (4 pulses at 100 Hz) repeated at 5–10 Hz (Larson et al., 1986), and can be induced by a single burst if another burst (Larson and Lynch, 1986) or single pulse (Rose and Dunwiddie, 1986) precedes that burst by 130–200 msec (i.e., at latencies that parallel frequencies of 5–7 Hz; "priming"). Finally, patterned stimulation is most effective in inducing LTP in dentate gyrus when delivered at the peak of the dentate theta rhythm (Pavlides et al., 1988). Together, these data suggest that brief episodes of high-frequency stimulation, when applied in the appropriate temporal relationship to prior activity and to the ongoing theta rhythm, can reliably enhance synaptic efficacy within the hippocampus.

A recent study by Otto et al. (1991b) found that all three of these characteristics of patterned stimulation, together optimal for hippocampal LTP induction, occur simultaneously, *selectively during episodes of mnemonic processing* (figure 17.5). Examination of the firing patterns of CA1 "complex spike" cells (Ranck, 1973) in rats engaged in learning hippocampal-dependent simultaneous odor discriminations revealed that these cells indeed discharged in high-frequency bursts, phase-locked to the positive peak of the dentate theta rhythm. Furthermore, these bursts were preceded by neural activity preferentially at intervals corresponding to the theta rhythm. Importantly, these patterns emerged only during significant behavioral events associated with likely periods of stimulus analysis, selection, or storage. Thus the optimal conditions for hippocampal LTP induction are indeed present when animals actively engage hippocampal processing for putative mnemonic functions.

The functional significance of these learning-related patterns of hippocampal unit activity remains to be determined; we can, however, offer two suggestive hypotheses regarding their potential involvement in memory. One possibility is that theta-bursting by CA1 pyramids results in the induction of LTP in efferent cortical targets of the hippocampus, and that the potentiation so induced represents a physiological substrate of memory encoding. Consistent with this hypothetical account are the neuropsychological and electrophysiological data presented previously, and the observation of LTP in parahippocampal cortex (Otto and Eichenbaum, 1993) and frontal cortex (Laroche et al., 1990) following stimulation of ipsilateral CA1. Alternatively, or perhaps in addition, theta-bursting in CA1 might reflect the induction of LTP in other CA1 cells innervated by axon collaterals (cf. Martin et al., 1992). Finally, these patterns could potentially reflect the induction of LTP within the very cells demonstrating theta-bursting, resulting from afferent activation by cells in CA3. The latter two accounts are consistent with the suggestion that LTP in CA1 may act as a temporary memory store, or "buffer," prior to permanent storage of that information elsewhere (Rawlins, 1985), as a working memory store (see above), or as an "index" to loci of information stored permanently in cortex (Teyler and DiScenna, 1987).

The relationship between cycles of olfactory information acquisition and its processing by the hippocampus are consistent with the hypothesis that the parahippocampal cortex is a likely site of learning-related, natu-

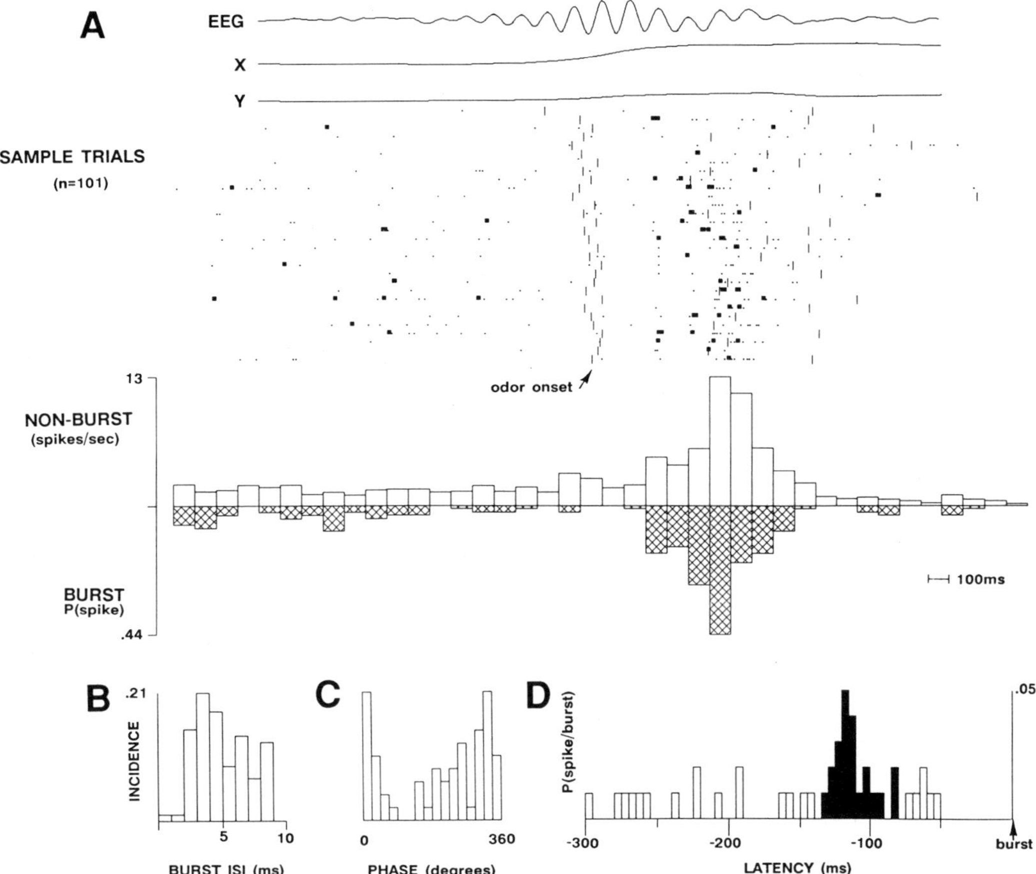

Figure 17.5 Analysis of the activity of a cue-sampling cell recorded during hippocampal-dependent simultaneous olfactory discrimination learning. Each of the three patterns of activation which together are optimal for the induction of hippocampal LTP are reflected in the firing of this cell. (*A*) EEG, movement, and hippocampal unit activity across trials time-locked to the first dentate theta peak after odor onset. EEG was first normalized then averaged across trials; movement was averaged across trials as well. EEG (upward deflections reflect positivity) was first normalized within the 4-sec sampling period, then averaged across trials. Separate X and Y movement traces respectively indicate averaged head movements toward the sampling ports and to the left (up) or right (down). Rasters show single spike (small dots) and burst (large dots) for a subset of analyzed trials. Note that spike and burst activity are aligned to the cycles of averaged EEG across trials. Histograms indicate the incidence of single spikes (open bars) and bursts (shaded bars). P(spike) refers to the probability of burst-related spike activity within each 100 msec bin. (*B*) Distribution of within-burst interspike intervals. (*C*) Distribution of the incidence of burst onsets across phases of the theta cycle. Zero and 360° indicate successive positive peaks of dentate theta rhythm. (*D*) Distribution of latencies of unit activity prior to identified bursts. Darkened bars indicate latencies corresponding to 7.0–12.5 Hz. P(spike/burst) refers to the probability of a spike occurring during each 10-msec bin prior to an identified burst. (From Otto et al., 1992b, by permission)

rally occurring synaptic enhancement. Specifically, during odor sampling, the sniffing cycle becomes precisely phase-locked with the hippocampal theta rhythm such that cycles of inhalation precede theta peaks by approximately 140–200 msec, or one theta cycle (Macrides et al., 1982). These data suggest that in some senses cycles of data acquisition may be considered to "drive" the cycles of its processing by the hippocampus. Moreover, during odor sampling, pyramidal cells in piriform cortex frequently discharge in complex-spike "bursts" similar to those found in CA1 (McCollum et al., 1991). When considered in light of the route of olfactory projections to and from the hippocampal formation, these data indicate that parahippocampal cortex likely receives convergent, highly entrained bursts of activity resulting from the coding properties of primary sensory cortex on the one hand and the hippocampal processing of the coded odor on the other. These tightly time-locked bursts of convergent activity are consistent with the conditions in which "associative" LTP is induced, and represent a highly speculative but nonetheless intriguing possibility concerning the role of theta-bursting in the encoding of hippocampal-dependent associations between multiple olfactory stimuli.

Hippocampal Projections to the Parahippocampal Cortex Support LTP

A test of the hypothesis that synaptic potentiation within the parahippocampal cortex serves a temporary storage function for hippocampal-dependent olfactory memories requires at a minimum that this area display LTP following electrical stimulation of the hippocampal formation. A recent study (Otto and Eichenbaum, unpublished) suggests that electrical stimulation of the hippocampus that mimics the endogenous, learning-related discharge patterns in CA1 can indeed induce LTP in the parahippocampal cortex. Briefly, a bipolar stimulating electrode was implanted in the dorsal hippocampus near the CA1-subiculum border, and a recording electrode was positioned in the deep layer of ipsilateral posterior perirhinal cortex so as to produce a small but reliable monosynaptic EPSP. Single-pulse stimuli of an intensity sufficient to evoke and EPSP of roughly half its maximum amplitude (approximately 0.5 mV at 250 μA) were delivered to the hippocampus every 15 sec, and the slope of the initial descending phase of the resultant cortical EPSP was measured. Following establishment of a stable baseline, "theta-burst" stimulation (2 bouts of 10 bursts [4 pulses at 100 Hz], bursts separated by 200 msec [5 Hz], 15 sec between bouts, stimulation intensity doubled) was delivered to the hippocampus. Theta-burst stimulation produced a stable LTP in perirhinal cortex

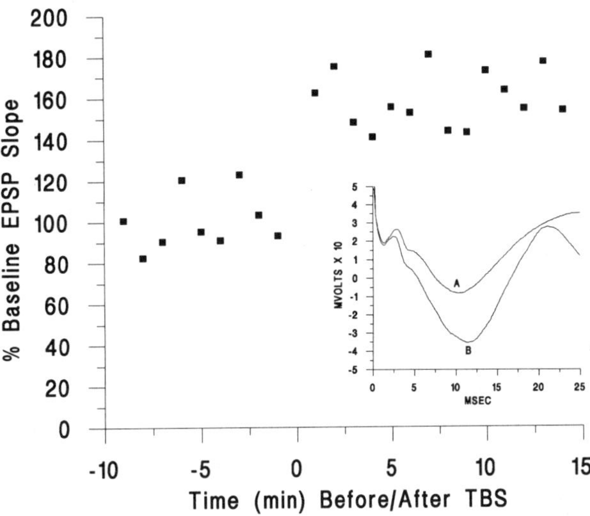

Figure 17.6 Long-term potentiation in perirhinal cortex following theta-burst stimulation (TBS) of the ipsilateral CA1 subfield of the hippocampus. *Scatterplot*: Each point represents the average of four consecutive EPSPs, and is plotted as a function of the percentage of the overall average baseline EPSP slope. *Inset*: (*A*) Baseline EPSP waveform averaged over the entire baseline recording period. (*B*) EPSP waveform averaged across the final five minutes of recording following TBS.

as indicated by a reliable increase in the slope of the EPSP evoked by subsequent single-pulse stimulation (figure 17.6).

A number of important issues remain to be resolved regarding both the physiological basis and functional significance of LTP in the hippocampal-parahippocampal pathway. These include examining whether LTP in this pathway involves cellular mechanisms similar to those critical to LTP induction in hippocampus and elsewhere, notably dependence on activation of the NMDA receptor, and whether naturally occurring enhancements of synaptic efficacy in this pathway accompany learning which is itself dependent on the integrity of the parahippocampal region. Consistent with the hypothesis that LTP in the hippocampal-parahippocampal pathway depends on activation of the NMDA receptor are data indicating high concentrations of this receptor in the parahippocampal cortical areas (Monaghan and Cotman, 1985). There are no data at present bearing on the functional significance of LTP within the parahippocampal area. However, the convergence of both primary olfactory and hippocampal formation outputs to parahippocampal cortex suggests, consistent with the neuropsychological data described previously and the notion advanced throughout the present treatment, that the parahippocampal cortex might serve as the site of preconsolidation memory storage for hippocampal-dependent olfactory memory.

Does LTP within the Hippocampal Formation Serve a Role in Information Storage?

The data outlined above are consistent with the hypothesis that memory storage within the hippocampal system may be subserved by a naturally occurring analog of experimentally induced LTP in the parahippocampal cortex. This interpretation does not, however, preclude the possibility that LTP within the hippocampal formation is *necessarily uninvolved* in the acquisition and temporary storage of information. Indeed, a recent study by Winocur (1990) suggests that in rats damage limited to the dorsal hippocampus results in a retrograde memory gradient. Thus it is possible that some forms of memory are temporarily stored within the hippocampal formation prior to their ultimate consolidation in other, presumably cortical, areas, and that the electrophysiological manifestation of this storage has escaped detection in previous electrophysiological studies of the firing properties of hippocampal neurons. Two approaches have been developed to address the issue of the specific involvement of the hippocampal formation in temporary memory storage; these are discussed below.

One experimental approach used in investigating the role of hippocampal LTP in memory storage involves parallel behavioral and electrophysiological assessments of the effect of pharmacological antagonism of the NMDA receptor, a subclass of the glutamate receptor known to play a critical role in the induction of LTP in hippocampus (Collingridge et al., 1983), using selective antagonists such as APV and MK-801. One particularly elegant example of a study of this type was reported recently by Morris (1989), who, using APV, found that the dose-response curves for blocking hippocampal LTP induction and learning in the hippocampal-dependent, spatially cued water maze task were virtually identical. However, while these studies suggest that both hippocampal LTP and acquisition of hippocampal-dependent learning may share a common chemistry, they do not demonstrate a necessary role for *LTP within hippocampus per se* in learning. NMDA receptors are found in relatively high concentrations not only within the hippocampus but throughout the brain, including the parahippocampal cortical areas and other areas outside the hippocampal system (Monaghan and Cotman, 1985). It is therefore plausible that the effects of NMDA antagonism on hippocampal LTP and on learning are parallel processes that are not causally linked, and that the behavioral effects are mediated by blockade of NMDA-dependent LTP in nonhippocampal circuits.

A second approach to addressing the role of hippocampal synaptic enhancement in information processing entails the search for naturally occurring, LTP-like changes in hippocampal circuits that accompany exploration. Sharp, McNaughton, and Barnes (1985, 1989; see also Green et al., 1990) reported that brief bouts of exploration in a novel environment resulted in short-term (approximately 15 min) enhancements of efficacy in the perforant path–dentate gyrus pathway, consistent with a role for changes of synaptic in information processing (but see Hargreaves et al., 1990). More recent data, however, indicate that these activity-dependent enhancements are likely due to changes in brain temperature (Moser et al., 1992). Perhaps more convincing are the findings indicating that in vitro hippocampal slices taken from rats reared in complex, enriched environments display significantly larger EPSPs and population spikes than those from rats reared in isolation (Green and Greenough, 1986). Equally interesting are the observations of naturally occurring changes in synaptic efficacy that accompany explicit bouts of nonspatial learning. Skelton and colleagues (1987) demonstrated long-lasting (10 days) enhancements of synaptic efficacy in the dentate gyrus following acquisition of an operant task. Although the interpretation of these data is complicated by the fact that these behaviors are unaffected by hippocampal lesions, they suggest that hippocampal synaptic enhancement may be involved in the acquisition and storage of specific information. Alternatively, the observed enhancement may reflect the effects of use-dependent mechanisms that are unrelated to information storage per se. Future studies examining naturally occurring potentiation phenomena within both intra- and extrahippocampal pathways and their relationship to hippocampal-dependent and -independent learning will help to clarify the role of synaptic enhancement in learning and memory.

SUMMARY AND FORMAL STATEMENT OF THE SPECULATIVE MODEL

Synthesis of the neuropsychological and electrophysiological data described herein leads to the following speculative conclusions regarding the sites and durations of various forms of olfactory memory, and the nature, locus, and duration of the potentiation phenomena on which these various forms of memory may depend. First, memory for relatively permanent assignments of significance for individual, isolated stimuli, such as in the successive-cue discrimination task, depends neither on processing within parahippocampal cortical areas nor within the hippocampus proper. Rather, these performance adaptations are mediated by stimulus representations stored in piriform cortex, or by interactions between these and other nonhippocampal representations. Hippocampal system–independent representations would likely take the form of enhancements of responsiveness of the specific areas of piriform cortex that code for the to-be-learned odor cue. By this account, the potentiation so induced should be learning-dependent, long-lasting, and limited to the area that was initially activated by the learned odor. Precise perceptual matching of a current stimulus to a representation of one presented earlier, such as in the cDNM task, might also occur at the level of piriform cortex provided a relatively short retention period; retention periods extending beyond the capacity of piriform cortex may require the additional, interactive processing of adjacent parahippocampal cortex. These short- and intermediate-term representations, encoded in the absence of participation by the hippocampus per se, might be supported by short-term potentiation phenomena within parahippocampal cortex, or alternatively by a "reverberating circuit" established within the reciprocal connections between piriform cortex and the lateral entorhinal cortex.

Second, associations between or organizations of individual stimuli depend on both the parahippocampal cortex and the hippocampal formation. Prior to their permanent consolidation, these relational representations are likely stored not within the hippocampal formation itself but rather within the parahippocampal cortical areas. These hippocampal-dependent relational memories might be encoded within the parahippocampal cortex by virtue of an interaction between direct sensory input and hippocampal output, such that a convergence of the two sources of information is required for lasting alterations of synaptic efficacy. Synaptic enhancement within this system would be limited anatomically to the specific connections between the piriform associational system and the parahippocampal cortex that were involved in the detection and relation of the remembered stimuli, and limited functionally to a role in the temporary, preconsolidation storage of hippocampal-dependent memories.

Finally, the permanent encoding of the relationships among multiple cues in piriform cortex might involve a repeated activation of the reciprocal projections between piriform and parahippocampal cortex, and rely on mechanisms conceptually similar to those putatively accounting for memory storage within parahippocampal cortex prior to consolidation. This final stage of memory formation is currently the least well understood and remains to be addressed experimentally.

These properties might be instantiated in the olfactory-hippocampal circuit in the following way. The encoding of an individual odor stimulus, as in hippocampal-independent successive discrimination learning, might occur within piriform cortex itself as a function of the convergence of primary sensory (bulbar) afferent activation and as yet unspecified afferent activity reflecting reinforcement. The encoding of the association among multiple odors, as in hippocampal-dependent simultaneous discrmination or paired associate learning, might occur at the level of the parahippocampal cortex as a function of precisely time-locked afferent activity originating in piriform cortex, within the parahippocampal cortex, and the CA1 field of hippocampus. By this scheme, a relatively rapid switching between sniffing two separate odors might result in strong activation among the neural ensembles coding for each odor within parahippocampal cortex. This activation, combined with the concommitant, time-locked theta-bursting afferent activation from hippocampal output (CA1) cells, might then result in potentiation of not only the primary sensory afferent connections themselves but additionally the associational connections between ensembles. This mechanism might then account for the ultimate psychological and electrophysiological association between stimuli observed during paired associate and simultaneous odor discrimination learing.

ACKNOWLEDGMENTS

Preparation of this manuscript was supported in part by ONR grant N0014-91-J-1881 to H. E. and T. O., and NIH grant NS26402 and NIA grant AG09973 to H. E. We wish to thank Michael Bunsey for helpful discussions of the material presented herein.

REFERENCES

Andersen, P., Sundberg, S. H., Sveen, O., and Wigström, H. (1977) Specific long-lasting potentiation of synaptic transmission in hippocampal slices. *Nature* 226:736–737.

Barnes, C. A. (1979) Memory deficits associated with senescence: A neurophysiological and behavioral study in the rat. *J. Comp. Physiol. Psychol.* 93:74–104.

Bliss, T. V. P., and Lømo, T. (1973) Long-lasting potentiation of synaptic transmission in the dentate area of the anaesthetized rabbit following stimulation of the perforant path. *J. Physiol.* 232:331–356.

Brown, M. W., Wilson, F. A. W., and Riches, I. P. (1987) Neuronal evidence that inferomedial temporal cortex is more important than hippocampus in certain processes underlying recognition memory. *Brain Res.* 409:158–162.

Bunsey, M., and Eichenbaum, H. (1992) Paired associate learning in rats: Critical involvement of the perirhinal-entorhinal cortex. *Soc. Neurosci. Abstr.* 18:1059.

Cahusac, P. M. B., Miyashita, Y., and Rolls, E. T. (1989) Responses of hippocampal formation neurons in the monkey related to delayed spatial response and object-place memory tasks. *Behav. Brain Res.* 33:229–240.

Chapman, P. F., Kairiss, E. W., Keenan, C. L., and Brown, T. H. (1990) Long-term synaptic potentiation in the amygdala. *Synapse* 6:271–278.

Collingridge, G. L., Kehl, S., and McLennan, H. (1983) Excitatory amino acids in synaptic transmission in the Schaffer collateral pathway of the rat hippocampus. *J. Physiol.* 334:33–46.

Diamond, D. M., Dunwiddie, T. V., and Rose, G. M. (1988) Characteristics of hippocampal primed burst potentiation in vitro and in the awake rat. *J. Neurosci.* 8:4079–4088.

Eichenbaum, H., Fagan, A., and Cohen, N. J. (1986) Normal olfactory discrimination learning set and facilitation of reversal learning after combined and separate lesions of the fornix and amygdala in rats: Implications for preserved learning in amnesia. *J. Neurosci.* 6:1876–1884.

Eichenbaum, H., Fagan, A., Mathews, P., and Cohen, N. J. (1988) Hippocampal system dysfunction and odor discrimination learning in rats: Impairment or facilitation depending on representational demands. *Behav. Neurosci.* 102:3531–3542.

Eichenbaum, H., Mathews, P., and Cohen, N. J. (1989) Further studies of hippocampal representation during odor discrimination learning. *Behav. Neurosci.* 103:1207–1216.

Eichenbaum, H., Otto, T., and Cohen, N. J. (1992) The hippocampus—What does it do? *Behav. Neural Biol.* 57:1–35.

Finch, D. M., Wong, E. E., Derian, E. L., and Babb, T. L. (1986) Neurophysiology of limbic system pathways in the rat: Projections from the subicular complex and hippocampus to the entorhinal cortex. *Brain Res.* 397:205–213.

Gaffan, D., Gaffan, E. A., and Harrison, S. (1984) Reversal learning by fornix-transected monkeys. *Q. J. Exp. Psychol.* 36B:223–234.

Gilbert, M. E., and Mack, C. M. (1990) The NMDA antagonist MK-801 suppresses long-term potentiation, kindling, and kindling-induced potentiation in the perforant path of the unanesthetized rat. *Brain Res.* 519:89–96.

Green, E. J., and Greenough, W. T. (1986) Altered synaptic transmission in dentate gyrus of rats reared in complex environments: Evidence from hippocampal slices maintained in vitro. *J. Neurophysiol.* 55:739–750.

Green, E. J., McNaughton, B. L., and Barnes, C. A. (1990) Exploration-dependent modulation of evoked responses in fascia-dentata: Dissociation of motor, EEG, and sensory factors and evidence for a synaptic efficacy change. *J. Neurosci.* 10:1455–1471.

Haberly, L. B., and Bower, J. M. (1989) Olfactory cortex: Model circuit for study of associative memory? *Trends Neurosci.* 12:258–264.

Haberly, L. B., and Price, J. L. (1978) Association and commissural fiber systems of the olfactory cortex of the rat. I. Systems originating in the piriform cortex and adjacent areas. *J. Comp. Neurol.* 178:711–740.

Hargreaves, E. L., Cain, P. P., and Vamderwolf, C. H. (1990) Learning and behavioral-long-term-potentiation: Importance of Controlling for motor activity. *J. Neuros.* 10:1472– 1478.

Jung, M. W., Larson, J., and Lynch, G. (1990) Long-term potentiation of monosynaptic EPSPs in rat piriform cortex in vitro. *Synapse* 6:279–283.

Kanter, E. D., and Haberly, L. B. (1989) APV dependent induction of long-term potentiation in piriform (olfactory) cortex slices. *Soc. Neurosci. Abstr.* 15:929.

Kosel, K. C., Van Hoesen, G. W., and West, J. R. (1981) Olfactory bulb projections to the parahippocampal area of the rat. *J. Comp. Neurol.* 198:467–482.

Laroche, S., Jay, T. M., and Thierry, A. M. (1990) Long-term potentiation in the prefrontal cortex following stimulation of the hippocampal CA1/subicular region. *Neurosci. Lett.* 114:184–190.

Larson, J., and Lynch, G. (1986) Induction of synaptic potentiation in hippocampus by patterned stimulation involves two events. *Science* 232:985–988.

Larson, J., Wong, D., and Lynch, G. (1986) Patterned stimulation at the theta frequency is optimal for the induction of hippocampal long-term potentiation. *Brain Res.* 368:347–350.

Luskin, M. B., and Price, J. L. (1982) The distribution of axon collaterals from the olfactory bulb and the nucleus of the horizontal limb of the diagonal band to the olfactory cortex demonstrated by double retrograde labeling techniques. *J. Comp. Neurol.* 209:249–263.

Luskin, M. B., and Price, J. L. (1983) The topographic organization of associational fibers of the olfactory system in the rat, including centrifugal fibers to the olfactory bulb. *J. Comp. Neurol.* 216:264–291.

Lynch, G. (1986) *Synapses, Circuits, and the Beginnings of Memory*. Cambridge, Mass.: MIT Press.

Lynch, G., Dunwiddie, T., and Gribikoff, V. (1977) Heterosynaptic depression: A postsynaptic correlate of long-term potentiation. *Nature* 266:737–739.

Macrides, F., Eichenbaum, H., and Forbes, W. B. (1982) Temporal relationship between sniffing and limbic theta rhythm during odor discrimination reversal learning. *J. Neurosci.* 2:1705–1717.

Martin, P. D., Lake, M., and Shapiro, M. L. (1992) Effects of burst stimulation on neighboring single CA1 neurons in rat hippocampus. *Soc. Neurosci. Abstr.* 18:1348.

McCollum, J., Larson, J., Otto, T., Schottler, F., Granger, R., and Lynch, G. (1991) Short-latency single unit processing in the olfactory cortex. *J. Cog. Neurosci.* 3:293–299.

McNaughton, B. L., Douglas, R. M., and Goddard, G. V. (1978) Synaptic enhancement in fascia dentata: Cooperativity among coactive afferents. *Brain Res.* 157:277–293.

Monaghan, D. T., and Cotman, C. W. (1985) Distribution of *N*-methyl-D-aspartate sensitive L-[^{3}H]-glutamate binding sites in rat brain. *J. Neurosci.* 5:2909–2919.

Morris, R. G. M. (1989) Synaptic plasticity and learning: selective impairment of learning in rats and blockade of long-term potentiation in vivo by the *N*-methyl-D-aspartate receptor antagonist AP5. *J. Neurosci.* 9:3040–3057.

Moser, E. I., Mathiesen, I., and Andersen, P. (1992) The short-term exploratory modulation (STEM) of perforant path synapses reflects increased brain temperature. *Science* 259:1324–1326.

Mouly, A. M., Vigouroux, M., and Holley, A. (1985) On the ability of rats to discriminate between microstimulations of the olfactory bulb in different locations. *Behav. Brain Res.* 17:45–58.

Murray, E. A. (1991) Medial temporal lobe structures contributing to recognition memory: The amygdaloid complex versus the rhinal cortex. In *The Amygdala: Neurobiological Aspects of Emotion, Memory, and Mental Dysfunction*, J. P. Aggleton (ed.). New York: Wiley, pp. 453–470.

Otto, T., and Eichenbaum, H. (1992a) Toward a comprehensive account of hippocampal function: Studies of olfactory learning permit an integration of data across multiple levels of neurobiological analysis. In *Neuropsychology of Memory*, 2nd Edition, L. R. Squire and N. Butters (eds.). New York: Guilford.

Otto, T., and Eichenbaum, H. (1992b) Complementary roles of orbital prefrontal cortex and the hippocampal system in an odor-guided delayed non-matching to sample task. *Behav. Neurosci.* 106:762–775.

Otto, T., and Eichenbaum, H. (1992c) Neuronal activity in the hippocampus during delayed non-match to sample performance in rats: Evidence for hippocampal processing in recognition memory. *Hippocampus* 2:323–334.

Otto, T., and Eichenbaum, H. (1993) Long-term potentiation in entorhinal cortex induced by theta-pattern stimulation of CA1. *Soc. Neurosci. Abstr. 19.*

Otto, T., Staubli, U., and Lynch, G. (1988) Theta stimulation of the lateral olfactory tract serves as a cue in a successive-cue olfactory discrimination task. *Soc. Neurosci. Abstr.* 14:568.

Otto, T., Schottler, F., Staubli, U., Eichenbaum, H., and Lynch, G. (1991a) The hippocampus and olfactory discrimination learning: Effects of entorhinal cortex lesions on learning-set acquisition and on odor memory in a successive-cue, go/no-go task. *Behav. Neurosci.* 105:111–119.

Otto, T., Eichenbaum, H., Wiener, S. I., Wible, C. G. (1991b) Learning-related patterns of CA1 spike trains parallel stimulation parameters optimal for inducing hippocampal long term potentiation. *Hippocampus* 1:181–192.

Pavlides, C., Greenstein, Y. J., Grudman, M., and Winson, J. (1988) Long-term potentiation in the dentate gyrus is induced preferentially on the positive phase of theta rhythm. *Brain Res.* 439:383–387.

Racine, R. J., Milgram, N. W., and Hafner, S. (1983) Long-term potentiation phenomena in the rat limbic forebrain. *Brain Res.* 260:217–231.

Ranck, J. B., Jr. (1973) Studies on single neurons in the dorsal hippocampal formation and septum in unrestrained rats. *Exp. Neurol.* 41:461–555.

Rawlins, J. N. P. (1985) Associations across time: The hippocampus as a temporary memory store. *Brain Behav. Sci.* 8:479–496.

Riches, I. P., Wilson, F. A. W., and Brown, M. W. (1991) The effects of visual stimulation and memory on neurons of the hippocampal formation and the neighboring parahippocampal gyrus and inferior temporal cortex of the primate. *J. Neurosci.* 11:1763–1779.

Robinson, G. S., Crooks, G. B., Shinkman, P. G., and Gallagher, M. (1989) Behavioral effects of MK-801 mimic deficits associated with hippocampal damage. *Psychobiology* 17: 156–164.

Roman, F., Staubli, U., and Lynch, G. (1987) Evidence for synaptic potentiation in a cortical network during learning. *Brain Res.* 418:221–226.

Rose, G.M., and Dunwiddie, T. V. (1986) Induction of hippocampal long-term potentiation using physiologically-patterned stimulation. *Neurosci. Lett.* 69:244–248.

Sakai, K., and Miyashita, Y. (1992). Neural organization for the long-term memory of paired associates. *Nature* 354:152–155.

Sakurai, Y. (1990a) Cells in the rat auditory system have sensory-delay correlates during the performance of an auditory working memory task. *Behav. Neurosci.* 104:856–868.

Sakurai, Y. (1990b) Hippocampal cells have behavioral correlates during the performance of an auditory working memory task in the rat. *Behav. Neurosci.* 104:253–263.

Sharp, P. E., McNaughton, B. L., and Barnes, C. A. (1985) Enhancement of hippocampal field potentials in rats exposed to a novel, complex environment. *Brain Res.* 339:361–365.

Sharp, P. E., McNaughton, B. L., and Barnes, C. A. (1989) Exploration-dependent modulation of evoked responses in fascia dentata: Fundamental observations and time course. *Psychobiology* 17:257–269.

Skelton, R. W., Scarth, A. S., Wilkie, D. M., Miller, J. J., and Phillips, A. G. (1987) Long-term increases in dentate granule cell responsivity accompany operant conditioning. *J. Neurosci.* 7:3081–3087.

Squire, L. R., Cohen, N. J., and Nadel, L. (1984) The medial temporal region and memory consolidation: A new hypothesis. In *Memory Consolidation*, H. Weingartner and E. Parker (eds.). Hillsdale, N.J.: Erlbaum.

Staubli, U., Ivy, G., and Lynch, G. (1984) Hippocampal denervation causes rapid forgetting of olfactory information in rats. *Proc. Nat. Acad. Sci. USA* 81:5885–5887.

Stripling, J. S., Patneau, D. K., and Gramlich, C. A. (1988) Selective long-term potentiation in the piriform cortex. *Brain Res.* 441:281–291.

Swanson, L. W., and Cowan, W. M. (1977) An autoradiographic study of the efferent connections of the hippocampal formation in the rat. *J. Comp. Neurol.* 172:49–84.

Switzer, R. C., de Olmos, J., and Heimer, L. (1985) Olfactory system. In *The Rat Nervous System. Vol. 1. Forebrain and Midbrain*, G. Paxinos (ed.). Orlando, Fa.: Academic Press.

Teyler, T. J. (1989) Comparative aspects of hippocampal and neocortical long-term potentiation. *J. Neurosci. Meth.* 28:101–108.

Teyler, T. J., and DiScenna, P. (1987) Long-term potentiation. *Annu. Rev. Neurosci.* 10:131–161.

VanGroen, T., and Wyss, J. M. (1990) Extrinsic projections from area CA1 of the rat hippocampus: Olfactory, cortical, subcortical, and bilateral hippocampal formation projections. *J. Comp. Neurol.* 303:1–14.

Watanabe, T., and Niki, H. (1985) Hippocampal unit activity and delayed response in the monkey. *Brain Res.* 325:241–254.

Wiener, S. I., Paul, C. A., and Eichenbaum, H. (1989) Spatial and behavioral correlates of hippocampal neuronal activity. *J. Neurosci.* 9:2737–2763.

Winocur, G. (1990) Anterograde and retrograde amnesia in rats with dorsal hippocampal or dorsomedial thalamic lesions. *Behav. Brain Res.* 38:145–154.

Witter, M. P., Groenewegen, H. J., Lopes Da Silva, F. H., and Lohman, A. H. M. (1989) Functional organization of the extrinsic and intrinsic circuitry of the parahippocampal region. *Prog. Neurobiol.* 33:161–253.

Zola-Morgan, S., Squire, L. R., and Amaral, D. G. (1989) Lesions of the hippocampal formation but not lesions of the fornix or mammillary nuclei produce long-lasting memory impairment in the monkey. *J. Neurosci.* 9:898–913.

V Synaptic Plasticity and Computational Neurobiology

18 A Hierarchical Model Derived from an n-Level Field Theory to Study the Effects of Long-Term Potentiation on System Properties of the Hippocampus

Gilbert A. Chauvet and Theodore W. Berger

One of the ultimate goals of studies of long-term potentiation (LTP) is to determine the consequence of synaptic plasticity on functional properties of the larger neural network in which the potentiated synapses are embedded. More specifically in the case of the hippocampus, the problem is to determine the relationship between the molecular events that modify the probability of glutamate-gated channel opening and the cognitive functions of learning and memory resulting from the collective, population activity of neurons making up the entorhinal, hippocampal, and subicular cortices. The essential issue is how to establish an analytical basis for relating the molecular mechanisms of LTP to the population dynamics of the global neuronal network when those mechanisms exert consequences only within a local domain, that is, modification of receptors and channels within the limited spatiotemporal region of activation. In the present chapter, we outline a theoretical basis for resolving this issue, and present results of a specific application to the hippocampal dentate gyrus. Models are proposed for representing the dynamics of each of the receptor-channel, synaptic, and neuronal levels of function using a common theoretical framework through which these different levels in the neuronal hierarchy can be related.

To describe how the physiological properties of each of these structural levels is expressed at each of the other levels (i.e., their functional organization), we have used an approach based on an n-level field theory developed by Chauvet (1988, 1993b). Central to the theory is the concept of functional interaction, which describes the elementary effect of one structural unit on another structural unit at the same, or at a different, level of organization. In the case of the dentate gyrus, an example of a functional interaction is the action of one fiber of the perforant path on the synapses of a granule cell, that is, the transformation of input activity into output activity (Chauvet and Berger, 1990, 1994).

With an n-level field theory, the dynamics of the global system result from the dynamics at each level of organization, represented by the variation in time and space of physiological variables that are the field variables. The first models of a neural network derived from a field theory considered only one level of organization, such as those proposed by Beurle (1956), Griffith (1963,1965), Wilson and Cowan (1972), Fischer (1973), Ermentrout and Cowan (1979), and Kishimoto and Amari (1979). In the present framework, each level of functional organization is defined by the physiological function, that is, the collective dynamics of a set of structural units defined relative to a given time scale. Thus, activity generated by neurons of a network is the state variable for the dynamics at the network level. Likewise, synaptic efficacy is a product of the synapses within one neuron, and therefore is the state variable for the dynamics at the synaptic level. Within a single synapse, the dynamics of the receptors and their associated channels generate the postsynaptic currents that lead to the synaptic events. The hierarchical system can be conceived as follows: receptors and channels constitute the lowest level, together with all the structures that contribute to a long-term modification of postsynaptic currents that depend on receptor-channel function. Synapses, together with the processes that contribute to modifying synaptic efficacy, constitute the second level. Neurons interconnected by the network structure so as to constitute the variation in time and space of system activity constitute the third level. Such a construction can be generalized to define the function produced by the set of neural networks (i.e., dentate gyrus, CA1, and CA3). In addition, because the locations of synapses, dendrites, and neurons in physical space are required by the formalism of partial derivative equations, a major advantage of an n-level field theory is the ability to take into account both the hierarchy and the geometry of a neural system.

The approach demonstrated here to developing a biologically based model of the hippocampus has been shaped by several different constraints, both biological and theoretical. In the following section, we will review successively the biological constraints: the hierarchical organization of the nervous system, the influence of neuronal geometry, and the molecular bases of synaptic plasticity, and then the related theoretical problems: the nature of the fields, the role of nonlocality, and the statistical formalism imposed by the large numbers of synapses. Finally, specific models for the postsynaptic currents produced by NMDA receptor-channel activation, synaptic efficacy, and population activity at each of the three levels of the functional organization, respectively, are presented, as well as their integration into a unique system to obtain the functional properties of the hippocampus.

CONSTRAINTS IN THE STUDY OF BIOLOGICAL NEURAL NETWORKS

Biological Constraints

The Relation between Network Architecture and Network Function A first constraint is the relation between network architecture and network function, which leads to two specific problems. (1) The precise role of neuronal geometry in determining network function. The level of detail in the geometry that must be considered in formulating a model of a real neural network is not known. Even if every anatomical characteristic of a neural system were known, the relative contribution of each characteristic to properties of the global network cannot now be determined. (2) Because the geometry of neural systems is specified by the continuous spatial densities of their component neurons, a continuous formalism is needed to represent the geometry. Including experimentally determined densities in a continuous representation constitutes a major advantage with respect to the coupling between geometry (the location of cells) and topology (the connectivity between cells) of the network. In contrast, models of neural networks based on automata theory can represent only the topology of the system.

Hierarchical Organization of Real Neural Networks A second constraint arises from the hierarchical organization of the nervous system, because the physiological dynamics of the network have to be described in temms of subcellular components, neurons, and populations of neurons (e.g., Ambros-Ingerson et al., 1990). The issue to be resolved is whether a formalism can be identified that describes and relates the physiological dynamics at each of these levels.

Learning Rules Consistent with Mechanisms of Synaptic Efficacy The third constraint is the biological relevance of learning rules incorporated into models of real neural networks. With few exceptions, learning rules are incorporated into formal neural networks in an arbitrary manner, as if learning rules and network dynamics or network structure are independent of each other. For example, in some formal neural networks, learning rules deduced from Hebb's principle (Hebb, 1949) express a statistical relation (principle of correlation), and not a physical one (deduced from molecular mechanisms of conductivity). However, the learning rule expressed by a biological neural network (i.e., activity-dependent changes in synaptic efficacy) must be the consequence of biochemical kinetics at the molecular level (Brown et al., 1990; Zador et al., 1990), as well as the changes in excitability due to synaptic interactions with other, distant neurons (Xie et al., 1992). We have shown that the physiological dynamics of the network can include biologically relevant learning rules by choosing an appropriate interaction operator that represents the dynamics at the lower, molecular level.

Theoretical Constraints

The Representation of the Geometry of Neuronal Elements The first biological constraint can be solved by using a field theory (i.e., a continous representation) where the state variables of the system are defined in real space (r) and time (T). Local partial differential equations and specific concepts included to give the biological system its particular features, lead to a field theory. In a field theory, a field variable, say ψ, can describe the functional interaction, that is, the action from a source (the neuron), to a sink (the postsynaptic site of another neuron): ψ is transported from a source, represented by Γ, to a sink under the action of a field operator H (figure 18.1). The formal equation which represents the variation in space and time of the field variable is:

$$H\psi = \Gamma \tag{1}$$

where H is a space and time differential operator, for example,

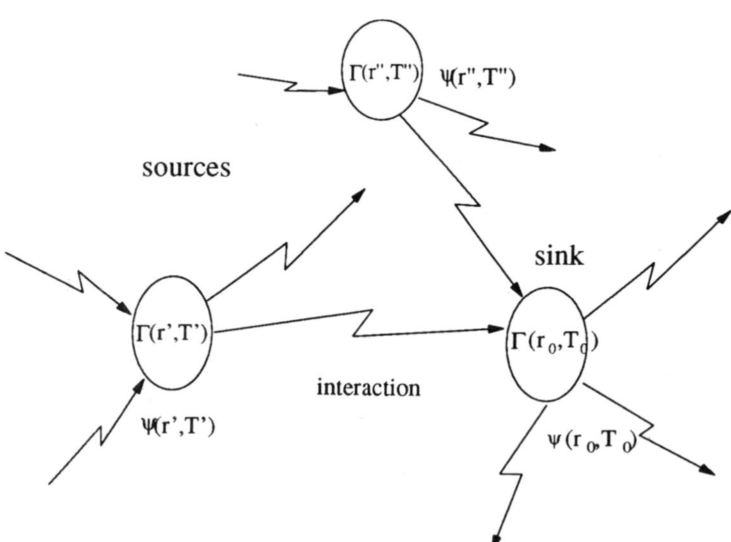

Figure 18.1 Representation of fields that act from neurons located at r', r'' to several synapses located in a neuron at r_0 in the neural network.

$$H \equiv \frac{\partial}{\partial T} - D\nabla^2 + H_I$$

with the diffusion operator, ∇^2, and the interaction operator, H_I. An important determination is that of the interaction operator. Several conceptual issues are encountered only when multiple levels of organization are included in a representation. An example of *nonlocality* is the propagation of neuronal activity from the emitting neuron (the source) to the postsynaptic target (the sink); the interaction operator describes the interaction between sources and sinks.

In summary, a field theory has the following characteristics: (1) it can represent the dynamics at several levels of organization within the nervous system, including subcellular, cellular, and multicellular; (2) a field theory can include nonlocality by means of a nonlocal interaction operator; (3) a field theory describes the space-time evolution of patterns, and relates topology and geometry because it integrates the connectivity described by the topology and the spatial densities dependent on geometry. The relation between topology and geometry is given by a "density-connectivity" function. The synaptic density-connectivity is the density of synapses at one point in the space of one neuron connected with one point of the n-neuron space; (4) the neural dynamic system represents the time evolution of state variables that have a physical meaning (i.e., neuronal soma membrane potential), which results in activity, and synaptic efficacy; (5) because the dynamic equations include the mechanisms of synaptic plasticity, learning rules can be deduced from the continuous description of the biological system (Chauvet, 1986).

The dynamics of a neural network, and thus the extent to which those dynamics mimic the cognitive processes of learning and memory, are a product of the properties of its elements, the neurons, including its topology and its geometry—and the physiological mechanisms of each cellular type, particularly the influence of molecular kinetics on synaptic modifiability, that is, the basis of learning rules. Incorporating both anatomy and physiology in a model provides for a better description of network dynamics than if only topology is included, as in the case of formal neural networks. In addition to allowing the study of geometry and molecular mechanisms on network function, a field theory approach also can provide a basis for relating formal neural network models (McNaughton and Morris, 1987; Levy et al., 1989; Treves and Rolls, 1992) and compartmental neuron models (Traub et al., 1985, 1988; Holmes and Levy, 1990; Spruston and Johnston, 1992; Wilson and Bower, 1992) of the same brain system. The field equations describing a biological neural network can be made equivalent to the automata equations typical of formal neural networks (Hopfield, 1982; Kohonen, 1972, 1978) by representing densities of neurons as Dirac functions, and can be made the equivalent of a compartmental neuron model by simplifying the interaction operator (Chauvet, 1988, 1993b).

The Hierarchical Organization of the System Is the Basis for "Nonlocal" Interactions The second biological constraint has an important consequence in the formulation of the model. Indeed, inside the structural unit (e.g., "cell") there is a set of transformations at different levels of organization. When a molecule or a signal is emitted by a structural unit, that molecule or signal acts either on a similar structural unit at the same level of organization, or on a similar structural unit at a lower level. In the case of the nervous system, it will be a neuron or synapses of another neuron. The sequence of at least two processes occurs: (1) the modification of soma depolarization at the level of neurons (the neural or network level) until the threshold is reached, then the generation and initiation of an action potential at the axon hillock of the neuron (the source), and the active transport of this action potential along the axon toward the synapses of another neuron; (2) at the level of the synapses (the synaptic or neuronal level) of the target neuron (the sink), the sequence of processes includes the presynaptic release of transmitter, its diffusion in the synaptic cleft and binding with postsynaptic receptors, the resulting generation of postsynaptic potentials, and the integration of all postsynaptic potentials for modification of the soma membrane potential at the neural level.

The system composed of synapses is included in the system composed of neurons. Because the dynamics of these two systems are dependent on each other, state variables for the entire system are (ψ, μ). This means that the dynamics of the entire system consists of the variation in space and time of the state variables considered as field variables. Thus, they are related to each point of the neuron space for ψ and to each point of the synapstic space for μ. Because the space of synapses is included in the space of neurons, and because, functionally, two time scales are attached to each of these spaces, a hierarchical functional system is defined from the field variables (Chauvet, 1993b). With the field theory, the field variable evolves from r to $r + dr$ in the continuous r-space, that is, between two infinitesimally close neurons considered to be points. The structural unit, the "neuron," *is not reducible to a point* because synapses (different structural units) exist in the neuron, and their dynamics determine the dynamics of soma depolarization. Therefore, at least two levels of organization can be

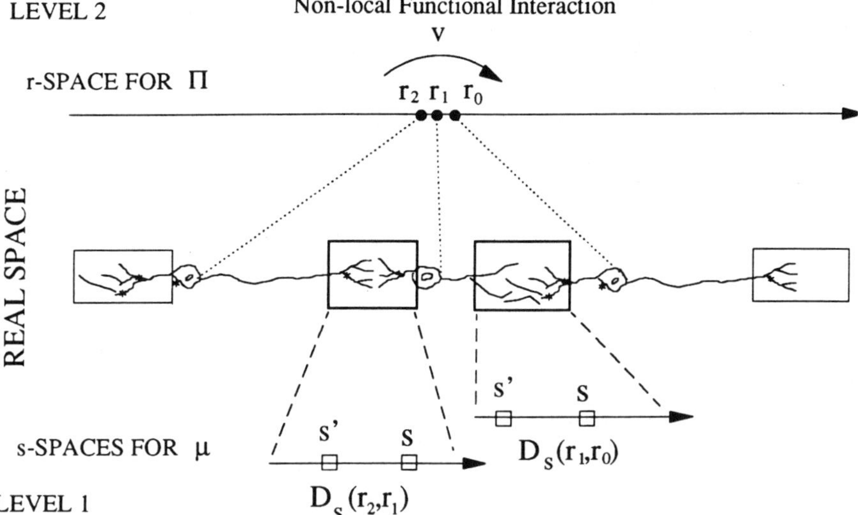

Figure 18.2 Non-locality in the space of neurons. The propagation of the field from a point r to another infinitesimally close $r + dr$ is, in the physical space over an extent that is the space of synapses $D_s(r, r_0)$.

accounted for. Let us denote the location of soma depolarization by x_r and the location of synapses by x_s (figure 18.2). Because the structural unit is not reducible to a point, x_r and x_s cannot be located precisely. Conceptual difficulties in the use of a field theory in biology arise from this characteristic. One solution is to represent the biological structure in "spaces of units" that are the abstract spaces where the functional process evolves. The formulation depends on the level of observation: in the real space, but with a different space scale, experimental data can provide the density of neurons, the density of synapses for a given neuron, and the density of receptors in a given synapse.

The Large Number of Synapses per Neuron Requires a Statistical Theory of the Field States Although the field equations give the values of soma membrane potential at each point of the neural network and the values of the synaptic efficacy at each point of the neuron for every neuron, some neuronal processes are not represented in the field equations, such as the modification of the internal state of a neuron as a consequence of action potential generation, or the return to resting membrane potential due to intrinsic regulatory mechanisms such as ionic pumps. An important theoretical difficulty is the mathematical formulation of the phenomena from the synaptic level to the neuronal level when the number of synapses is considerably larger than the number of neurons.

Our method to identify these phenomena is, first, to mathematically describe their effects by introducing specific parameters, and, second, to determine these parameters from the observed curves, for example, from the waveforms of the extracellular evoked field potential. We have introduced a function, the *distribution function f of the states of the fields* (Chauvet, 1990, 1993a; Chauvet and Berger, submitted) which describes the number of synapses that are in a given *synaptic state*. In the present case, the *synaptic state* is defined by the postsynaptic potential (PSP) $\Phi(\psi)$ and the synaptic efficacy μ. Therefore, the state of a synapse at s_0 in r_0 is defined by the field variables and the *distribution function f of the states of the fields* is the function $f(s_0, t; \psi(r_0, t), \mu) = f_0$. Thus, the large number of synapses per neuron which are in a given synaptic state depends on soma potential, synaptic efficacy, and postsynaptic potential. Given that large number of small variations, a statistical "activity" of synapses, generally called the synaptic activation, is observed even if the postsynaptic cell is not active.

It has been shown that the behavior of a population of units can be deduced from the fields at all levels, that is, the behavior of elements. In the following, the distribution function of synaptic states is taken at points s_0 and r_0, and is thus denoted as f_0. Because of the different time scales, the time variation of the distribution function includes three time-independent variations that can be written as the sum:

$$(\Delta f_0) = (\Delta f_0)_{\text{ext}} + (\Delta f_0)_\psi + (\Delta f_0)_\mu \qquad (2)$$

representing, from left to right: (1) the fraction $Q(t)$ of synapses that modify their state under the influence of an external stimulus intensity below or over the threshold; (2) the modification of the internal state of the cell and the corresponding synaptic states because of a variation of the soma membrane potential, either as a consequence of a stimulation that leads to firing (feedback from the action potential to the emitting cell) or a spontaneous "relaxation," in other words, a modification of potential without external stimulation but with internal modification $\Phi(\psi)$; (3) the modification of the internal state of the cell and the corresponding synaptic states after any long-term variation in synaptic efficacy (e.g., LTP).

The meaning of the distribution function can be expressed as follows: the two variables, soma membrane potential, ψ, and synaptic efficacy, μ, define the distribution function, $f_0 \equiv f(s_0, t; \Phi[\psi(r_0, t)], \mu)$, which represents the probability of having an (s_0, t)-synapse in the state defined by $(\Phi(\psi), \mu)$. Because of the statistical nature of the problem, new parameters (e.g., Q) characterize the population of synapses and have to be determined experimentally. An important issue remaining to be resolved is whether it is possible to express the relation between the distribution function of synaptic states and known neurobiological processes that are other variables of the system, and thus, to identify the effects of synaptic activation on the activity of the network.

MODELS FOR THE RECEPTOR-CHANNELS, SYNAPSES, AND NEURONS

A Model of the Kinetics of the NMDA Receptor-Channel Complex

We recently have developed a mathematical model of the kinetics of the NMDA receptor-channel complex (Urban et al., 1992). This model describes the action of glutamate, glycine, and magnesium on excitatory postsynaptic currents (EPSCs) as observed in outside-out patch-clamp experiments for a variety of experimental conditions. It has been used to determine the importance of several physiologically observed elements (specifically, rate of channel opening, predominance of the desensitized state, and channel open-time) on the kinetics of the NMDA response. We have proposed an extension of the idea that a separate desensitized channel state accounts for the desensitization of the response as proposed by Katz and Thesleff (1957). This model of the NMDA receptor accounts for the desensitization by allowing for multiple states to be entered after the binding of glycine and glutamate. While the model does not yet account for the number of distinct states apparent in the work of Gibb and Colquhoun (1991), it can be extended to do so.

The model is based on several assumptions: (1) ligands bind specifically to the receptor sites and the binding follows first-order kinetics while the off-binding follows zeroth-order kinetics; (2) the complex passes through states and the conductance of the channel is determined by the state of the complex; (3) the rates of the transitions between the states for a given complex are functions of the ligands bound to that complex and these rates are independent of the state of other complexes and also independent of the ligands bound to other complexes; (4) the transitions are dependent only on the current state of the receptor-channel complex and not on any of its previous states.

One means for representing the desensitization of the NMDA response is with the inclusion of a long-lasting desensitized channel state. If a channel is in a desensitized state then it will not open even if the agonist is bound to the appropriate site. Thus after opening, the NMDA channel would enter a state in which it was glutamate and glycine insensitive. Thus the model includes three channel states, *open, closed and desensitized,* and binding sites for three ligands, glutamate, glycine, and magnesium (figure 18.3). Various numbers of binding sites for glutamate and glycine have

Zero or one glutamate site bound

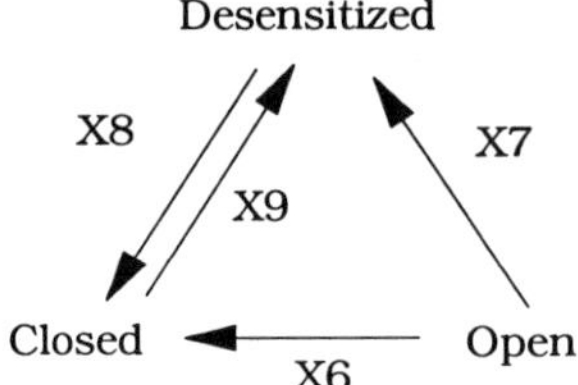

Two glutamate sites bound

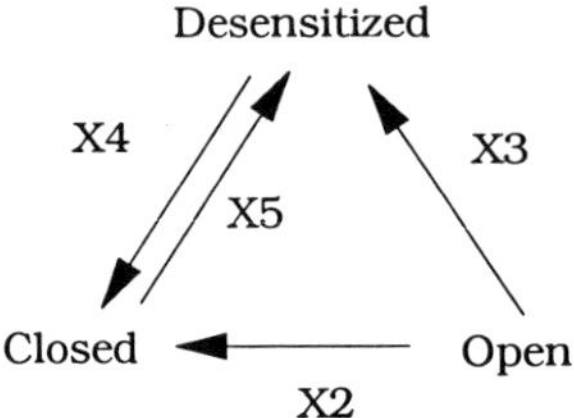

Two glutamate and one glycine site bound

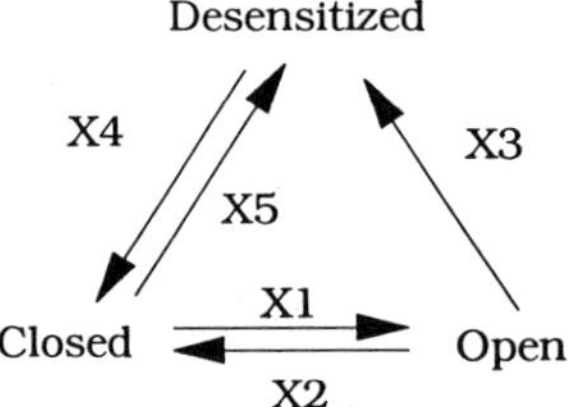

Figure 18.3 Description of states and rate constants in the NMDA receptor-channel complex model.

been modeled, but for the simulations presented here, the number of sites for both ligands was assumed to be two. Thirteen parameters are used to describe the kinetics of the system. These parameters all represent rate constants for either ligand binding or unbinding, or for transitions between channel states.

Formulation of the Model This kinetic model is based on the simple mass action laws which govern the binding and unbinding of ligands and the transitions between conformation states. In the simple case of two successive chemical reactions with rate constants k_1 and k_3 of ligand binding in the order 1, 3, and k_2 and k_4 of ligand unbinding in the order 2, 4, at equilibrium, the series of reactions is represented by the matrix $\mathbf{P}$ such that $\mathbf{P} = \mathbf{N} - \mathbf{D}_{ij}$:

$$\begin{bmatrix} -k_1 & k_2 & 0 \\ k_1 & -(k_2+k_3) & k_4 \\ 0 & k_3 & -k_4 \end{bmatrix} = \begin{bmatrix} 0 & k_2 & 0 \\ k_1 & 0 & k_4 \\ 0 & k_3 & 0 \end{bmatrix} - \begin{bmatrix} k_1 & 0 & 0 \\ 0 & k_2+k_3 & 0 \\ 0 & 0 & k_4 \end{bmatrix} \tag{3}$$

where $\mathbf{D}_{ij}$ is defined by:

$$\left\{ \begin{matrix} \sum_{j=1}^{n} \mathrm{Col}_j N: & \text{if } i = j \\ 0: & \text{otherwise} \end{matrix} \right\} \tag{4}$$

For a larger set of reactions, the dynamics of the kinetic system is determined by a matrix equation $\mathbf{X}' = \mathbf{PX}$, where $\mathbf{X}$ is the state vector and $\mathbf{P}$ is the matrix which describes all the chemical reactions as in eqn. (8). This matrix includes the matrix $\mathbf{N}$ of state transitions where each element N_{ij} is the rate of the transition from x_i to x_j. Each state of the configuration is defined by each possible variation of receptor conformation and set of bound ligands. The variables in the dynamical system are the number of receptors. Thus, a receptor-channel complex with an open channel and only glutamate bound defines one state. The number of receptors in this state is, for example, x_i, and the number of complexes that have a desensitized channel and no ligands bound (another state) is, say, x_j. The rate constants $k_1 \ldots k_n$ are then rates of ligand binding and unbinding and also rates of conformational changes. If a system of equations is to describe the kinetics of a receptor-channel system with 3 channel conformations and 3 ligand sites the number of states, and thus the number of variables and equations, is $n = 3(2^3) = 24$. The number of possible transitions, and thus the number of rate constants, is then $n^2 - n = 552$ for the case $n = 24$. Fortunately, many of these transitions will be disallowed by certain physiologically based assumptions described below while others will have identical rates.

Consider a receptor channel system with k ligand binding sites and c channel conformations. For such a system the total number n of receptor-channel states to consider is $2kc$. Let $\mathbf{X}$ be an n-element vector with elements x_i the number of receptor channel complexes in each state where a state is defined by a ligand configuration and a channel conformation. The individual elements of $\mathbf{X}$ can be written descriptively as

$$\begin{pmatrix} \alpha \\ \beta \end{pmatrix} = \begin{pmatrix} \alpha \\ b_1, b_2, \ldots, b_k \end{pmatrix} \tag{5}$$

where $\alpha \in \{0, 1, \ldots, c-1\}$ indicates the channel conformation, open, closed or desensitized in the case of NMDA, and β is a binary vector

Table 18.1 Variables and notation

Variable Name	Description
k	Number of ligand binding sites per receptor channel complex
c	Number of channel conformations allowed
n	Total number of states. $n = 2^k \cdot c$
α	Variable for channel conformation. $\alpha = (1, 2, \ldots, c-1)$
b_i	Elements of β; each describes state of a single receptor
β	Variable for ligand configuration. Often β is read as a binary c-vector
N	Matrix of rates for transitions. N_{ij} = rate of transition of x_j to x_i
P	Matrix used for overall reaction, $X' = PX$. P is determined by N as in eqn. 4
X	n-Vector of state variables $(x_1, \ldots, x_n)$
T_β	c by c submatrix of N which describes channel conformation changes for ligand configuration β
M_{ij}	c by c submatrix of N which gives rates for ligand configuration i changing to ligand configuration j

indicating ligand configuration such that $\beta = (b_1 b_2, \ldots, b_k)$. It can be taken as a convention that $b_i = 0$ indicates that the ith receptor site it unoccupied and that otherwise $b_i = 1$. β can also be written as a decimal integer which corresponds to the appropriate binary number. For example, if $\beta =$ (1001) this may be written as $\beta = 9$ without ambiguity.

The structure of the matrix **N** is determined by the order in which the states are represented in the vector **X**. The states can be ordered so that **N** is composed of submatrices $\mathbf{T}_\beta$ and $\mathbf{M}_{ij}$. The matrix $\mathbf{T}_\beta$ describes conformation changes for a given ligand configuration β and the matrix $\mathbf{M}_{ij}$ describes the change from ligand configuration i to j (see table 18.1). With one logical ordering of **X** described elsewhere (Urban et al., 1992), the matrix **N** can be written generally for any number of ligands and any number of channel conformations. For $k = k_0$ we have:

$$N_{k_0} = \begin{bmatrix} T_0 & M_{1,0} & M_{2,0} & \cdots & M_{2^{k_0-1},0} \\ M_{0,1} & T_2 & M_{2,1} & \cdots & M_{2^{k_0-1},1} \\ M_{0,2} & M_{1,2} & T_3 & \cdots & M_{2^{k_0-1},2} \\ \vdots & \vdots & \vdots & \ddots & \vdots \\ M_{0,2^{k_0-1}} & M_{1,2^{k_0-1}} & M_{2,2^{k_0-1}} & \cdots & T_{2^{k_0-1}} \end{bmatrix} \tag{6}$$

There are three different channel conformations, single open, closed, and desensitized in the case of NMDA with binding sites for glutamate, glycine, and magnesium. Thus, β is a three-vector and **T** and **M** are three by three matrices. As an example, if b_1 of β represent the magnesium site, b_2 represent the glycine site and b_3 represent the glutamate site and let an

α of 0, 1 and 2 indicate closed, open and desensitized respectively, then the reaction:

$$\begin{pmatrix} 1 \\ 001 \end{pmatrix} \Rightarrow \begin{pmatrix} 1 \\ 011 \end{pmatrix} \tag{7}$$

will represent the binding of a molecule of glycine to a receptor channel complex with an open channel at which glutamate, but not magnesium is already bound. With this very general method, the ordinary differential matrix equation $\mathbf{X}' = \mathbf{PX}$ can be solved for $\mathbf{X}$, that is, gives the number of receptor channel complexes in each state for given rate constants in matrix $\mathbf{P}$.

Description of Experiments The model was used to simulate the behavior of the NMDA receptor-channel complex under three sets of experimental conditions found in the literature. For each type of experiment the simulated and the experimental (observed) results were compared by computing the error (given by [[observed − simulated]/observed]) for certain characteristics of the EPSC relevant to the given experiment (*e.g.*, tau — the decay constant, time to peak, and ratio of steady state to peak response). In brief, the experimental conditions and measured characteristics of the EPSC are given below and in figure 18.4.

Simulation one In this experiment (based on Sather et al., 1990) a saturating solution of glutamate is added to a bath in which glycine is present at saturating levels. The concentrations of glutamate and glycine remain constant throughout the experiment (for 800 msec). The characteristics measured are time to peak, decay time constant, and ratio of steady state to peak response.

Simulation two In this experiment (based on Sather et al., 1990) a saturating solution of glutamate is added to a bath in which glycine is present at saturating levels. The concentration of glutamate is then reduced after 750 ms and returned to saturating levels after an additional 750 msec. In addition to the characteristics measured in simulation one, the ratio of the second peak to the first is measures for comparison.

Simulation three In this experiment (based on Lester et al., 1990) a saturating solution of glutamate is added to a bath in which glycine is present at saturating levels. The concentration of glutamate is then reduced after a period of about 5 msec. In this experiment the time to peak, the decay time constant, and the value of the response at 800 msec are measured for comparison.

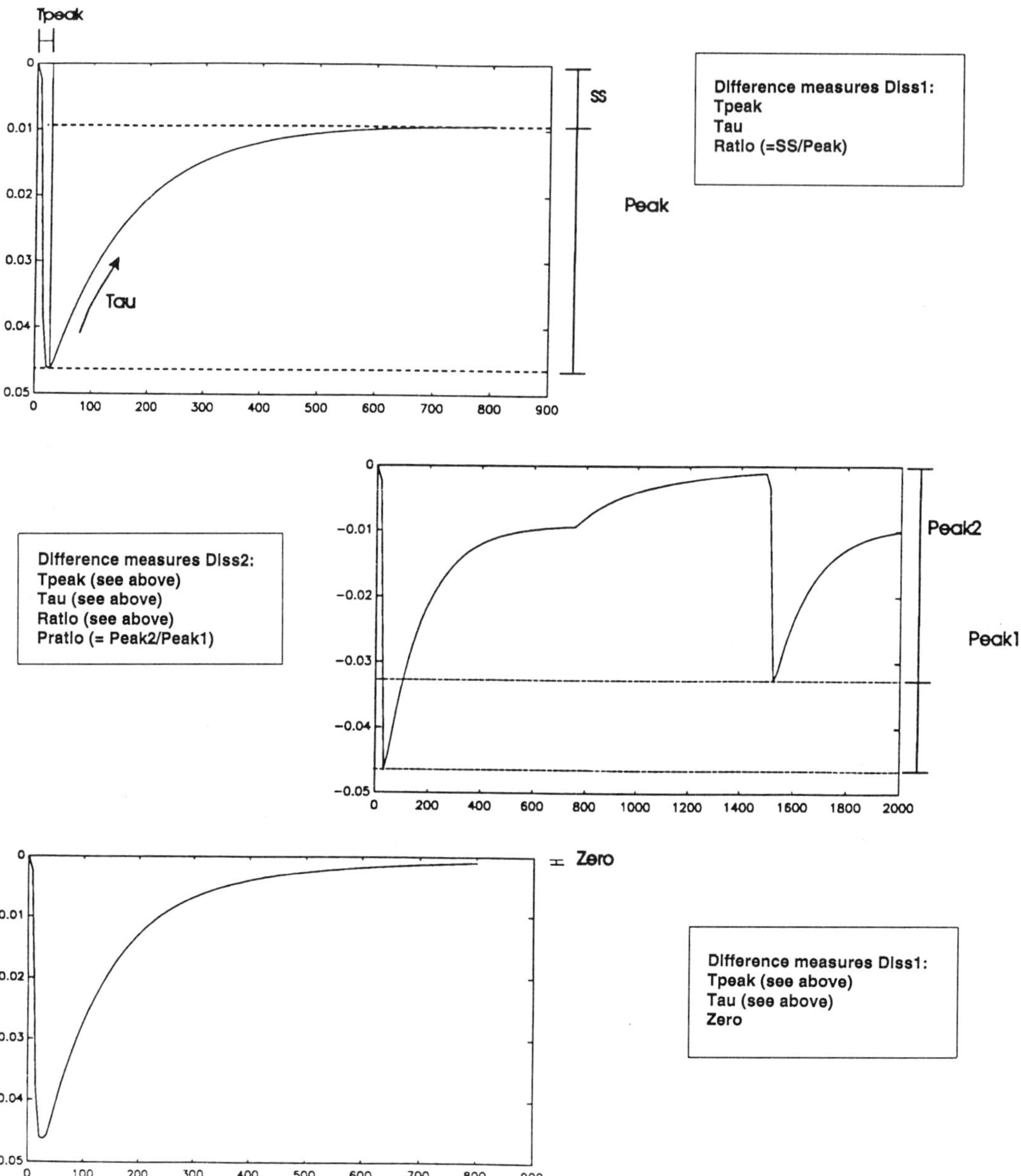

Figure 18.4 Results of the simulation for three experiments at the molecular level (NMDA receptor-channel model). The characteristics are: *Tpeak,* the amplitude of the minimum of the curve; *Tau,* the decay constant; *Ratio,* the ratio between *SS,* the asymptotical value of the EPSC, and *Peak,* which equals *Tpeak* minus *SS*; *Pratio* equals the ratio of the two peaks when there are two drug administration times. (*Top*) A saturating solution of glutamate is added to a bath in which glycine is present at saturating levels (Sather et al., 1990). (*Middle*) The same experiment, but the concentration of glutamate is then reduced after 750 msec and returned to saturating levels after an additional 750 msec (based on Sather et al., 1990). (*Bottom*) A saturating solution of glutamate is added to a bath in which glycine is present at saturating levels. The concentration of glutamate is then reduced after a period of about 5 msec (based on Lester et al., 1990).

Comparison of Simulations to Experiments The parameters of the model were varied (constrained by experimental data where available) and the total error (as calculated by taking the square root of the sum of the squares of all the differences in individual curve characteristics) between the experimental results and the simulations was minimized using the simplex method of nonlinear optimization. The parameter set that resulted in this minimum was taken as a standard.

Then, to determine the influence of changes in various parameters of the model, individual parameters were varied from the standard and the value of $(D_{\text{varied}} - D_{\text{standard}})/D_{\text{standard}}$ (where D can be any characteristic of the EPSC) was calculated. The results for the variation of parameters X1 (rate of channel opening), X3 (rate of open channel desensitization), and X4 (rate of channel resensitization) are given in figure 18.5. Next we varied pairs of parameters in order to investigate the independence of the various parameters as well as to look for other sets of parameters that might minimize the error (other local minima of the total error in the parameter space). These results for the pairs $\{X1, X3\}$ and $\{X1, X4\}$ are displayed in figure 18.6. These figures display the mesh diagram over the contour plot of the same data and show that the influence of any one of these parameters depends significantly on the value of the other two.

The three-dimensional plots given in figure 18.6 are plots of the error space of the system in the chosen parameters. From these plots information about the independence of certain parameter pairs can be gained, and further, other local minima can be detected and examined for physiological relevance. In the cases investigated above—the error space plot of parameters X1 (rate of channel opening with glutamate and glycine sites full) and X4 (rate of channel resensitization with glutamate bound)—we can see that as parameter X1 was increased, the effect of increasing parameter X4 was negated. In the plot, this is indicated by the back left side of the plot being flatter than the front right. Physiologically, this means that if the rate of channel opening is slow enough, then the rate of resensitization becomes less important.

In the error space of parameters one and three, we observed a long valley as parameter X1 increased and parameter X3 decreased or vice versa. Along this valley, as parameter X1 increases the peak height (maximum number of channels open during the response) also increases while the error shows little variation above the minimum at the center of the plot. It is significant to note that while the error as described above remains relatively unchanged along this valley, the number of open channel at the peak of the response does not. Increasing X1 while decreasing X3 produces larger simulated EPSCs without significantly altering their shape.

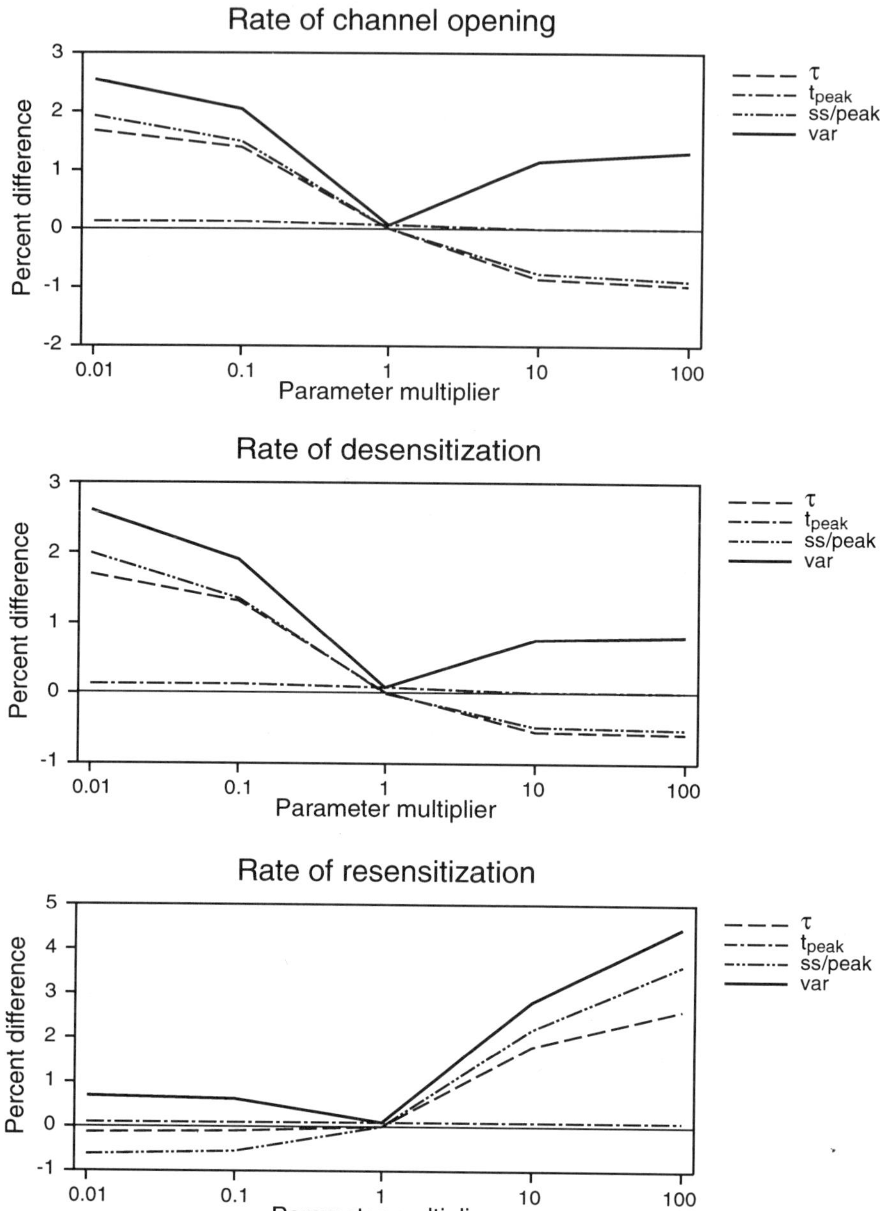

Figure 18.5 Study of the influence of single rate constants in the NMDA model by minimizing the total error (square root of the sum of the squares of all the differences in individual curve characteristics) between the experimental and the simulated curve in three cases: (*top*) variation of parameter X1 (rate of channel opening); (*middle*) X3 (rate of open channel desensitization); (*bottom*) X4 (rate of channel resensitization).

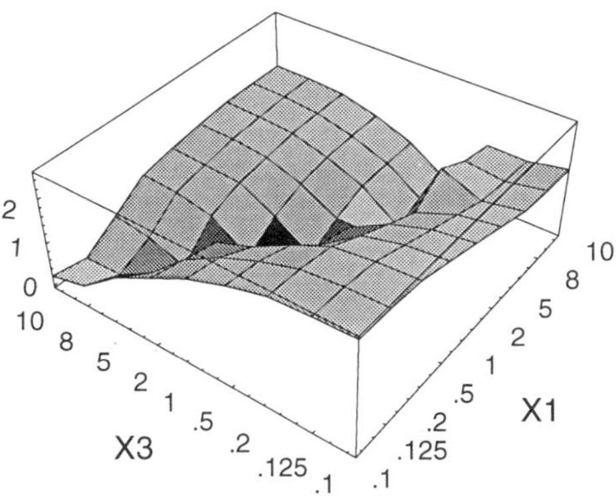

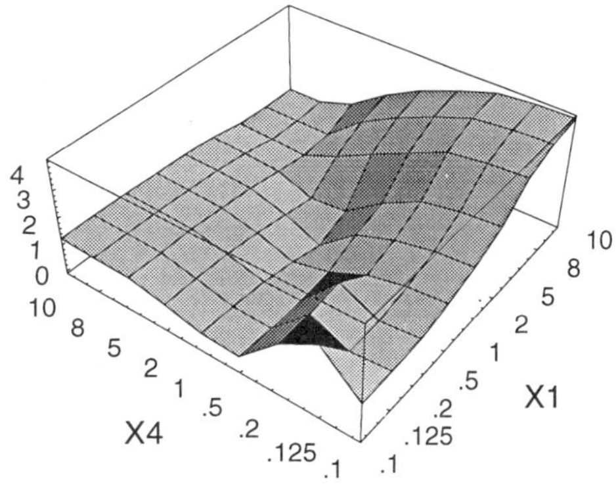

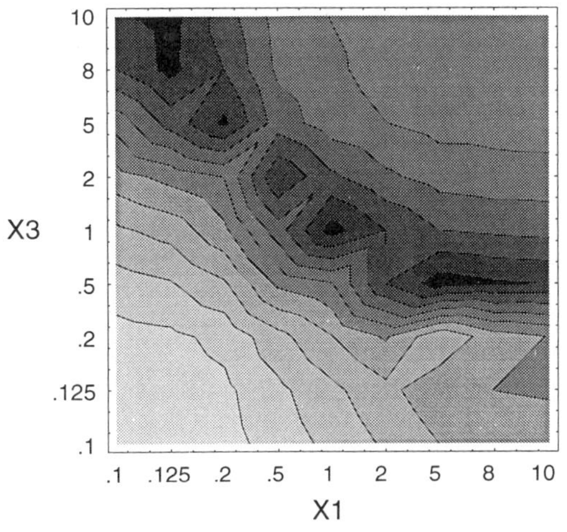

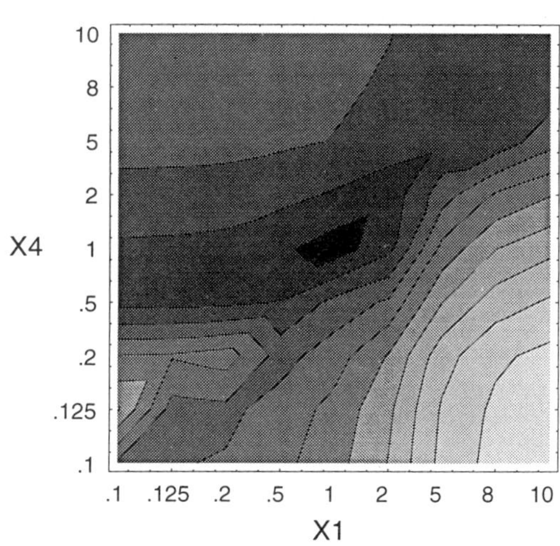

Figure 18.6 Study of the dependence of rate constants pairs in the NMDA model by minimizing the total error in the space of parameters in two cases: (*left*) X1 and X3; (*right*) X1 and X4. (X1 = rate of channel opening); X3 = rate of open channel desensitization; X4 = rate of channel resensitization).

Formulation of the n-Level Field Model If ψ denotes the field variable for soma depolarization, the local equation for ψ has to describe the variation in time of ψ from one "point" r (the axon hillock or the soma) to another "point" $r + dr$ in the space of body cells. The second process consists in the modification of the synaptic efficacy μ due to a variation of activity at any other synaptic input, and is described by the field variable μ. The field equations that describe these two processes can be written:

$$\frac{\partial\psi(r_0, T_0)}{\partial t} = \nabla_r(D^r\nabla_r\psi(r_0, T_0)) + \int_{D_R(r_0)} \rho_1(r)\psi(r, T) \int_{D_s(r, r_0)} B(r_0, T_0, r, T)\pi_0(s, r; r_0)\mu(s, t)\, ds\, dr + \Gamma_\psi[\psi(r_0, T_0^-), \psi_{\text{refr}}] \tag{8.1}$$

$$\frac{\partial\mu(s, t)}{\partial t} = \nabla_s[D^s\nabla_s\mu(s, t)] + \int_{D_R(r_0)} \rho_1(r') \int_{D_s(r', r_0)} \mu_0(s', t')\pi_0(s', r'; r_0)A(s, s')\, ds'\, dr' + \Gamma_\mu(s, t) \tag{8.2}$$

$$s \equiv s(r, r_0)$$

$$D_R(r_0) = \bigcup_{r'} D_s(r', r_0)$$

where $T = T_0 - d/v_\psi$, $d = \|r - r_0\|$, v_ψ is the transport velocity of the ψ-interaction, i.e., the neural activity, and $t' = t - \|s - s'\|/v_\mu$, with v_μ being the velocity of the μ-interaction.

Equations (8.1) and (8.2) are the two neural 2-level field equations for the activity of neural tissue at two corresponding points $(s(r, r_0), t)$ and (r, T) in space-time level of organization. Their coupling is imposed by synaptic efficacy μ, and the neural tissue is characterized by two geometrical functions: the density of presynaptic neurons, ρ_1, and the density-connectivity π of synapses. $D_s(r, r_0)$ is the space of synapses in neurons localized in r_0, which correspond to neurons in r. Space $D_R(r_0)$ is the neural integration space which is the union of subspaces $D_s(r, r_0)$ (figure 18.7).

The first term on the right of the nonlocal equation for the soma potential represents the local diffusion due to the biophysical ionic transport in the extracellular space from r_0 to infinitesimally close points $r_0 + dr$, which leads to a variation of the membrane potential. The second term integrates all the nonlocal neuronal influences on the given neuron, and results from

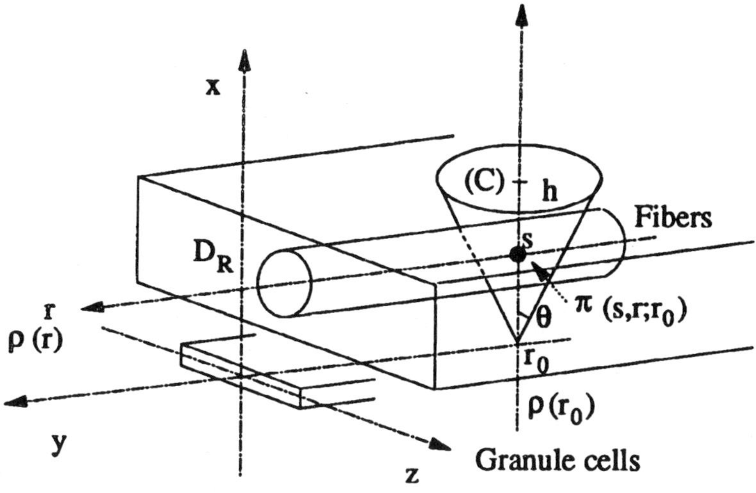

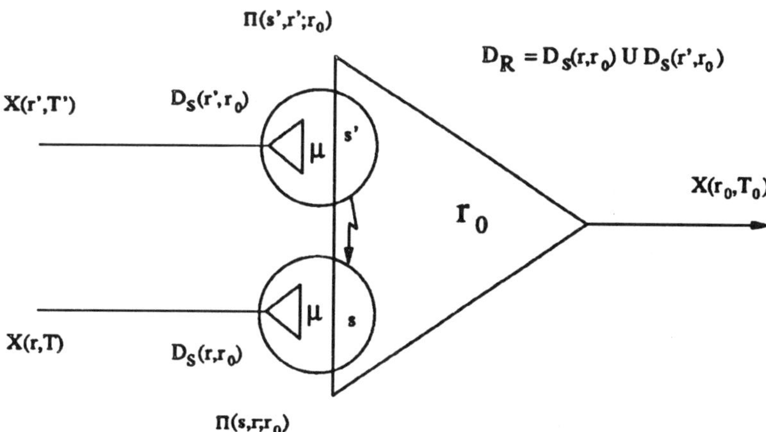

Figure 18.7 Notations used in the n-level field theory equations. $X(r, T)$ is the activity at point r and at time T. The space $D_s(r, r_0)$ contains all the synapses that are in the neuron at r_0, and correspond to the presynaptic neurons at r. The space $D_R(r_0)$ is the union of all the space of synapses for a neuron at r_0.

the space and time variations of the potentials in the afferent neurons at r. When the resulting potential reaches the threshold, then the source emits an action potential.

The first term on the right of the nonlocal equation for the synaptic efficacy represents the local space diffusion from s to infinitesimally close points $s + ds$. The second term is the nonlocal interaction operator which represents the influence of all distant synapses on the given synapse at s. The third term represents the local effect of the source, that is, the local transformation that allows μ to modify its value under a local influence after a time dt (e.g., some modification of the conformation of receptors).

Meaning of the Concepts Introduced in the n-Level Field Model
The variations in time and space of the related field variables are described by the field equations. The nonlocal term included in the field equation at a given level can be simply interpreted as resulting from the integration of all the system-external inputs at this level to increase the value of the related state variable. The source term represents a strictly local contribution to that increase.

When the dynamics inside the source is coarsely represented by a linear increase that only allows the firing, there are two parameters in the field equation for the soma membrane potential ψ: the *source coefficient of action potential* p, in other words, the rate constant of ψ, and the *time period of action potential* ΔT_{ap} between threshold and peak of the action potential for both the rising and falling phases. They can be interpreted as follows. The nonlocal term, which integrates all the postsynaptic inputs to increase the soma potential membrane up to the threshold value, depends in this model only on the geometrical distribution of the synapses at the lower level. Then, the soma membrane potential increases sharply due to the local transformations inside the axon hillock. For a better description, a model of all the action potential would be necessary. But, in the present case, we are interested only in the generation of active cells.

In the same way, with linear dynamics of the source to describe coarsely the increase in the source production, and not its relaxation, there are two parameters in the field equation for the synaptic efficacy μ: the *source coefficient of synaptic efficacy* m (i.e., the rate constant of μ), and the constant h that represents the degree of correlation (i.e., the *pre-postsynaptic correlation factor* between pre- and postsynaptic activity). The interpretation is the same as above: the nonlocal term integrates all the effects of the synaptic transmission at a distance and depends only on the geometrical distribution of the receptor channel complexes. Then, a local transformation follows due to the dynamics at the molecular level.

These parameters are important, and specially h because they are related to the events at the molecular level. The general case has been studied by Chauvet (1990, 1993b) as follows: (1) Presynaptic mechanisms at any point (s, t) in a neuron at (r_0, T_0) and corresponding to a presynaptic neuron at (r, T) are described by terms that can interpret the kinetic properties of neurotransmitter release (Magleby and Zengel, 1982). The presynaptic phenomenology is included in ξ, *presynaptic efficacy*, which describes the *probability of release of the neurotransmitter.* (2) The postsynaptic mechanisms are included in the *postsynaptic efficacy* η that describes the *proportion of channels in the open state,* and thus is similar to a conductivity. They include transmitter-receptor binding, transitions between opened and closed channel states, conductance variation of voltage-dependent channels, the resulting variation of postsynaptic potentials (PSP), and the action of neuromodulators. (3) The interaction between ξ and η that results in μ_0 (the local synaptic efficacy) is assumed to be multiplicative:

$$\mu_0(s, t) = \langle \xi(s, t)\eta(s, t)\rangle \tag{9}$$

The solutions ξ and η are given either as solutions of the complex and nonlinear local dynamics which partly is presented in a next section as concerns the dynamics of the NMDA receptor-channel complex, or by the average of pre- and postsynaptic activities, $\langle X(r, T)\rangle$ and $\langle X(r_0, T_0)\rangle$, respectively, at two points r and r_0:

$$\xi(s, t; \xi^0(\langle X\rangle)) = a(\xi^0(s); r)\langle X(r, T)\rangle(t) \tag{10}$$

$$\eta(s, t; \eta^0(\langle X\rangle)) = b(\eta^0(s_0, s); r)\langle X(r_0, T_0)\rangle(t) \tag{11}$$

Assumptions (4), (5), and (6) lead to the learning rules of the biological neural network, because they drive the variation of μ in time. These assumptions, and particularly these learning rules, can be included in the field equation (8.2) to give the particular μ-field equation:

$$\begin{aligned}\frac{\partial \mu(s, t)}{\partial t} &= \nabla_s(D^s\nabla_s\mu(s, t)) \\ &\quad + \int_{D_R(r_0)} \rho(r') \int_{D_s(r', r_0)} a(s'; r')b(s'; r_0)A(s, s')\langle X(r', T')\rangle(t') \\ &\quad \times \langle X(r_0, T_0)\rangle(t')\pi(s', r'; r_0)\, ds'\, dr' + \Gamma_\mu(s, t)\end{aligned} \tag{12}$$

As discussed further, for integrating the molecular dynamics in the field equations, a molecular model is needed that results in the variation in time of pre- and postsynaptic efficacies. It is clear that, within this interpretation, they are explicitly related with the pre- and postsynaptic activities. In contrast, the molecular mechanisms which are independent on the input or output activities will be included in the source term of the equation. Therefore, the complete model will result from the analysis of the pre- and postsynaptic molecular mechanisms.

Synaptic Population Activation Described by a Statistical Distribution Function

As shown earlier, the theoretical constraint of using two different formalisms to describe the neurons and the large number of synapses per neuron leads to introducing two kinds of parameters. The first kind corresponds to the neurons and the synapses. The variations in time and space of the related field variables are described by the field equations. The nonlocal term included in the field equation at a given level can be simply interpreted as resulting from the integration of all the system-external inputs at this level to increase the value of the related state variable. The source term represents a strictly local contribution to that increase. The second kind of parameter corresponds to the population of synapses that is described by the distribution function of synaptic states. We give now an interpretation of these parameters.

Regarding the parameters for the population of synapses, we can define three contributions to the statistical distribution of the field states. The third one is studied in the next section because it concerns the neuron-internal "inputs" after a variation of the synaptic efficacy due to, for example, LTP.

Synaptic States Influenced by Neuron-External Inputs The first contribution is $(\Delta f_0)_{ext} = Q(t) f_0(t)\Delta t$, which represents the neuron-external inputs and is determined by two parameters: Q_0, the *initial fraction of synaptic states*, and Q_1, the *first-order fraction of synaptic states*, that is, the rate of change of rate of variation in time of f. They are such as $Q(t) = Q_0 + Q_1(t)$, where the factor Q represents the fraction of states that enter, if $Q(t) > 0$, or escape, if $Q(t) < 0$. We can recognize here a variation similar to a first-order chemical reaction with a rate "constant" Q. We have shown that a linear dependence in time gives correct results.

Synaptic States Influenced by Neuron-Internal Input Potentials The second contribution is $(\Delta f_0)_\psi$ which represents the neuron-internal "inputs" after a variation of the soma membrane potential, namely, the internal state of the neuron which can be considered as a transition of ψ from one state to another. We have shown that four parameters, ΔT_f and a_ψ, b_ψ, a_f, are sufficient to describe the synaptic activation due to the neuron-internal inputs. They correspond to three time periods, pre-firing, firing, and post-firing. The time interval $[0, T]$ during which the process is observed is decomposed into three intervals $[0, t_a]$, $[t_a, t_a + \Delta T_f]$, $[t_a + \Delta T_f, T]$ where ΔT_f, called the *time period of population discharge,* is the duration of the feedback caused by firing on the distribution of synaptic states onset, starting at time t_s when $\psi = \psi^a$. Parameters a_ψ and b_ψ determine the expression of $(\partial f/\partial t)_\psi$ when there is no firing (i.e., when there is only a spontaneous transition of ψ). They can be called respectively the *first-order spontaneous* ψ-transition and the *initial spontaneous* ψ-transition. Parameter a_f determines the *first-order stimulated* ψ-transition, that is, a *firing coefficient* that determines the transitions of the soma membrane potential due to a modification of the internal state of the neuron after firing. This firing coefficient gives expression of the *stimulated* ψ-transition, because it has the same meaning as a_ψ, except that it describes what happens during the firing period, from t_a to $T_f = t_a + \Delta T_f$. Coefficient b_f is calculated from a_ψ, a_f, b_ψ. After firing (i.e., for $t \geq t_a$), the internal state of the neuron is submitted, as before firing, to a spontaneous relaxation described by the same parameters $a_3 = a_\psi$ and b_3 deduced from ΔT_f.

Synaptic States Influenced by Neuron-Internal Input Synaptic Efficacies. Neuronal activation has been described above based on two mechanisms: (1) the increase in membrane potential until a threshold due to the integration of presynaptic activities, and (2) the generation of an action potential due to the fast local transformation in the source (i.e., the axon hillock). Synaptic activation is the second phenomenon due to the effect of the population of synapses, and, for this reason, is a statistical effect. We have shown that this statistical effect is due to two contributions: first, the neuron-external inputs determined by the factor Q, and, second, the neuron-internal inputs (i.e., the internal state of the neuron) determined by the parameters a_ψ, a_f, b_ψ following a modification of the soma membrane potential, whatever the cause. Now, we examine the consequences of a modification of the second field variable μ. Because LTP modifies the synaptic efficacy, the synaptic state, represented by the postsynaptic potential, is also modified, and therefore the internal state of the neuron. But, because such a modification is not realized on the same time scale as a modification of the soma membrane potential, its contribution to the statistical distribution can be added to the contribution of the soma membrane potential.

Therefore, the specific contribution to synaptic activation of, say, LTP, is obtained using the same techniques as in the previous section. We have shown that the contribution $(\Delta f_0)_\mu$ of the neuron-internal "inputs" (i.e., due to a modification of the internal state of the neuron, which can be considered as a transition of μ from one state to another) is determined by two parameters, a_μ and b_μ such that:

$$(\Delta f_0)_\mu = (a_\mu t + b_\mu)\Delta t \qquad t \in [0, T] \tag{13}$$

It is clear that, with this interpretation in terms of non-local fields, the coefficient a_μ describes the *first-order stimulated* μ-transition in the distribution f, as a_ψ described previously the first-order stimulated transition of ψ-states, and b_μ. is the initial stimulated μ-transition.This coefficient represents the statistical population effect of LTP on synaptic activation. If a model of the receptor-channel complex is coupled with the present model, then we will obtain an expression of μ in terms of kinetic and molecular parameters.

Interpretation of the Synaptically Evoked Population Waveform in a Monosynaptic Pathway

For the values shown in table 18.2, a typical simulated waveform is presented in figure 18.8. We have verified the formal interpretation of the previous parameters. Each waveform is represented for three values of a parameter around its default value (see table 18.2) that gives the best fitting with the experimental data.

First, we consider synaptic activation parameters: Q_0, Q_1 and a_ψ, b_ψ, a_f. Figure 18.8, top row shows the effect of parameters Q_1 and Q_0 on the extracellular field potential waveform. The two following numerical observations—(1) the variance and the average of the distribution of synaptic

Table 18.2 Parameters of the n-level field model, meaning and default values

	Meaning	Default value
Population Parameters		
Q_1	First-order fraction of synaptic states	-0.1
Q_0	Initial fraction of synaptic states	1.0
a_ψ	First-order spontaneous transition of ψ-states	0.2
b_ψ	Initial spontaneous transition of ψ-states	-20.0
a_f	First-order stimulated transition of ψ-states	15.0
ΔT_f	Time period of population discharge	15.0
a_μ	First-order stimulated transitions of μ-states	0.0
b_μ	Initial stimulated transition of μ-states	1.0
Element Parameters		
p	Source coefficient of action potential	$-.0005$
m	Source coefficient of synaptic efficacy	.0100
h	Pre-postsynaptic correlation factor	10^{-12}
ΔT_{ap}	Time period of action potential	3.0
$\mu(0)$	Steady-state synaptic efficacy	$1.5 * 10^{-5}$

states are largely modified, linearly with Q_1 and Q_0, and (2) the location of the population spike remains fixed—are an argument to confirm that the variation in time of subthreshold synaptic activation due to external-neuron inputs is represented by parameters Q_1 and Q_0. The delay for the population of synapses to be maximum decreases when the parameters are increased (i.e., the rate constant of the synaptic transformations) has an influence on the statistical dispersion and location of the distribution. In other terms, if we apply higher excitation to the neuron, the average delay of synaptic activation decreases.

Figure 18.8, second row shows that, during the period of relaxation before firing, the distribution of synaptic states does not change significantly, in agreement with the interpretation of coefficients a_ψ, b_ψ. This means that the internal state of the neuron does not change significantly. However, there is an influence of the initial value b_ψ of the rate of distribution to the right of the waveform. In contrast, during firing, coefficient a_f has a sharp influence on the distribution, in accordance with the interpretation of the internal state varying subsequently to firing. The parameter ΔT_f strongly changes the queue of the distribution, that is, the distribution comes back to the initial baseline after a time varying directly according to the value of ΔT_f. Thus, the meaning of this parameter is the duration of the influence of firing on synaptic states: a reaction to firing is visible as the larger number of synapses in the opposite direction to the main number of synapses after a certain period. The number of synapses that react in the "opposite" state increases with the "intensity" of the internal state (i.e., in relation to the duration of the influence of firing on the internal state). In this case, as shown in figure 18.8, third row, parameters a_f and ΔT_f have no influence on the spike population, that is, these parameters are related only to the statistical distribution of synaptic states regarding the internal state of the neuron, and not the field variables, namely, the soma membrane potential and the synaptic efficacy.

Second, we consider the effect of a variation of the source coefficients p for the field equations (figure 18.8, bottom left panel). We observe an earlier onset of population spikes. This effect can be obtained with any of the following parameters that have a similar role according to the structure of field equation (2): p, m, a, $\mu(0)$, $\{t\}$. The effect of a variation of m is represented on figure 18.8, bottom right panel that can be compared with second row, left panel.

The qualitative deformation of the extracellular waveform obtained with numerical simulations agrees with the physiological mechanisms introduced in the original equations. Experiments are now made to test the model in order to confirm these results. The extracellular field potential will be recorded for each of these parameters in the specified conditions of

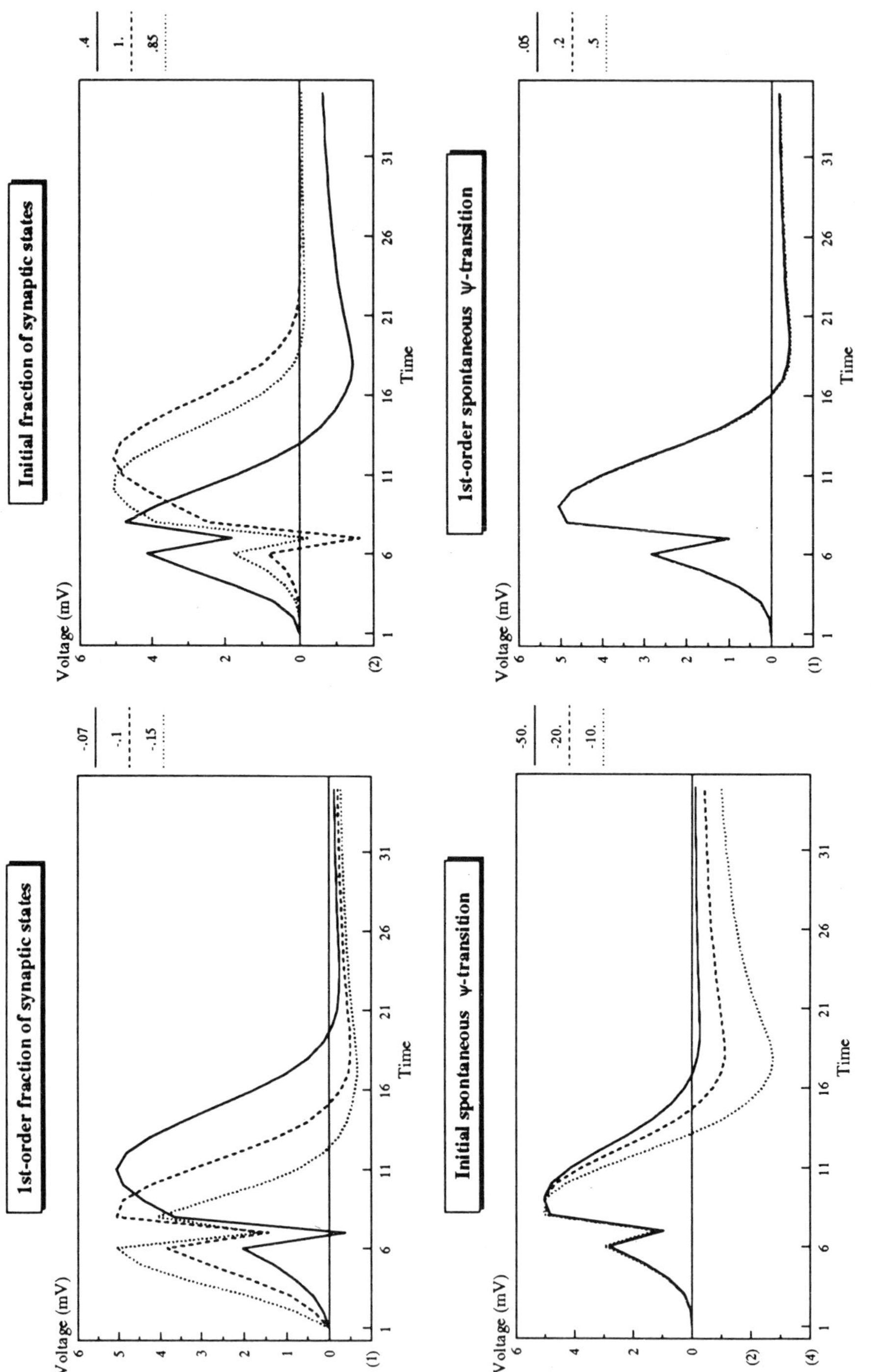
1st-order fraction of synaptic states
-.07
-.1
-.15
Initial fraction of synaptic states
.4
1.
.85
Initial spontaneous ψ-transition
-50.
-20.
-10.
1st-order spontaneous ψ-transition
.05
.2
.5
Voltage (mV)
Time

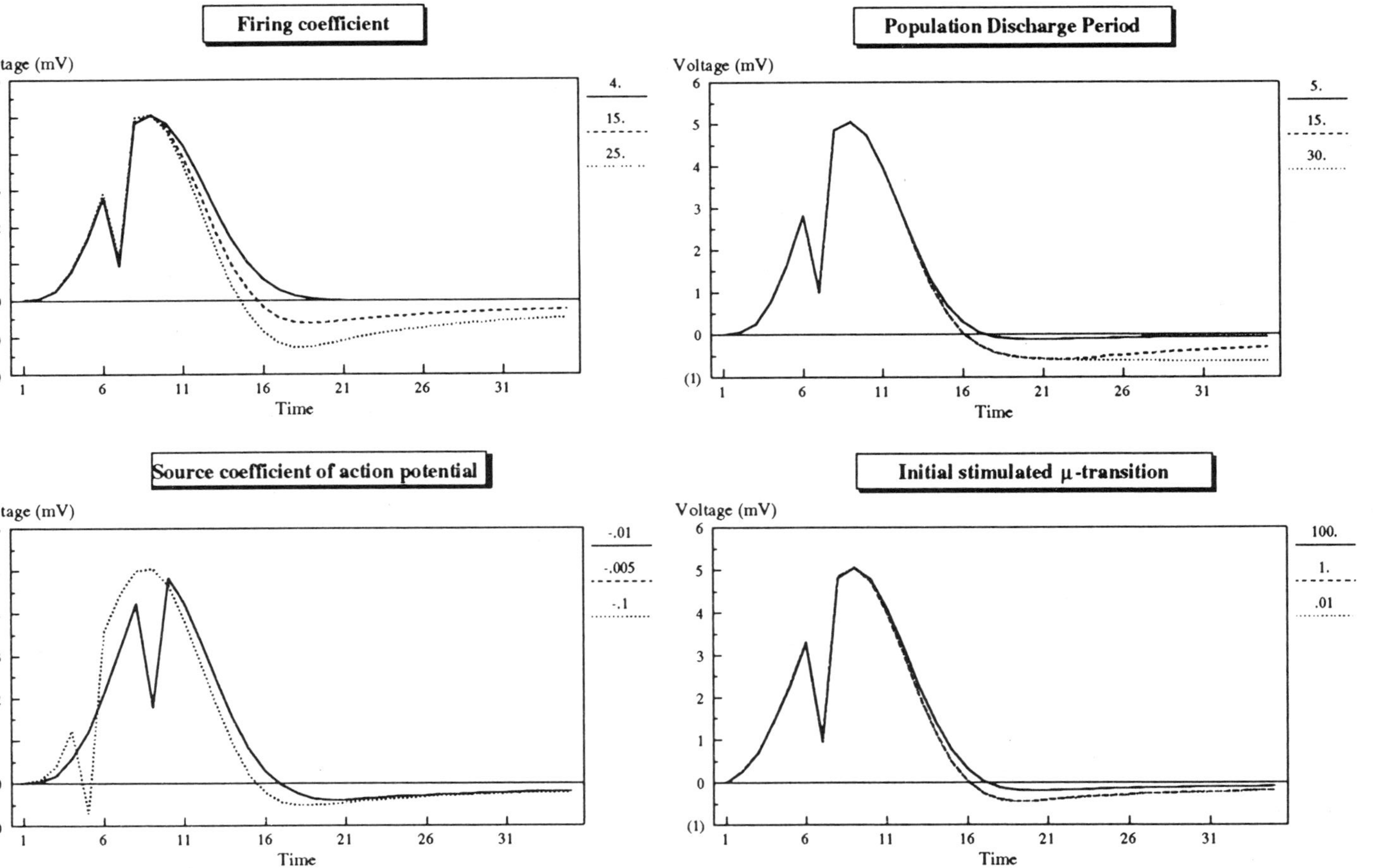

Figure 18.8 Study of the role of parameters occurring in the 2-level field model for the monosynaptically evoked granule cell population activity. Legends indicate which parameter is concerned in the study. Three values of the considered parameter around a basic standard value (obtained to fit the experimental curve obtained in usual conditions) give the three curves in each schema.

experiment. If the qualitative deformation of the recorded waveform is the same, it will be possible to identify the parameters with the mechanisms included in the proposed experiments.

AN INTEGRATED HIERARCHICAL MODEL OF THE DENTATE GYRUS

The present models at different levels of the dentate gyrus, molecular, synaptic and neuronal, can be hierarchically integrated to predict the space-time evolution of patterns of activity under the effect of LTP. Each level of the functional organization corresponds to a set of structures. From the bottom to the top, we can define: (1) modification of the conductance of channels after the release of neurotransmitters; (2) modification of synaptic efficacy propagated from one synapse to another; (3) activity propagated in the neural network.

To incorporate the molecular mechanisms in the n-level model, we can first determine the presynaptic efficacy using a phenomenological expression of these kinetic properties as a function of time in the neuron space. As did Magleby and Zengel (1982) it is obtained by fitting the observed curve with the product of the four functions that describe facilitation, potentiation, and depression (decreasing) rates:

$$\xi(s,t) = \xi^0(s,t) \prod_{i=1}^{4} (1 + d^{(i)}(s,t)) \tag{14}$$

This expression gives the probability of neurotransmitter release when the four functions $d^{(i)}(s,t)$ are known. The basal level of presynaptic efficacy $\xi^0((s,t); \langle X \rangle)$ is a function of activity $\langle X \rangle(t)$. A simplified expression including only two terms was used by Finkel and Edelman (1987) to describe presynaptic facilitation and synaptic depression. Second, following Finkel and Edelman (1987), Changeux and Heidmann (1987), a simple two-state kinetic model can be chosen to describe the binding of transmitter to postsynaptic receptor sites. The instantaneous postsynaptic efficacy is defined as a conductivity. In a simplified model, it is a sum of two conductance terms: (1) the product of the number $(1 - f_l(\Phi))$ of non-modified (but modifiable) channels with conductance g with the number of non-activated receptors $1 - R^*$; and (2) the product of the number $(1 - f_l^*(\Phi))$ of modified channels with conductance g^* with the number of activated receptors R^*:

$$\eta = \eta^0[(1 - R)(1 - f_I(\Phi))g + R^*(1 - f_I^*(\Phi))g^*] \tag{15}$$

The postsynaptic potential (PSP) is obtained using the Hodgkin-Huxley equation (Hodgkin and Huxley, 1952):

$$\frac{\partial^2 \Phi(s,t)}{\rho \partial s^2} = K\left[\left(C\frac{\partial \Phi}{\partial t}\right) + \sum_k g_k(\Phi - \Phi_k) - G(\Phi(s,t) - \Phi_\infty)\right] \tag{16}$$

where the g's and $G = Rg$ are the conductances that depend on the postsynaptic potential and on the synaptic and ionic currents; K is a coefficient, and C is the membrane capacity. This reaction-diffusion equation expresses the local conservation of the number of ions during their transport accross the membrane. Because of the symmetrical space diffusion term, it cannot represent the transport of action potential from one neuron to one another.

The present method can be used to integrate the model of the NMDA receptor-channel complex into the field equations with the following steps: (1) to determine an expression for presynaptic efficacy from a model which includes the dynamics of neurotransmitter release, uptake, etc.; (2) to deduce an expression for postsynaptic efficacy from the present model of the NMDA receptor-channel combined with a corresponding model of the AMPA receptor-channel; (3) these expressions being given, they can be included in equation (4): $\mu_0(s,t) = \langle \xi(s,t)\eta(s,t) \rangle$.

Because of the definition of the source term in the field equation, the last problem consists in determining the identity of the local transformation in the source. That is to say, what is the part of the μ-variation related to transformations at a distance, and what is the part related to only local transformations. Numerical simulations could give an answer, specifically by testing the phenomenon of saturation of LTP.

Experimental Verification of the Integrated Model

Experimental verification of the integrated model is complicated by the very fact that multiple processes from several levels of neuronal function are included in the representation. The nonlinearities inherent in the processes underlying the activity of individual neurons, and the complex functional organization imposed by the network structure represent significant barriers to achieving a model of a real, biological neural network. These obstacles can be overcome, however, using an application of nonlinear systems analysis (Berger et al., 1989, 1991; Sclabassi et al., 1988, 1989). The essential components of this approach (see Marmarelis and Marmarelis, 1978) involve electrically stimulating the perforant path with a series of electrical impulses having a randomly determined distribution of inter-impulse intervals, to generate a wide range of interactions among the neural elements and processes contained in the preparation. During the stimulation, evoked responses are recorded electrophysiologically from dentate granule cells. The functional properties of granule cells expressed in response to such a broad-band stimulus can be represented as the kernels of a functional power series, that is, as a set of input/output functions. Several procedures can be used to estimate the linear and higher-order nonlinear terms of the series, that is, for estimating the set of kernel functions that best describes the relationship between the randomized parameter of the input and granule cell output. Together, the kernels constitute a complete characterization of the composite dynamics of all stimulus-evoked processes at the molecular, cellular, and network levels. Thus, the dynamic properties of the network of neurons forming the hippocampus can be observed directly, and the contribution of all processes—both those that are known and observable and those that are unknown or unobservable—are included in the input/output functions. The general nature of this analytic approach also allows its application to at all levels of network organization: population activity, single cell activity, and the activity of subcellular processes (Berger et al., in press). In addition, we recently have developed procedures for estimating kernels for conditions of nonstationarity, that is, during the induction of synaptic plasticity, or in terms used above, during the transitions between different distributions of synaptic states (Krieger et al., 1992).

The same impulses sequence can be used as the stimulus conditions for simulating the generation of an extracellular waveform, the parameters of which can be identified from the field equation for neural activity ψ, and from the distribution of synaptic states restricted to the two first terms. In addition, a series of impulses with a frequency sufficient to induce a modification of synaptic efficacy, can be used and, therefore, an identification of the parameters of the two field equations and the complete statistical distribution function.

ACKNOWLEDGMENTS

The authors are grateful to the Office of Naval Research (N00014-90-J-4000), NIMH (MH00343 and MH45156), the NIH Biomedical Simulations Resource (P41 RR01861), and to the Conseil Général de Maine-et-Loire for their support of this research.

REFERENCES

Ambros-Ingerson, J., Granger, R., and Lynch, G. (1990) Simulation of paleocortex performs hierarchical clustering. *Science* 24:1344–1348.

Berger, T. W., and Sclabassi, R. J. (1988) Long-term potentiation and its relation to changes in hippocampal pyramidal cell activity and behavioral learning during classical conditioning. In *Long-term Potentiation: From Biophysics to Behavior*, P. W. Landfield and S. A. Deadwyler (eds.). N.Y.: Alan R. Liss, pp. 467–497.

Berger, T. W., and Yeckel, M. F. (1991) Long-term potentiation of entorhinal afferents to the hippocampus: Enhanced propagation of activity through the trisynaptic pathway. In

M. Baudry and J. Davis (Eds.), *Long-Term Potentiation: A Debate of Current Issues*. Cambridge, Mass.: MIT Press, pp. 327–356.

Berger, T. W., Harty, T. P., Barrionuevo, G., and Sclabassi, R. J. (1989) Modeling of neuronal networks through experimental decomposition In *Advanced Methods of Physiological System Modeling*, Vol. II, V. Z. Marmarelis (ed.). New York: Plenum, pp. 113–128.

Berger, T. W., Barrionuevo, G., Levitan, S. P., Krieger, D. N., and Sclabassi, R. J. (1991) Nonlinear systems analysis of network properties of the hippocampal formation. In *Neurocomputation and Learning: Foundations of Adaptive Networks*, J. W. Moore and M. Gabriel (eds.). Cambridge, Mass.: MIT Press, pp. 283–352.

Berger, T. W., Barrionuevo, G., Chauvet, G., Krieger, D. N., Levitan, S. P., and Sclabassi R. J. (in press). Theoretical and experimental strategies for a biologically constrained model of the hippocampus. In *Synaptic Plasticity: Molecular, Cellular, and Functional Aspects*, R. F. Thompson, M. Baudry, and J. Davis (eds.). Cambridge, Mass.: MIT Press.

Beurle, R. L. (1956) Properties of a mass of cells capable of regenerating pulses. *Phil. Trans. R. Soc. (Lond.)* 240:8–94.

Brown, T, H., Kairiss, E. W., and Keenan, C. L. (1990) Hebbian synapses: Biophysical mechanisms and algorithm. *Annu. Rev. Neurosci.* 13:475–511.

Changeux, J. P., and Heidmann, T. (1987) Allosteric receptors and molecular models of learning. In *Synaptic Function*, G. M. Edelman, W. E. Gall, W. M. Cowan, (eds.). New York: Wiley.

Chauvet, G. (1986) Habituation rules for a theory of the cerebellar cortex. *Biol. Cyber.* 55:1–9.

Chauvet, G. (1988) Correlation principle and physiological interpretation of synaptic efficacy. In *Systems with Learning and Memory Abilities*, J. Delacour and J. C. S. Levy (eds.). Amsterdam: Elsevier.

Chauvet, G. (1990) *Traité de Physiologie Théorique, Vol. III: Physiologie intégrative: Champ et organisation fonctionnelle.* Paris: Masson [*Theoretical Systems in Biology, Vol. III: Integrative Physiology: Field and Functional Organization.* Translation in progress, Pergamon Press]

Chauvet, G. (1993a) Hierarchical functional organization of a formal biological system: A dynamical approach. III. The concept of non-locality leads to a field theory describing the dynamics at each level of organization of the (D-FBS) sub-system. Phil. Trans. B R. Soc. London, 339:463–481

Chauvet, G. (1993b) An n-level field theory of biological neural network. *J. Math. Biol.* 31:771–795.

Chauvet, G., and Berger, T. W. (1990) Two-level field theory interpretation of hippocampal extracellular field potentials. *Soc. Neurosci. Abstr.* 16:739.

Chauvet, G., and Berger, T. W. (1994) A mathematical model of synaptically-evoked granule cell population activity in hippocampus. *Biol. Cyber.*, submitted.

Ermentrout, G. B., and Cowan, J. D. (1979) Temporal oscillations in neuronal nets. *J. Math. Biol.* 7:265–280.

Finkel, L. H., and Edelman, G. M. (1987) Population rules for synapses in networks. In *Synaptic Function*, G. M. Edelman, W. E. Gall, and W. M. Cowan (eds.). New York: Wiley.

Fisher, B. (1973) A neuron field theory: Mathematical approaches to the problem of large numbers of interacting nerve cells. *Bull. Math. Biol.* 35:345.

Gibb, A. J., and Colquhoun, D. (1991) Glutamate activation of a single NMDA receptor-channel produces a cluster of channel openings. *Proc. R. Soc. Lond.* 243:39–45.

Griffith, J. (1963) A field theory of neural nets I. *Bull. Math. Biophys.* 25:111–120.

Griffith, J. (1965) A field theory of neural nets II. *Bull. Math. Biophys.* 27:87–195.

Hebb, D. O. (1949) *The Organization of Behavior: A Neuropsychological Theory*. New York: Wiley.

Hodgkin, A. L., and Huxley, A. F. (1952) A quantitative description of current and its application to conduction and excitation in nerve. *J. Physiol.* 117:500.

Holmes, W. R., and Levy, W. B. (1990) Insights into associative long-term potentiation from computational models of NMDA receptor-mediated calcium influx and intracellular calcium concentration changes. *J. Neurophysiol.* 63:1148–1168.

Hopfield, J. J. (1982) Neural networks and physical systems with emergent collective computational abilities. *Proc. Natl. Acad. Sci. USA* 79:2554–2558.

Katz, B., and Thesleff, S. (1957) A study of the "desensitization" produced by Acetylcholine at the motor end plate. *J. Physiol.* 138:63–80.

Kohonen, T. (1972) Correlation matrix memories. *IEEE Trans. Comput.* C-21:353–359.

Kohonen, T. (1978) *Associative Memory: A System-Theoretical Approach*. Berlin: Springer-Verlag.

Kishimoto, K., and Amari, S. (1979) Existence and stability of local excitations in homogeneous neural fields. *J. Math. Biol.* 7:303–318.

Krieger, D. N., Berger, T. W., and Sclabassi, R. J. (1992) Instantaneous characterization of time-varying nonlinear systems. *IEEE Trans. Biomed. Eng.* 39:420–424.

Lester, R. A. J., Clements, J. D., Westbrook, G. L., and Jahr, C. E. (1990) Channel kinetics determine time-course of NMDA receptor-mediated synaptic currents. *Nature* 346:565–567.

Levy, W. B., Colbert, C. M., and Desmond, N. L. (1989) Elemental adaptive processes of neurons and synapses: A statistical/computational perspective. In *Neuroscience and Connectionistic Theory*, M. Gluck and D. Rumelhart (eds.). Hillsdale, N.J.: Lawrence Erlbaum Associates, pp. 187–235.

Magleby, K. L., and Zengel, J. E. (1982) A quantitative description of stimulation induced changes in transmitter release at the frog neuromuscular junction. *J. Gen. Physiol.* 80:613–638.

Marmarelis, P. Z., and Marmarelis, V. Z. (1978) *Analysis of Physiological Systems: The White Noise Approach*. New York: Plenum.

McNaughton, B. L., and Morris, R. G. M. (1987) Hippocampal synaptic enhancement and information storage within a distributed memory system. *Trends Neurosci.* 10:408–415.

Sather, W., Johnson, J. W., Henderson, G., and Ascher, P. (1990) Glycine-insensitive desensitization of NMDA responses in cultured mouse embryonic neurons. *Neuron* 4:725–731.

Sclabassi, R. J., Krieger, D. N., and Berger, T. W. (1988) A systems theoretic approach to the study of CNS function. *Ann. Biomed. Eng.* 16:17–34.

Sclabassi, R. J., Krieger, D. N., Solomon, J., Samosky, J., Levitan, S., and Berger, T. W. (1989) Decomposition through modeling of neuronal networks. In *Advanced Methods of Physiological System Modeling, Vol. II*, V. Z. Marmarelis (ed.). New York: Plenum, pp. 129–146.

Spruston, N., and Johnston, D. (1992) Perforated patch-clamp analysis of the passive membrane properties of three classes of hippocampal neurons. *J. Neurophysiol.* 67:508–529.

Traub, R. D., Dudek, F. E., Snow, R. W., and Knowles, W. D. (1985) Computer simulations indicate that electrical field effects contribute to the shape of the epileptiform field potential. *Neuroscience* 15:947–958.

Traub, R. D., Miles, R., and Wong, R. K. S. (1988) Large scale simulations of the hippocampus. *IEEE Eng. Med. Biol.* 7:31–37.

Treves, A., and Rolls, E. T. (1992) Computational constraints suggest the need for two distinct input systems to the hippocampal CA3 network. *Hippocampus* 2:189–200.

Urban, N., Berger, T. W., and Chauvet, G. A. (1992) A model of the kinetics of the NMDA receptor-channel complex. *J. Neurophysiol.*, submitted.

Wilson, M, and Bower, J. M. (1992) Cortical oscillations and temporal interactions in a computer simulation of piriform cortex. *J. Neurophysiol.* 67:981–995.

Wilson, H. R., and Cowan, J. D. (1972) Excitatory and inhibitory interactions in localized populations of model neurons. *Biophys. J.* 12:1–24.

Xie, X., Berger, T. W., and Barrionuevo, G. (1992) Isolated NMDA receptor-mediated synaptic responses express both LTP and LTD. *J. Neurophysiol.* 67:1009–1013.

Zador, A., Koch, C., and Brown, T. H. (1990) Biophysical model of a Hebbian synapse. *Proc. Natl. Acad. Sci. USA* 87:6718–6722.

19 Intrinsic and Emergent Corticohippocampal Computation

Richard Granger, Richard Myers, Eric Whelpley, and Gary Lynch

DIFFERENT FORMS OF POTENTIATION IN THE CORTICOHIPPOCAMPAL PATHWAY

Figure 19.1 schematically illustrates the components of the hippocampus and its interfaces with cortex. We take as a starting assumption that learning, in the hippocampus and elsewhere in telencephalon, is mediated by synaptic change. Physiologically induced synaptic change has been demonstrated in every link in each of these components; however, the characteristics of synaptic change vary from circuit to circuit. Hippocampal field CA1 shows long-lasting changes that have been measured for weeks without measurable decrement via chronically implanted electrodes (Staubli and Lynch, 1987). Potentiation in dentate gyrus, in contrast, has been reported to decrement over the course of days, eventually returning to baseline (Green et al., 1990). Both of these forms of potentiation have been shown to depend differentially on two prevalent subtypes of telencephalic glutamate receptor: they both are induced by the NMDA receptor and mediated or expressed by the AMPA receptor (Muller et al., 1988; Kauer et al., 1988). A form of potentiation exhibiting these same receptor dependencies has recently been demonstrated to occur in piriform cortex (Kanter and Haberly, 1990; Jung et al., 1993); however, this cortical potentiation, in contrast with hippocampal potentiation, is difficult to elicit in in vitro slices and is dependent on the behavioral set of the animal in vivo (Roman et al., 1987).

Potentiation in the mossy fiber synapses in hippocampal field CA3 has been shown to be independent of NMDA activation, thereby classifying it as a biochemically distinct form of synaptic change (Staubli et al., 1990; Zalutsky and Nicoll, 1990). The long-lasting nature of NMDA-mediated long-term potentiation (LTP) is one of the primary reasons for its being considered a candidate substrate for long-term memory; another key reason is the fact that LTP is synapse-specific, rather than a whole-cell or whole-axon change, which would be far too restrictive in terms of capacity

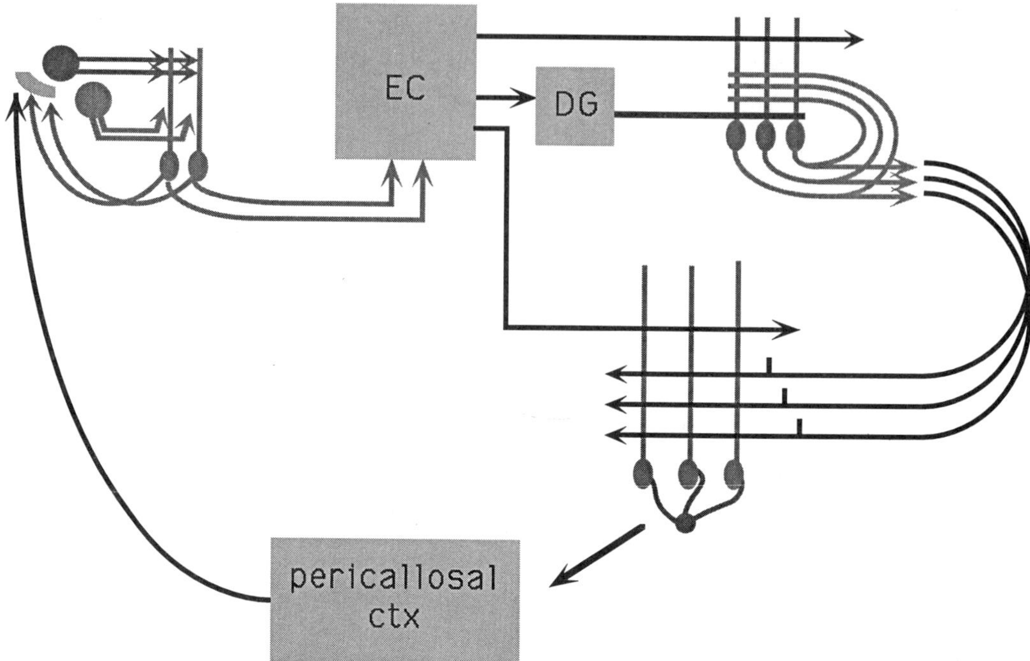

Figure 19.1 Hippocampus: its constituent circuitry and interfaces with cortex. Cortical circuitry (including olfactory paleocortex and polysensory association cortices) project to entorhinal cortex (EC), which in turn innervates all three constituents of hippocampus: dentate gyrus (DG), field CA3, and field CA1. DG in turn generates the mossy fiber projection to CA3, and CA3 provides the Shaffer/collateral projection to CA1. CA1 projects out, to deep layers of entorhinal cortex, to subiculum, and to pericallosal cortex, all of which in turn project back to association cortex.

for a long-term memory system (Lynch and Granger, 1992). The duration of mossy-fiber potentiation (MFP) in field CA3 is still unknown but initial findings seem to indicate that it is relatively transient, perhaps lasting for only hours; moreover, synapse specificity of MFP has yet to be demonstrated, and we conjecture that MFP is a whole-axon effect.

We have individually approached each of the constituent networks comprising the corticohippocampal pathway in our program of research, and have discussed distinguishing anatomical and physiological features and their possible differential contributions to memory processing (Lynch and Granger, 1991, 1992; Lynch et al., 1991). For each such circuit the anatomical architecture and physiological operating and plasticity rules have been deduced from experiments in our and others' laboratories and subjected to simulation and theoretical analysis. These studies have identified distinct functionality emergent from the different circuitries. Perhaps not surprising for a highly evolved system addressing the processing of complex time-varying signals, each individual component provides useful

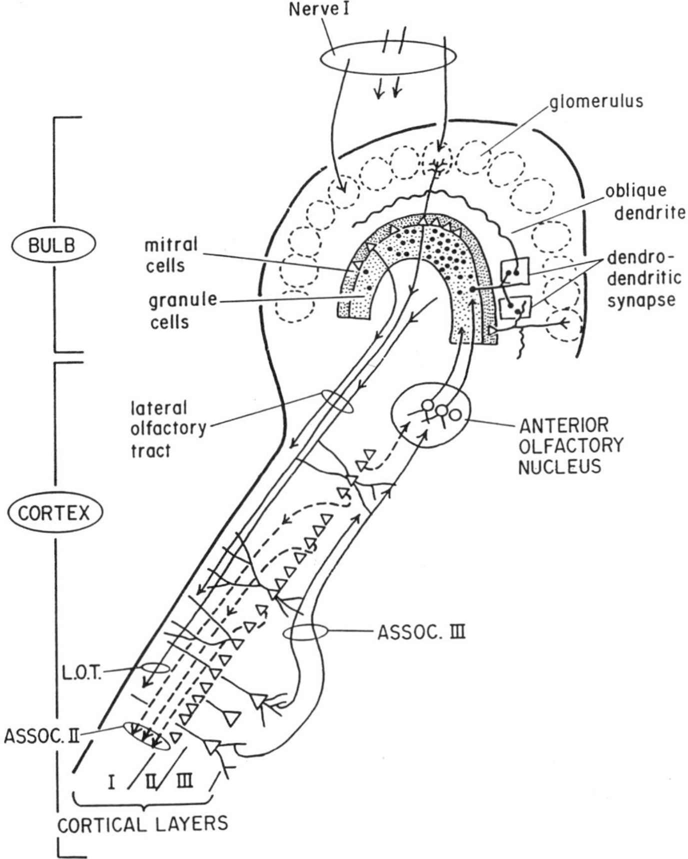

Figure 19.2 Olfactory system anatomy: receptor cells in the nasal epithelium form a topography-preserving projection via the first cranial nerve to the excitatory mitral cells in the olfactory bulb; these cells in turn project sparsely and nontopographically to layer I of olfactory cortex, contacting apical dendrites of layer II and III cells. Cortical cells project back, both directly and via the anterior olfactory nucleus, to the inhibitory granule cells of bulb, which dendrodendritically inhibit mitral cells.

computational functionality, and combined action of multiple circuits adds significant new functions not present in the constituents alone. We here outline the findings in two selected constituent circuits and some combined circuit operations leading to the beginnings of a hypothesis of differential contributions of hippocampal components to perceptual learning.

PRIMARY OLFACTORY CORTEX

Anatomy

Figure 19.2 schematically illustrates the major anatomical features of the primary olfactory (piriform) cortex and its primary subcortical input structure, the olfactory bulb. A key architectural feature is the presence of a loop consisting of (1) excitatory feedforward fibers from bulb mitral cells to cortical layer II/III cells, and (2) an inhibitory feedback projection from cortex to bulb. In detail, the primary bulb excitatory cells, mitral/tufted cells, project monosynaptically to layer I of olfactory (piriform) cortex, where they synapse sparsely and nontopographically onto apical dendrites of layer II and III cortical cells (Price, 1973). Layer II cells emit collateral axons which flow predominantly caudally, contacting more caudal layer II and III cells, and forming the primary input to lateral entorhinal cortex (Zimmer, 1971; Wyss, 1981). The axons of layer III cells in caudal piriform cortex flow predominantly in the opposite direction, mostly passing via layer III back to bulb, to synapse not on the excitatory mitral/tufted cells, but rather on bulb granule cells, which in turn inhibit mitral/tufted cells (Mori, 1987).

Synchronous Cyclic Activity

Unrestrained animals actively exploring an environment exhibit synchronous EEG activity at roughly the 4- to 8-Hz theta rhythm in hippocampus (Hill, 1978). In an olfactory learning task, the entire olfactory-hippocampal pathway becomes synchronized to the theta rhythm (Komisaruk, 1970; Macrides, 1975; Eichenbaum et al., 1987), and the latency of peaks of this theta wave in each of the constituent circuits is roughly what would be expected due to synaptic transmission delays between the circuits (Macrides, 1975).

It has been demonstrated that this rhythm is the optimal stimulation pattern for the induction of LTP in hippocampal field CA1: when afferents to a target are activated in brief (4-pulse) high-frequency (100 Hz) bursts separated by 200 msec (i.e., the 5-Hz theta rhythm), LTP is robustly induced, in vitro or in vivo, without the need for GABA blockers, voltage clamps, or physiologically implausible long (seconds) high-frequency bursts of activity (Larson and Lynch, 1986; Larson et al., 1986). The mechanism underlying the optimality of the 200-msec interval is understood: the initial afferent burst activates both excitatory and inhibitory postsynaptic potentials (PSPs), with their characteristic different time courses, EPSPs lasting 10–20 msec and IPSPs lasting 100–300 msec. At 200 msec after this first burst, the initial IPSP has roughly returned to baseline, but the inhibitory synapse has become refractory, due to presynaptic $GABA_B$ inactivation of $GABA_A$ transmission (Mott and Lewis, 1991). Thus the second burst activates predominantly only excitatory currents. These therefore summate with each other over the duration of the brief burst, enabling them to surpass the threshold of the voltage-sensitive NMDA receptor channel (Collingridge et al., 1983), allowing it to open and pass Ca^{2+}, initiating the cascade of events inducing LTP (Lynch et al., 1983).

Emergent Computation

LTP in olfactory cortex (Roman et al., 1987; Kanter and Haberly, 1990; Jung et al., 1993) is induced via stimulation patterns (brief bursts at the theta rhythm) matching those found during learning in behaving animals (Komisaruk, 1970; Otto et al., 1991). This argues for investigations of cortical memory function based on this synchronized operating mode together with the induction and expression rules for LTP. Some computational models of the olfactory system have been offered based on hypothetical synaptic plasticity rules other than LTP (Haberly and Bower, 1989; Hasselmo et al., 1990; Freeman, 1991). Hypotheses and computational analysis of the olfactory system that do incorporate these features of LTP induction and expression have led to findings suggesting not just encoding of sensory cues, but organization of the resulting memories into structures not typically seen in neural network models. Implementation of a repetitive sampling feature meant to represent the cyclic sniffing behavior of mammals (Komisaruk, 1970) produced a system that exhibited successively finer-grained encodings of learned cues over sampling cycles. Each sampling cycle includes feedforward activity from bulb to cortex followed by feedback from cortex to bulb. Because this feedback activates long-lasting inhibition in the bulb, the next sample of the input arrives against an inhibitory background in bulb, effectively masking part of the input, and thereby resulting in different activity patterns in bulb and in cortex. Thus, resampling a fixed cue generates different cortical responses with each new sampling cycle.

Learning in this model results in the initial cortical responses becoming nearly identical to sufficiently similar inputs from bulb. This phenomenon, called "clustering," has been predicted based on Hebbian learning rules by many researchers (von der Malsburg, 1973; Grossberg, 1976; Rumelhart and Zipser, 1985). Essentially, the resulting cortical response corresponds to general families or clusters of inputs; a given response signals membership of the input in a given cluster (e.g., "fruit" odors versus "meat" odors versus "floral" odors), thereby coarsely partitioning the input space.

The feedback from cortex to the bulb granule cell layer selectively inhibits those portions of bulb response giving rise to the cortical firing pattern; this inhibition in bulb lasts for hundreds of milliseconds (Nicoll, 1969). Resampling then causes new bulb and cortical activity against the background of this long-lasting inhibition. The resulting cortical response corresponds to odor components not shared across category members;

that is, the first sample response to a set of flowers will all be identical, signifying that these odors are all members of a single category; subsequent samples correspond to differences among different flowers, thereby effectively distinguishing among subcategories of floral odors. Thus learning via LTP in the model generates a multilevel hierarchical memory that uncovers statistical relationships inherent in collections of learned cues, and during retrieval sequentially traverses this hierarchical recognition memory (Ambros-Ingerson et al., 1990).

Predictions from Modeling

These findings from computer simulations and theoretical analyses lead to a specific hypothesis of paleocortical function: repetitive sampling and learning combine to cause the cortex to organize its memories into a hierarchical tree which is traversed during recognition. Specific findings from the models give rise to specific and testable predictions at both behavioral and physiological levels: behaviorally, animals should learn similarity-based categories; physiologically, cell spiking in piriform cortex should be sparse and odor-specific during olfactory learning and recognition, in contrast to models with extensive global activity. More specifically, different cortical cells should discharge over successive sampling cycles with progressively more selective tuning.

A recent set of behavioral experiments has tested one primary prediction of the model, namely that rats will spontaneously encode and use similarity-based categories. In the model, cortical responses to category members become identical only after a number of similar cues are learned by the model. This behavior is distinguishable from stimulus generalization, which predicts that choice behavior should, without learning, arise from physical similarity of stimuli alone, rather than from the existence of a sufficient number of similar stimuli. The experimental results supported the computer prediction, providing the first evidence that rats build unsupervised similarity-based categories, and that they can do so with widely spaced learning sessions (Granger et al., 1991).

Different computational models have generated distinct predictions about the physiological responses of cortical cells during learning of novel olfactory stimuli: our published modeling results have predicted that spiking responses to odors will be sparse (Lynch and Granger, 1989; Granger et al., 1989), whereas modeling by Freeman and associates predicts extensive excitatory activity (Freeman, 1975, 1991). Surprisingly few experimental studies have focused on the responses of single, primary sensory cortical units during acquisition of novel olfactory stimuli in behaving, freely moving animals. Behavioral studies have shown that in olfactory learning tasks mammals are capable of very rapid locomotor responses, within 0.5 sec, to olfactory cues, even though typical odor delivery systems may take up to 100 msec from odor onset to time of reception at the epithelium (Staubli et al., 1987; Eichenbaum et al., 1987). Recent experiments were performed to study the processing that occurs within this relatively narrow time period, during which odors are detected, recognized, and appropriate responses organized. The results indicated that the majority of cells in piriform cortex do not respond to most odors, and a small number of cells exhibited odor specificity of response, that is, cortical coding is extremely sparse (McCollum et al., 1991).

The further prediction from this computational model is that different cortical cells should discharge over successive sampling cycles, with progressively more selective tuning; further experiments will be required to test this prediction.

HIPPOCAMPAL FIELD CA1

Structure and Activity

Primary excitatory pyramidal cells in field CA1 of hippocampus receive afferent contacts from entorhinal cortex (on their distal apical dendrites) and from field CA3 (on proximal apical dendrites). Both afferent systems make sparse contact and exhibit little topographic organization, although some regularities exist in projection patterns from CA3 to CA1 along the septal-temporal extent of the hippocampus. Also contacted by these two afferent systems are inhibitory interneurons whose axons arborize within a small local radius, such that each inhibitory cell makes contact with roughly 100 excitatory pyramidal cells, which in turn outnumber the interneurons by roughly two orders of magnitude. "Local circuits" in field CA1 therefore consist of several excitatory neurons jointly innervating and receiving feedback from a smaller number of shared inhibitory interneurons. As mentioned, the hippocampus exhibits relatively synchronous EEG activity at the 5-Hz theta rhythm during exploration and learning, and individual units have been shown to fire preferentially at or near peaks of the theta wave envelope (Otto et al., 1991). Simulations have been investigated based on this arrangement in which excitatory cells receive input stimulation probabilistically as a function of the theta rhythm. Simulated neurons receiving the most input activation are the first to reach their spiking threshold. Activation of the inhibitory cell(s) in each local circuit causes suppression of further excitatory cell activity. Thus only those cells in a local circuit that are most strongly activated can spike or burst before

Computational evaluation of this non-Hebbian, temporal LTP induction rule raises the question of how the resultant learning might be expressed during subsequent performance. Analysis with biophysical simulations using the SPICE program (figure 19.3) predicts that sequentially arriving inputs with different synaptic potencies will evoke the greatest depolarization in a target cell when they are activated in the order of their strengths (most-potentiated input first). Given the order-dependence of the LTP rule, this means that a trained cell reacts most strongly to the same temporal sequence as it was trained on. Thus potentiation of an input sequence "codes" a cell to recognize that sequence subsequently, causing the cell to act as a form of "sequence detector." The predicted effect occurs robustly across a wide range of simulated conditions, one of which is illustrated in figure 19.3, middle rows.

Initial tests of the prediction using two equal-sized independent sets (S1 and S2) of Schaffer-commissural projections to a common dendritic locus in field CA1 of the hippocampal slice showed that the amount of LTP expressed in response to the input sequence on which the target cell had been previously trained was larger than the amount expressed in response to input sequences on which the cell was not trained.

Emergent Computation of Field CA1

These physiological characteristics of LTP in field CA1, incorporated into a network of simplified cells derived from the SPICE and local-circuit models described above, enabled investigation of the computational properties of this set of non-Hebbian learning (induction) and performance (expression) rules. Intuitively, the resulting learning algorithm processes a temporal input sequence by performing order-dependent potentiation of the synapses of those cell(s) that survive lateral inhibition in the competitive patches. The synapses of surviving cells are potentiated to a level commensurate with the order in which their inputs were activated: the contact with the first-arriving input is potentiated the most and the contact with the last-arriving input is potentiated the least.

The performance algorithm preferentially activates only those cells whose synapses are appropriately potentiated at each input step. If any patch in the network contains no cells that survive this "honing" process through the set of temporal input steps, the input is rejected (not recognized); if each patch contains a surviving cell, the input is accepted (recognized). The resulting "accept/reject" or "match/mismatch" response to any recognized input X will be unique within a specified error tolerance (Granger et al., 1994).

During performance, voltage summation uses potentiated weights, whereas during learning, summation is performed using only naive weights, implementing the distinction between changes to AMPA versus NMDA receptor conductance with LTP induction (Muller et al., 1988; Kauer et al., 1988). In other words, during episodes of learning (induction), the NMDA receptor–mediated voltage dominates cell responses, and therefore predominantly determines which of the target cells will "win" the competition and be potentiated. Yet these NMDA receptor–mediated voltages change little or not at all as a result of prior learning episodes: the changes are, rather, predominantly to the AMPA component. One computational consequence of this is notable: in contrast to the great majority of learning algorithms in neural network studies, prior learning via LTP has little or no effect on subsequent LTP, preventing the formation of "attractor" cells, that is, cells that tend to respond to any of a number of inputs similar to those previously trained.

The resulting extreme selectivity of cells, in combination with the order-dependency of the LTP-based learning and performance rules, confers unusually large capacity to the network. The capacity of the network can be cast in terms of its errors as a function of the number of sequences stored in a network of a given size. Two types of errors can be distinguished: errors of collision, in which a particular set of cells respond to more than one temporal string or "word" during training, and errors of commission, in which a target cell responds at testing (performance) time to a string on which it has not been trained. Theoretical analysis of the values of these error rates were derived by Granger et al. (1994), who showed that these rates remained extremely low even with heavy "loading" of the network, that is, training of large numbers of sequences into a relatively small network. It was found, for example, that a network of 1000 cells could be trained to recognize 10,000 sequences of length 10, with an error rate of 0.0001. It was also shown that the capacity of the network scaled extremely well; larger networks could learn comparably larger numbers of sequences, in contrast with many neural network approaches in which the network must be exponentially larger than the number of items it is to learn. In particular, given a network consisting of M competitive local circuits of C cells apiece, each with A synapses, the number n of sequences of length S that can be learned without exceeding error rate E is

$$n = \frac{C\log(1 - E^{1/SM})}{\log(1 - (1/A))} \tag{1}$$

Figure 19.4 is a contour graph of the relationship among the number of patches in the network, the number of cells in each patch, and the number

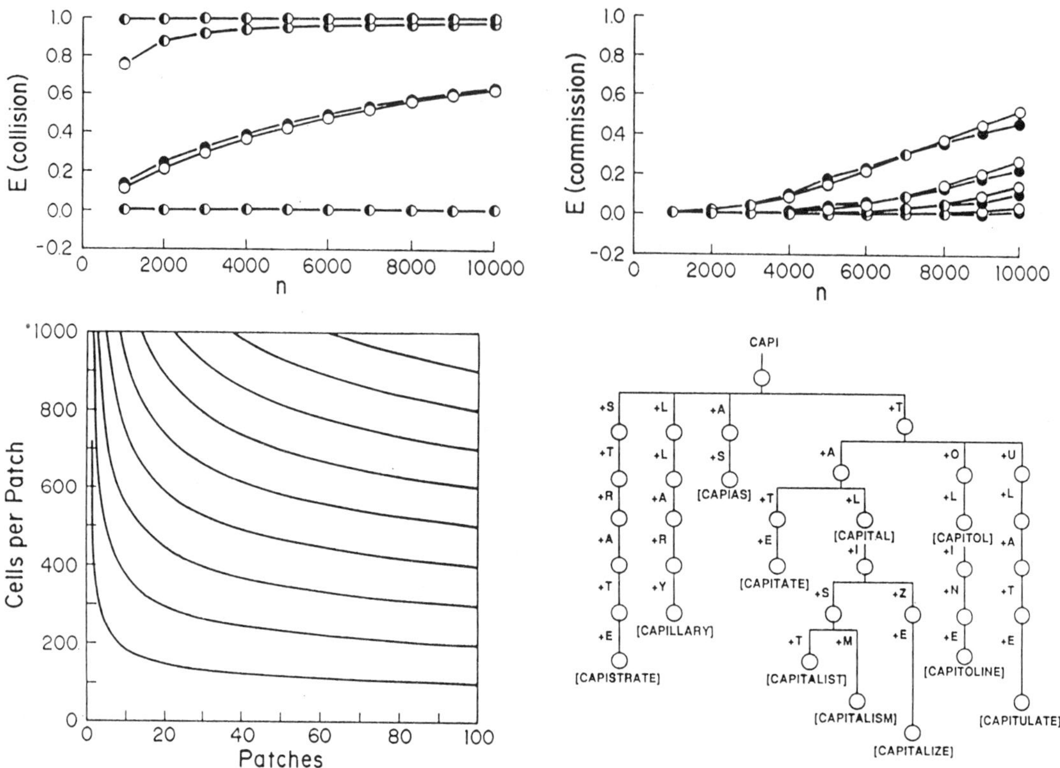

Figure 19.4 Capacity and scaling properties of the sequence-recognition network. Contour lines indicate the number n of sequences of length 10 that can be learned without exceeding a fixed error probability $E = 0.001$, for cells with 10,000 synapses each. The abscissa gives the number of patches in the network, and the ordinate denotes number of cells per patch. Each contour curve corresponds to a specific number n of learned sequences: the lowest curve denotes 5 million learned sequences: following the curve shows that this many sequences can be learned by a network of 5 patches (x-axis) and about 250 cells per patch (y-axis), or by a network of 40 patches of 125 cells per patch, etc. The contour curves increase by 5 million per curve; the highest curve shown corresponds to 45 million learned sequences (accomplished via 60 patches of 1000 cells each ranging to 100 patches of 900 cells each). The largest network for which theoretical values are plotted corresponds to the point at the upper right corner of the contour graph: this network consists of 100,000 cells (100 patches of 1000 cells each) and has a capacity of 50 million sequences, with error rate $E \leq 0.001$.

of sequences that can be stored without exceeding a fixed recognition error of p(commission) = 0.001.

If cells are assumed to receive 10,000 synaptic contacts on their dendrites, then the point at the upper right corner of the contour graph corresponds to a 100,000-cell network (100 patches of 1000 cells each) storing approximately 5×10^7 (50 million) random sequences of length 10, with a recognition error rate of 0.001. It is worth noting that 100,000 cells with 10,000 synaptic contacts apiece constitutes a relatively small network in real terms (less than the size of one field CA1 in a rat), and yet 50 million learned sequences is equivalent to one novel sequence learned every 10 sec for 8 hr each day for 50 years, with a 0.001 recognition error rate. This probably exceeds the rat's capacity by orders of magnitude; allowing for possible additional errors of transmission, contact and noise, it still provides a scheme capable of accounting for the dramatic capacity of our own memory systems. Standing (1973) found that subjects exposed for only seconds apiece to 10,000 pictures nonetheless exhibited 90% recognition rates for those stimuli when tested weeks later. Neural network models do not exhibit the capability for this kind of rapid learning, long retention and large capacity; rather, they typically require extensive training and cannot store many items without using very large networks. Surprisingly, what might have been thought to be excessively low-level biological detail of LTP induction and expression physiology give rise to networks that directly address the problematic question of how large memories can be implemented in brain circuitry.

HIPPOCAMPAL LOOP INTERACTIONS: CORTEX AND FIELD CA1

Supervised Clustering

Circuit-specific computational powers arise in simulations of piriform-entorhinal cortex and in hippocampal field CA1 as a consequence of the distinct synaptic plasticity rules and embedding architectures specific to each of these networks. Simulation of piriform-entorhinal cortex performs efficient hierarchical clustering (Ambros-Ingerson et al., 1990) and simulation of non-Hebbian LTP-based learning rules in field CA1 yields match/mismatch recognition with high capacity (Granger et al., 1994). Independently, these provide computationally efficient implementations of two functions of known utility and application. These two circuits are connected via the olfactory corticohippocampal loop schematized in figure 19.1: dentate gyrus projects to field CA3, which in turn projects to field

CA1; entorhinal cortex (superficial layers) projects to all three hippocampal fields with decreasing respective density, and field CA1 projects back to entorhinal cortex (deep layers, which in turn project vertically to superficial layers) via the subiculum. Viewing cortex and field CA1 in isolation led surprisingly to identification of individual contributed computational functionality; somewhat more quixotic is the hope that the interactions of these two circuits, and thus their combined contribution to the corticohippocampal loop, might be characterized. Our efforts in this regard take as their starting point the modeled outputs of cortex as input to CA1 and the output from CA1 as the input to cortex.

At the outset it is notable that the two circuits produce outputs that convey extremely different quantity of information. Simulated cortical responses carry rich cue-specific information indicating category membership, subcategory membership, and so on. In contrast, simulation indicates that CA1 output is simply of the match/mismatch variety, carrying very little differential information beyond the binary categories of recognized versus not recognized. An additional anatomical fact is that cortical pathways corresponding to all sensory modalities converge (by way of association cortical regions) onto hippocampus via entorhinal/perirhinal cortex; the presence of these afferents and their effect on field CA1 can be assumed without loss of generality. We have then a skeleton pathway containing two circuits: the cortical network receives sensory information, identifies category information, and passes it to CA1. CA1 receives sensory information from this and other cortical modalities, decides whether the coincidence of these inputs is recognized (match) or not (mismatch) and passes this binary signal back to cortex. This signal is impoverished with respect to information, and cannot possibly be used to direct cortical encoding in any significant fashion. We conjecture that this hippocampocortical feedback signal serves only the binary purpose of triggering the cortical network to iterate an additional activity cycle, thereby producing one further hierarchical layer of encoding of its sensory input: a mismatch signal triggers an additional output from cortex; a match triggers no further cortical response. Reflection shows that from this seemingly trivial corticohippocampal interaction blossoms a significant new capability: that of supervised learning.

As a simple example, imagine that the general category of "sweet" odors includes flowers and nuts. A single sniff of an object (i.e., a single iteration cycle of olfactory cortex) will produce only the information that the object is sweet. At the same time, other sensory modalities may operate on the object: it may taste edible or inedible; feel smooth, rough, sharp; look attractive or unattractive, large, small, colored, etc. A flower smells sweet, looks brightly colored, and tastes bad; a nut smells sweet, looks dull, and tastes good. The sweet smell is not predictive of edibility, but the olfactory system is quite capable of further resolution of "sweet" into "flowery," "nutty," etc., which are predictive in this instance. Coincidence or consistent sequence of features can be learned (matched) by the CA1 system. If, as conjectured, its match/mismatch signal triggers further cycling of the cortical network, resulting in finer resolution of the odor, then the combined cortical-CA1 system can find the appropriate level of resolution for predictive classification in a given context.

In the above example, it can be assumed that the intended "supervised" classification is that of edible versus inedible substances. For a particular olfactory input, the cluster information provided by cortex can be tested by CA1 for a match with the appropriate supervised category (edible or inedible). There are only three possible results for the cortical signal: (1) it has already been learned to uniquely match the intended classification; (2) CA1 has not learned that this cortical signal is a member of this classification; (3) it matches the intended classification but not uniquely. If (1), then the job of both CA1 and cortex is done: CA1 returns a "match" response, and cortex iterates no further. If condition (2) prevails, then the signal is to be learned, that is, CA1 is trained that this signal matches this class. If (3), then CA1 returns a "mismatch" signal, triggering another cortical operation cycle, and the process then recurs: the resulting finer-grained cortical cluster representation is tested for classification recognition in CA1, with the same three possible outcomes.

It can be seen that the cortical algorithm remains unchanged: it still is performing purely unsupervised clustering of the input. The operation of field CA1 likewise is as it was: it is matching or mismatching, recognizing or rejecting, a sequence (an item and its intended category). The new emergent supervised-learning faculty arises from CA1's recognition control of cortical processing. Now the hierarchical tree will be built out to precisely the depth needed to distinguish between members of different classifications. For easy distinctions, coarse-grained cluster information suffices and a single cortical iteration uniquely identifies the sensation as a member of the relevant class. For difficult discriminations, initial responses will be recognized as ambiguous, and further cortical processing is triggered to render the input with finer granularity. The result is variable-sized tiling of the set of inputs in the input space, so that tightly circumscribed regions are demarcated in those areas where similar-smelling inputs fall into different classes, but broad regions are marked when all inputs residing in that region are members of the same class.

Application and Efficiency of the Combined Cortex-CA1 Algorithm

The resulting combined algorithm has significant advantages over standard statistical and neural-network approaches, which can be illustrated via an example in the signal-processing domain. Simulations of this simplified cortical-CA1 interactive algorithm have been applied to complex signal-processing tasks and the results compared against those of other statistical algorithms and standard artifical neural-network approaches. The performance of the biological algorithm routinely exceeds that of the standard contenders, for reasons that will be made clear. Databases of preprocessed sonar signals compiled by the Navy have been used as a testbed for benchmark data on signal-processing algorithms; one such database contains processed sonar signals purportedly denoting signature sonar responses for two classes of undersea objects: rocks and mines. Each entry consists of a 60-dimensional vector corresponding to discrete Fourier transforms of fixed-time slices of a sonar response to a rock or a mine. A multilayer perceptron network consisting of 86 cells and trained via the extended delta or backpropagation learning rule took 300 epochs (each consisting of one training trial on each of the 104 training vectors) to converge to a solution with 100% correct classification of those training vectors; its generalization performance, that is, its classification score on the test vectors (those vectors on which it had not been trained) was 89.2% (Gorman and Sejnowski, 1988). The cortex-CA1 combined algorithm achieved 100% correct classification rate on the training vectors after only 20 training epochs, and its classification on the test vectors was 94.23%.

Figure 19.5 illustrates the type of partitioning performed by backpropagation (BP) versus the corticohippocampal algorithm: BP, and most other network approaches, partition the input space via placement of separator hypersurfaces such that members of one category reside on one side of the surface and vice versa. Much attention is paid in the literature to speed of movement of these surfaces, different surface shapes, etc., but the whole class of approaches nonetheless shares the attribute of constructing a single-level description of the space. The corticohippocampal algorithm, by contrast, constructs a hierarchical partitioning, consisting of division of the space into (relatively large) clusters, subdivision of those clusters into subclusters, etc. The figure illustrates the nature of the difference between these two approaches. In the depicted example (corresponding to another, simpler sonar database, obtained from the Naval Underwater Systems Center in Rhode Island) members of two categories are denoted by circles and squares. It can be seen that there are a few "outliers," that is, squares that seem more similar to most circles than to other squares, and circles

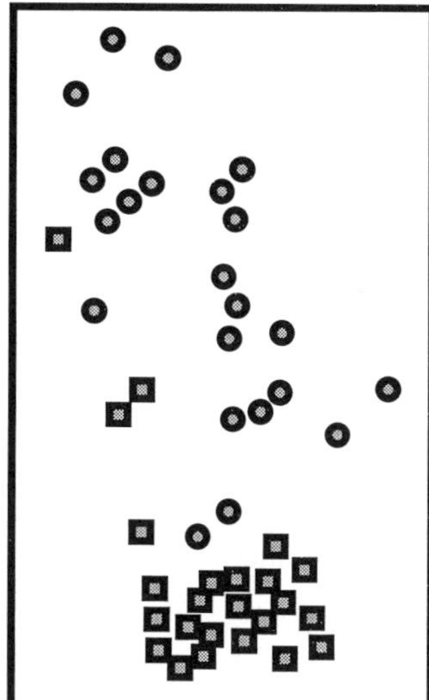
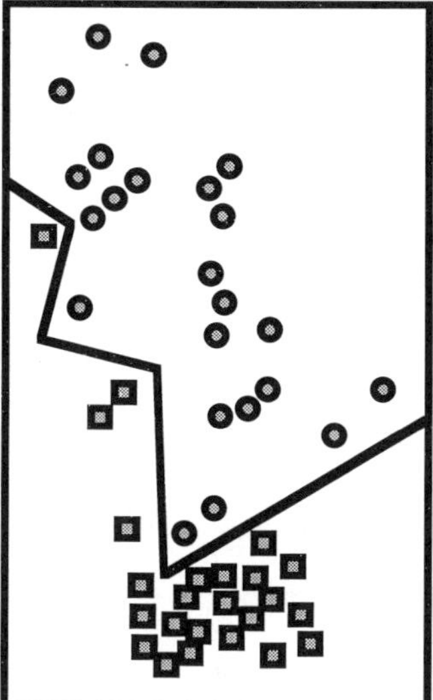
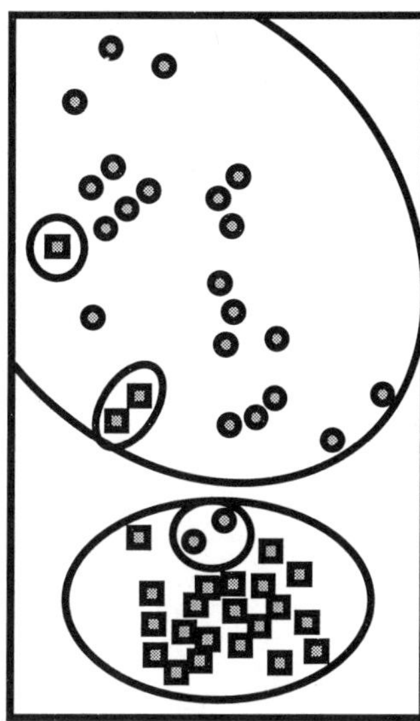

Figure 19.5 Partitioning methods via separators versus neighborhoods. (*Left*), Planar projection of two classes (circles/squares) of multidimensional data. (*Middle*) ,Separator surface partitioning identifies surfaces (in this example, piecewise-linear hyperplanes) such that all members of one category on one side of the surface; the presence of any difficult discriminations in the data typically renders such surfaces complex. (*Right*) Hierarchical partitioning identifies first coarse regions of input space each roughly characterizing the bulk of the training data, followed by identification of subregions (and sub-subregions, etc.) corresponding to outliers and exceptions.

that seem more similar to most squares than to other circles. These outliers are examples of sonar signatures that should be difficult to classify due to the difficulty of discriminating the slight differences between their signatures and those of their neighbors of the other category. Hyperspatial partitions identify category membership, although it cannot be determined which points are good category members versus which are outliers. Supervised hierarchical clustering first identifies general regions of the space, each of which in this instance contain most members of a single category, plus outliers from the other category; the next hierarchical level identifies those subregions of each cluster containing outliers. Thus the biological network conveys information beyond that provided by the backpropagation network, and does so with significantly less computational cost in terms of network size and required training time. These constitute general advantages of the approach: hierarchical decomposition of data conveys more information than single-level descriptions, and the network operates very efficiently in terms of space and time costs (Granger et al., 1989; Ambros-Ingerson et al., 1990; Granger et al., 1994).

Figure 19.6 contains an excerpt from the large hierarchical tree constructed by the biological network for the rocks/mines sonar data. As in

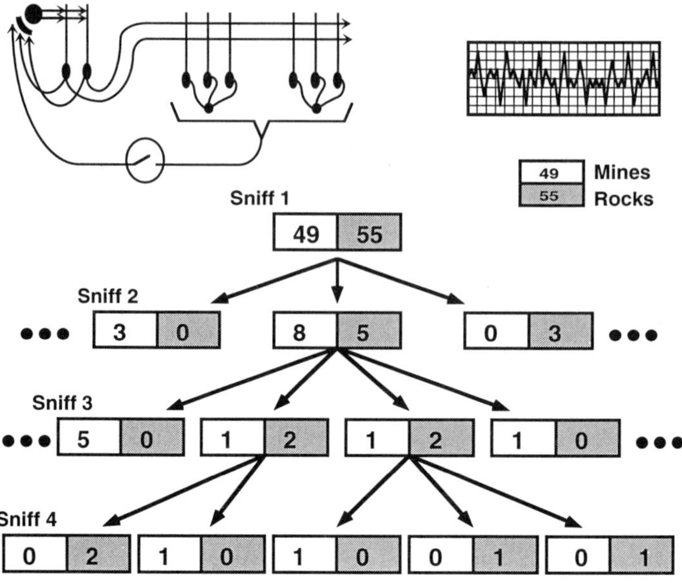

Figure 19.6 Application of the combined corticohippocampal supervised network algorithm to a standard benchmark database of naval sonar signals. The network is trained on members of two classes, rocks versus mines, and iterates until a level in the hierarchy is found at which all category members are from a single class, either rock or mine. The effect is that of an iterative-deepening beam search, with deeper levels in the tree being generated only when necessary for finer-grained descriptions of difficult discriminations between similar members of the two different classes. Shown is a portion of the tree constructed after training on 104 rocks and mines. The top category does not discriminate between classes. Those subordinate levels containing only rocks or only mines remain leaf nodes. Ambiguous subordinate levels cause generation of sub-subordinate levels, until unambiguous leaves are arrived at.

the above example, the hierarchical tree successively subdivides the space into regions containing similar items. At any given level, if not all of the members of a particular such region are either rocks or mines, then the CA1 response will be ambiguous, triggering the cortical tree to be built out one level deeper. Thus the tree is shallow where similarity-based regions of the space correspond well to the supervised categories (rocks and mines), and it is deeper where members of the two different categories yield sonar signatures that are very similar to each other and hence difficult to distinguish.

One key reason for the relatively low computational cost of the performance of the biological network is that, in constrast to the case for backpropagation, it does not need extensive feedback signalling from the network receiving supervised information (CA1) to the network partitioning the space (cortex). In other words, precisely what is absent from the biological model is extensive backwards-propagation of information, in contrast to the very process from which the backpropagation algorithm takes its name. Discussion in the literature about the biological plausibility of the backpropagation algorithm has raised the issue of whether a "weakened" form of backpropagation, with somewhat less stringent requirements on its backpropagated information, might exist in the brain (Zipser and Andersen, 1988; Zipser, 1991). In contrast, the operation of the cortex-CA1 algorithm intrinsically arises from and takes advantage of the simplicity of the information fed back from CA1 to cortex; in contrast, this level of feedback information is impoverished with respect to the computational needs of the backpropagation algorithm, and would prevent it from successful training. This is an instance in which the lack of an expensive computational resource (dense feedback connections) required by one algorithm nonetheless gives rise to a more powerful algorithm which computes a richer family of results.

DISCUSSION

It has long been noted that hippocampal damage can cause anterograde amnesia without significant retrograde amnesia (Scoville and Milner, 1957); it has therefore been proposed that the hippocampus serves as a "waiting platform" through which memories must pass, but upon which they reside only transiently, en route to their ultimate long-term destination, presumably neocortex (Squire, 1992). Yet long-term potentiation was identified in the hippocampus (Bliss and Lømo, 1973) and has been robustly replicated by scores of researchers for two decades. Moreover, the longest duration of LTP yet demonstrated has been in hippocampus (Staubli and Lynch, 1987). If hippocampal LTP is a substrate of long-term memory, what, if anything, might it be storing?

The theoretical findings discussed here constitute a proposal for a kind of "memory" that could reside in hippocampus. Learned match/mismatch information in hippocampus can play a crucial role in extrahippocampal cortical encoding, as described; loss of hippocampus would impair this encoding, without causing an obvious loss of prior "data" memories of the kind usually searched for with behavioral and verbal probes. Thus these results provide one potential resolution of the issue of long-term information storage in hippocampus, consistent both with the existence of LTP in hippocampus and the lack of retrograde amnesia with hippocampal damage. We put forward the hypothesis that hippocampus operates on memories, and that it learns while doing so, but that the information learned is not of a type typically tested for with neuropsychological damage. Indeed, it is difficult even to name this type of memory, without verbose reference to the computational analysis of its purported role in memory processing. This difficulty points up simultaneously the impoverishment of descriptive language for memorial phenomena, and yet our reliance on such language for characterization of behavioral and neuropsychological findings. It is hoped that further elucidation of emergent computation from specified brain circuitry may enrich the relevant descriptive language of psychology. Initial work on a central hippocampal structure, field CA3, has suggested its role in processing of temporal phenomena (Taketani et al., 1992), possibly providing, in combination with dentate gyrus and CA1, a set of specialized mechanisms for the differential treatment of recently experienced information over a range of time scales, from seconds to hours to days (Lynch and Granger, 1992). Thus possibly a form of memory for time and recency may be stored in hippocampus, and used in processing the stream of information that occurs over these different time scales.

REFERENCES

Ambros-Ingerson, J., Granger, R., and Lynch, G. (1990) Simulation of paleocortex performs hierarchical clustering. *Science* 247:1344–1348.

Bliss, T. V. P., and Lømo T. (1973) Long-lasting potentiation of synaptic transmission in the dentate area of the anaesthetized rabbit following stimulation of the perforant path. *J. Physiol.* (*Lond.*) 232:334–356.

Collingridge, G. L., Kehl, S. L., and McLennan, H. (1983) Exctitatory amino acids in synaptic transmission in the schaffer-commissural pathway of the rat hippocampus. *J. Physiol.* (*Lond.*) 334:33–46.

Coultrip, R., Granger, R., and Lynch, G. (1992) A cortical model of winner-take-all competition via lateral inhibition. *Neural Networks* 5:47–54.

Eichenbaum, H., Kuperstein, M., Fagan, M., and Nagode, J. (1987) Cue-sampling and goal approach correlates of hippocampal unit activity in rats performing an odor-discrimination task. *J. Neurosci.* 7:716–732.

Freeman, W. (1975) *Mass Action in the Nervous System*. New York: Academic Press.

Freeman, W. (1991) The physiology of perception. *Sci. Am.* 264:78–85.

Gorman, R. P., and Sejnowski, T. J. (1988) Analysis of hidden units in a layered network trained to classify sonar targets. *Neural Networks* 1:75–89.

Granger, R., Ambros-Ingerson, J., and Lynch, G. (1989) Derivation of encoding characteristics of layer II cerebral cortex. *J. Cog. Neurosci.* 1:61–87.

Granger, R., Staubli, U., Powers, H. A., Otto, T., Ambros-Ingerson, J., and Lynch, G. (1991) Behavioral tests of a prediction from a cortical network simulation. *Psychol. Sci.* 2:116–118.

Granger, R., Whitson, J., Larson, J., and Lynch, G. (1994) Non-hebbian properties of LTP enable high-capacity encoding of temporal sequences. *Proc. Natl. Acad. Sci.*, in press.

Green, E., McNaughton, B., and Barnes, C. (1990) Exploration-dependent modulation of evoked responses in fascia dentata. *J. Neurosci.* 10:1455–1471.

Grossberg, S. (1976) Adaptive pattern classification and universal recoding: I. Parallel development and coding of neural feature detectors. *Biol. Cybern.* 23:121–134.

Haberly, L. B., and Bower, J. M. (1989) Olfactory cortex: Model circuit for study of associative memory? *Trends Neurosci.* 17:258–264.

Hasselmo, M., Wilson, M., Anderson, B., and Bower, J. (1990) Associative memory function in piriform (olfactory) cortex: Computational modeling and neuropharmacology. *Cold Spring Harbor Symp. Quant. Biol.* 55:599–610.

Hill, A. (1978) First occurrence of hippocampal spatial firing in a new environment. *Exp. Neurol.* 62:282–297.

Jung, M., Larson, J., and Lynch, G. (1990) Long-term potentiation of monosynaptic EPSPs in rat piriform cortex in vitro. *Synapse* 6:279–293.

Kanter, E. D., and Haberly, L. B. (1990) NMDA-dependent induction of long-term potentiation in afferent and association fiber systems of piriform cortex in vitro. *Brain Res.* 525:175–179.

Kauer, J. A., Malenka, R., and Nicoll, R. (1988) A persistent postsynaptic modification mediates long-term potentiation in the hippocampus. *Neuron* 1:911–917.

Komisaruk, B. R. (1970) Synchrony between limbic system theta activity and rhythmical behavior in rats. *J. Comp. Physiol. Psychol.* 70:482–492.

Larson, J., and Lynch, G. (1986) Induction of synaptic potentiation in hippocampus by patterned stimulation involves two events. *Science* 232:985–988.

Larson, J., and Lynch, G. (1989) Theta pattern stimulation and the induction of LTP: The sequence in which synapses are stimulated determines the degree to which they potentiate. *Brain Res.* 489:49–58.

Larson, J., Wong, D., and Lynch, G. (1986) Patterned stimulation at the theta frequency is optimal for induction of long-term potentiation. *Brain Res.* 368:347–350.

Lynch, G., and Granger, R. (1989) Simulation and analysis of a cortical network. *Psychol. Learn. Motiv.* 23:205–241.

Lynch, G., and Granger, R. (1991) Serial steps in memory processing: Possible clues from studies of plasticity in the olfactory-hippocampal circuit. In *Olfaction: A Model System for Computational Neuroscience,* H. Eichenbaum, and J. L. Davis (eds.). Cambridge, Mass.: MIT Press.

Lynch, G., and Granger, R. (1992) Variations in synaptic plasticity and types of memory in cortico-hippocampal networks. *J. Cog. Neurosci.* 4:189–199.

Lynch, G., Larson, J., Kelso, S., Barrionuevo, G., and Schottler, F. (1983) Intracellular injections of EGTA block induction of hippocampal long-term potentiation. *Nature* 305: 719–721.

Lynch, G., Larson, J., Staubli, U., and Granger, R. (1991) Relating variants of synaptic potentiation to different types of memory operations in hippocampus and related structures. In *Memory: Organization and Locus of Change,* L. Squire, N. Weinberger, G. Lynch, and J. McGaugh (eds.). New York: Oxford University Press.

Macrides, F. (1975) Temporal relationships between hippocampal slow waves and exploratory sniffing in hamsters. *Behav. Biol.* 14:295–308.

McCollum, J., Larson, J., Otto, T., Schottler, F., Granger, R., and Lynch, G. (1991) Short latency single unit processing in olfactory cortex. *J. Cog. Neurosci.* 3:293–299.

Mori, K. (1987) Membrane and synaptic properties of identified neurons in the olfactory bulb. *Prog. Neurobiol.* 29:275–320.

Mott, D., and Lewis, D. (1991) Facilitation of the induction of long-term potentiation by $GABA_B$ receptors. *Science* 252:1718–1720.

Muller, D., and Lynch, G. (1988) Long-term potentiation differentially affects two components of synaptic responses in hippocampus. *Proc. Natl. Acad. Sci.* 85:9346–9350.

Muller, D., Joly, M., and Lynch, G. (1988) Contributions of quisqualate and NMDA receptors to the induction and expression of LTP. *Science* 242:1694–1697.

Nicoll, R. (1969) Inhibitory mechanisms in the rabbit olfactory bulb: Dendrodendritic mechanisms. *Brain Res.* 14:157–172.

Otto, T., Eichenbaum, H., Weiner, S., and Wible, C. (1991) Learning-related patterns of CA1 spike trains parallel stimulation parameters optimal for inducing hippocampal long-term potentiation. *Hippocampus* 1:181–192.

Price, J. L. (1973) An autoradiographic study of complementary laminar patterns of termination of afferent fiber to the olfactory cortex. *J. Comp. Neurol.* 150:87–108.

Roman, F., Staubli, U., and Lynch, G. (1987) Evidence for synaptic potentiation in a cortical network during learning. *Brain Res.* 418:221–226.

Rumelhart, D. E., and Zipser, D. (1985) Feature discovery by competitive learning. *Cog. Sci.* 9:75–112.

Scoville, W., and Milner, B. (1957) Loss of recent memory after bilateral hippocampal lesions. *J. Neurol. Neurosurg. Psychiatry* 201:11–21.

Squire, L. (1992) Memory and the hippocampus: A synthesis from findings with rats, monkeys and humans. *Psychol. Rev.* 99:195–231.

Standing, L. (1973) Learning 10,000 pictures. *Q. J. Exp. Psychol.* 25:207–222.

Staubli, U., and Lynch, G. (1987) Stable hippocampal long-term potentiation elicited by "theta" pattern stimulation. *Brain Res.* 435:227–234.

Staubli, U., Fraser, D., Faraday, R., and Lynch, G. (1987) Olfaction and the "data" memory system in rats. *Behav. Neurosci.* 101:757–765.

Staubli, U., Larson, J., and Lynch, G. (1990) Mossy fiber potentiation and long-term potentiation involve different expression mechanisms. *Synapse* 5:333–335.

Taketani, M., Ambros-Ingerson, J., Myers, R., Granger, R., and Lynch, G. (1992) Is field CA3 a reverberating short term memory system? *Soc. Neurosci. Abstr.* 18:1211.

von der Malsburg, C. (1973) Self-organization of orientation sensitive cells in the striate cortex. *Kybernetik* 14:85–100.

Wyss, J. (1981) Autoradiographic study of the efferent connetions of entorhinal cortex in the rat. *J. Comp. Neurol.* 199:495–512.

Zalutsky, R. A., and Nicoll, R. A. (1990) Comparison of two forms of long-term potentiation in single hippocampal neurons. *Science* 248:1619–1624.

Zimmer, J. (1971) Ipsilateral afferents to the commissural zone of the fascia dentata demonstrated in decommissurated rats by silver impregnation. *J. Comp. Neurol.* 23:393–416.

Zipser, D. (1991) Identification models of the nervous system. *Neuroscience* 47:853–862.

Zipser, D., and Andersen, R. (1988) A back-propagation programmed network that simulates response properties of a subset of posterior parietal neurons. *Nature* 331:679–684.

20 From Neurobiology to Neurocomputation

John M. Sarvey

It has been 20 years since the publication of the first two papers describing the phenomenon of long-term potentiation (LTP) revolutionized the way we study the molecular and cellular mechanisms of mammalian learning and memory (Bliss and Lømo, 1973; Bliss and Gardner-Medwin, 1973). Those two papers demonstrated for the first time that it was possible for brief events of neuronal activity on the millisecond to second scale to bring about long-lasting changes in neuronal function that can persist for days or weeks. It had seemed logical that learning and memory must function this way, although up to that time no one had succeeded in developing an experimental model. The development in the following years of the hippocampal slice preparation and the demonstration that LTP could be induced in this in vitro preparation in fields CA1 and CA3, as well as in the dentate gyrus, not only established the generality of the phenomenon, but also made it more amenable to biochemical, biophysical, and pharmacological analyses of the underlying mechanisms. Nevertheless, for several years a nagging problem remained: if LTP was the mechanism of learning and memory, why did attempts to produce it in other obvious mnemonic regions, such as the neocortex, generally meet with failure? In spite of these early misgivings, pioneering work by Lee (1982) and more recent contributions from others presented in this volume and elsewhere have clearly shown that LTP can be induced in other cortical regions of mammalian brain.

Although the concept of modeling neuronal function at the level of the synapse or up to the whole brain is not new, the development of realistic models, based on an understanding of how the brain actually processes information, is still in its infancy. Attempts at modeling learning, generally based on the concept of LTP, have been overwhelmed by experimental findings of the last 10 years, beginning, perhaps, with the demonstration of a role for NMDA receptors in LTP (Collingridge et al., 1983). As up to date as this volume's predecessor (Baudry and Davis, 1991) was, none of the chapter titles mentioned either nitric oxide or long-term depression (LTD); both are major topics of this volume.

Today, better computers and programs and increased interest in LTP, learning, and cognition have made it likely that modeling can begin to catch up with empirical breakthroughs. Commercially available programs that run on microcomputers, such as those reviewed in the November 1992 issue of *Trends in Neuroscience* (De Schutter, 1992), put modeling within virtually everyone's reach.

Currently, an appealing approach is to construct a model based on real neuronal systems, which is the subject of three chapters that I shall discuss in this chapter. Two of these, one by Granger and colleagues and the other by Chauvet and Berger, discuss computer models developed from experimental data collected from the mammalian limbic system.

The chapter by Charpier and colleagues describes a "model" system, the Mauthner cell in the goldfish reticulospinal system. Yet even this "simple" non-mammalian system has proven itself to be quite complex. Not only can LTP be induced in both the electrical and chemical excitatory synapses on the Mauthner cell, but both types of synapses also display LTD. The chapter focuses on yet another complexity of the system: LTP of an additional synapse, an inhibitory glycinergic synapse that can be induced by stimulation of the eighth nerve.

The Mauthner cell is an important, and in several ways, a rather unique model. Generally, inhibitory synapses in mammalian forebrain do not appear to exhibit LTP, although IPSPs can increase in amplitude secondary to LTP at excitatory synapses (Yamamoto and Chujo, 1978; Misgeld et al., 1979). Convincing evidence for LTP at an inhibitory synapse is provided by Charpier and colleagues in the form of (1) induction of LTP of the Mauthner cell IPSP during blockade of chemical transmission and LTP at the eighth nerve-interneuron synapse and (2) paired recordings from a single interneuron and a Mauthner cell during LTP. In the first experiment, stimuli reached the presynaptic (to the Mauthner cell) inhibitory interneuron by electrotonic synapses. In the second, the electrode in the interneuron served to elicit a test stimulus before and after induction of LTP.

The concept of "silent cells" that could become active during LTP has been proposed by a number of authors and could be involved in the apparent LTP of inhibitory synapses on the Mauthner cell. In particular, lateral diffusion of glycine following physiological stimulation could be the mechanism of activation of silent inhibitory cells. Also, cyclic AMP produced in the postsynaptic cell could diffuse to neighboring postsynaptic sites and thereby act at all inputs. Recent publications have shown that cyclic AMP-dependent protein kinase can phosphorylate and enhance the activity of AMPA receptors; such a mechanism has been proposed to underlie LTP at excitatory synapses (Raymond et al., 1993; Wang et al., 1993). Related phenomena in the dentate gyrus, norepinephrine-induced long-lasting potentiation and depression, have been shown to act through β-adrenergic receptors linked to adenylyl cyclase (Dahl and Sarvey, 1989, 1990; Sarvey et al., 1989). Thus, involvement of cyclic AMP in synaptic plasticity may be widespread. At the Mauthner cell, glycine could also be acting through enhancement of NMDA receptor efficacy, as it is now known that glycine acts as a coagonist with glutamate of the NMDA receptors. However, testing this hypothesis would be experimentally difficult, and the available data indicate that all the synaptic events and kinetics can be accounted for by diffusion alone. Furthermore, the glycine effect is blocked by strychnine, whereas the glycine site on the NMDA receptor is insensitive to strychnine.

Another observation that links LTP of inhibitory synapses on the Mauthner cell to LTP found in other systems is the finding that intracellular injection of BAPTA prevents LTP of the IPSP, just as it blocks LTP of the EPSP in field CA1 hippocampal pyramidal cells. However, it is not known whether hyperpolarization of the Mauthner cell while stimulating the inhibitory glycinergic neurons prevents the induction of LTP, as it does in mammalian hippocampal or cortical neurons. Such an experiment might be difficult to realize, because the Mauthner cell is a large cell with a low membrane resistance. Finally, these inhibitory synapses also exhibit LTD, which is produced by glycine released from the collaterals or applied from a pipette.

It also will be interesting to determine the generality of the observation that inhibition (in the contralateral Mauthner cell) is potentiated by weaker stimuli than the (ipsilateral) excitatory step of this simple network (see figure 10.1 in Charpier and colleagues' chapter). This wiring conveys the "advantage" of activating one side selectively to give the appropriate escape reflex. On the ipsilateral (to the activated VIIIth nerve) side, both electrotonic and chemical EPSPs exhibit LTP. Low stimulus intensity produces LTP of the contralateral IPSP. So, the reflex is potentiated in both cases.

Many other questions are raised by these findings and remain unanswered, including the question of the existence of a retrograde messenger for LTP of the Mauthner cell IPSP, and most importantly of the generality of LTP at inhibitory synapses. Finally, one can also ask what the phylogenetic significance of these results is.

It is interesting to watch the work of Chauvet and Berger in progress, like watching a house being built. Chauvet and Berger are attempting to model the perforant path–dentate granule cell synapse. Their goal is that their model, derived from field theory, will eventually be able to explain the interaction of neurons in the hippocampal network. They have thus far created a very "lifelike" field EPSP and population spike.

Because their model includes connectivity and geometry, it has an advantage over connectionist models, which must ignore geometry. A second advantage of using the field theory to model the system is that the same set of equations can represent the dynamics of the variables at different levels of organization, namely, the receptor level, the neuron level, or the circuit level. On the other hand, the relationship between some of the parameters used in the mathematical description of the model and the biological variables the model incorporates is sometimes obscure. In particular, the meaning of the parameter p (the source coefficient of the action potential) is not easy to understand, although it is clear that it is related to the generation of the field potential (EPSP + population spike). The field equation used for generation of action potentials does include a term that represents the integration of all external inputs. However, at the present stage of development of the model, the threshold value for the action potential is not a variable parameter. It is clearly possible to include additional elements, such as channels, that would confer variable threshold, as the behavior of the system at the lower level can be studied and included later in the solution at the upper level. So one step variable at the lower level is a parameter for the upper level and is included in the system. In this approach, the study of the mathematical properties of the field equation is used to understand the physiological characteristics of the real neural network it represents. Interestingly, changing the value of p actually *does* affect the initial slope of the EPSP and produces a shift in the population spike (see figure 18.8), a result which was certainly not intuitively obvious. On the other hand it is quite puzzling that changing the value of μ, the source coefficient of synaptic efficacy, produces only small changes in the EPSP/population spike waveform. The only change in the field potential occurs *after* the population spike and most of the synaptic wave, although one would have predicted that synaptic efficacy determines the initial slope of the EPSP. A considerable amount of work has been done to model the NMDA receptors, and the results, which are presented clearly, indicate that the modeling approach will be useful to determine which of the characteristics of these complicated receptors are critical for their roles in epsp properties and in LTP.

Richard Granger and colleagues have modeled two separate regions: hippocampal field CA1 and olfactory cortex. Their efforts to model the interaction between the two cortical areas (hippocampal and olfactory) appear to be extremely promising. In particular, they propose a totally new hypothesis concerning the role of the hippocampus in learning and memory processes. In their view, the hippocampus learns match/mismatch information which would play a critical role for extrahippocampal cortical encoding and storing of specific information. This could account for a number of experimental observations concerning the physiological properties of hippocampal circuitries as well as more traditional psychological views of hippocampal function. The question of whether the output from CA1 really conveys only "match-mismatch" information remains to be demonstrated in vivo, but the hypothesis makes a number of testable predictions, at both the physiological and behavioral level.

The two goals of modeling are (1) to interpret and (2) to predict. In general, it appears that modelers are becoming reasonably good at the first, but that they have a long way to go in the second. Nevertheless, Granger and coworkers have demonstrated that their computer model can *predict* the behavior and physiological activity of olfactory cortical neurons in an olfactory learning task. There is a good review of why stimulus trains delivered at the frequency of the theta rhythm effectively activate NMDA receptors to induce LTP at "Hebbian" CA1 hippocampal pyramidal cell synapses. The concept that patterns of activity in paired inputs producing LTP become coded and that the pyramidal cell becomes a "non-Hebbian" sequence detector is intriguing. The application of the model to the correct identification of sonar signals indicates that further analyses of the computational properties of different neuronal circuitries will generate new hypotheses concerning higher cognitive functions of the brain.

In summary, several models for long-term potentiation have been presented. One major test of the mathematical models is the question of whether LTP (or LTD) is presynaptic, postsynaptic, or both (Lisman and Harris, 1993). Will Chauvet and Berger's model, for example, be able to predict the locus of LTP (or LTD), or is that a question that can only be approached experimentally? Possibly, the field theory model, because it can look at various levels and because it includes geometry in its formalism, may be able to model and eventually to predict the interaction between pre- and postsynaptic elements at a particular synapse.

The hope for modeling is that eventually learning rules (derived from experimental results) will stop driving models and models will begin driving experimental searches for novel biochemical and physiological mechanisms. But, at least for the time being, if LTP of IPSPs turns out to be a general phenomenon that occurs in mammalian cortical neurons as well as the Mauthner cell, the other modelers in this book will be forced to add yet another set of parameters to their models. Nevertheless, the next big breakthrough could result from modelers telling experimentalists to look for a process or mechanism predicted by their computers.

REFERENCES

Baudry, M., and Davis, J. L., eds. (1991) *Long-term Potentiation: A Debate of Current Issues.* Cambridge, Mass.: MIT Press.

Bliss, T. V. P., and Gardner-Medwin, A. R. (1973) Long-lasting potentiation of synaptic transmission in the dentate area of the unanaesthetized rabbit following stimulation of the perforant path. *J. Physiol.* 232:357–,374.

Bliss, T. V. P., and Lømo, T. (1973) Long-lasting potentiation of synaptic transmission in the dentate area of the anaesthetized rabbit following stimulation of the perforant path. *J. Physiol.* 232:331–356.

Collingridge, G. L., Kehl, S. J., McLennan, H. (1983) Excitatory amino acids in synaptic transmission in the Schaffer collateral-commissural pathway of the rat hippocampus. *J. Physiol.* 334:33–46.

Dahl, D., and Sarvey, J. M. (1989) Norepinephrine induces pathway-specific long-lasting potentiation and depression in the hippocampal dentate gyrus. *Proc. Natl. Acad. Sci. USA* 86:4776–4780.

Dahl, D., And Sarvey, J. M. (1990) β-Adrenergic agonist-induced long-lasting synaptic modifications in hippocampal dentate gyrus require activation of NMDA receptors, but not electrical activation of afferents. *Brain Res.* 526: 347–350.

De Schutter, E. (1992) A consumer guide to neuronal modeling software. *Trends Neurosci.* 15:462–464.

Lee, K. S. (1982) Sustained enhancement of evoked potentials following brief, high-frequency stimulation of the cerebral cortex in vitro. *Brain Res.* 239:617–623.

Lisman, J. E., and Harris, K. M. (1993) Quantal analysis and synaptic anatomy—Integrating two views of hippocampal plasticity. *Trends Neurosci.* 16:141–147.

Misgeld, U., Sarvey, J. M., and Klee, M. R. (1979) Heterosynaptic postactivation potentiation in hippocampal CA3 neurons: Long-term changes of the postsynaptic potentials. *Exp. Brain Res.* 37:217–229.

Raymond, L. A., Blackstone, C. D., and Huganir, R. L. (1993) Phosphorylation and modulation of recombinant GluR6 glutamate receptors by cAMP-dependent protein kinase. *Nature* 361:637–641.

Sarvey, J. M., Burgard, E. C., and Decker, G. (1989) Long-term potentiation: Studies in the hippocampal slice. *J. Neurosci. Meth.* 28:109–124.

Wang, L.-Y., Taverna, F. A., Huang, X.-P., MacDonald, J. F., and Hampson, D. R. (1993) Phosphorylation and modulation of a kainate receptor (GluR6) by cAMP-dependent protein kinase. *Science* 259:1173–1175.

Yamamoto, C., and Chujo, T. (1978) Long-term potentiation in thin hippocampal sections studied by intracellular and extracellular recordings. *Exp. Neurol.* 58:242–250.

Contributors

Alain Artola
Department of Neurobiology
Swiss Federal Institute of Technology Zurich
Zurich, Switzerland

Attila Baranyi
Centre National de la Recherche Scientifique
Institut Alfred Fessard
Gif-sur-Yvette, France

Carol A. Barnes
Arizona Research Laboratories
Division of Neural Systems, Memory and Aging
University of Arizona
Tuscon, Arizona

Michael F. Barry
Department of Physiology
University College London
London, England

Michel Baudry
Neuroscience Program
University of Southern California
Los Angeles, California

Theodore W. Berger
Department of Biomedical Engineering
Program in Neuroscience
University of Southern California
Los Angeles, California

Lynn J. Bindman
Department of Physiology
University College London
London, England

Georg Andreas Böhme
Rhône-Poulenc Rorer
Centre de Recherche
Vitry-Alfortville, France

Stephen R. Bolsover
Department of Physiology
University College London
London, England

Christelle Bon
Rhône-Poulenc Rorer
Centre de Recherche
Vitry-Alfortville, France

David S. Bredt
Department of Neurosciences
Johns Hopkins School of Medicine
Baltimore, Maryland

Franck A. Chaillan
Neurobiologie des Comportements
Université de Provence Aix Marseille
Marseille, France

Stéphane Charpier
Laboratory of Cellular Neurobiology
Institut Pasteur
Paris, France

Gilbert A. Chauvet
Institut de Biologie Théorique
Université d'Angers
Angers, France

Geri Christofi
Department of Physiology
University College London
London, England

Francis Crépel
Laboratoire de Neurobiologie et Neuropharmacologie du Développement
Université Paris-Sud
Orsay, France

Hervé Daniel
Laboratoire de Neurobiologie et Neuropharmacologie du Développement
Université Paris-Sud
Orsay, France

Dominique Debanne
Centre National de la Recherche Scientifique
Institut Alfred Fessard
Gif-sur-Yvette, France

Adam Doble
Rhône-Poulenc Rorer
Centre de Recherche
Vitry-Alfortville, France

Howard Eichenbaum
Center for Behavioral Neuroscience
State University of New York
Stony Brook, New York

Christopher D. Ferris
Department of Neurosciences
Johns Hopkins School of Medicine
Baltimore, Maryland

Yves Frégnac
Centre National de la Recherche Scientifique
Institut Alfred Fessard
Gif-sur-Yvette, France

Kohji Fukunaga
Department of Pharmacology
Kumamoto University
Kumamoto, Japan

Richard Granger
Center for the Neurobiology of Learning and Memory
University of California, Irvine
Irvine, California

Lewis B. Haberly
Department of Anatomy and Neuroscience Program
University of Wisconsin
Madison, Wisconsin

Thomas G. Hedberg
Departments of Neuroscience and Neurology
Albert Einstein College of Medicine
Bronx, New York

Nathalie Hémart
Laboratoire de Neurobiologie et Neuropharmacologie du Développement
Université Paris-Sud
Orsay, France

Danielle Jaillard
Laboratoire de Neurobiologie et Neuropharmacologie du Développement
Université Paris-Sud
Orsay, France

Evan D. Kanter
Department of Anatomy and Neuroscience Program
University of Wisconsin
Madison, Wisconsin

Kevin L. Ketchum
Department of Anatomy and Neuroscience Program
University of Wisconsin
Madison, Wisconsin

Henri Korn
Laboratory of Cellular Neurobiology
Institut Pasteur
Paris, France

Dimitri M. Kullmann
Department of Pharmacology and Physiology
University of California, San Francisco
San Francisco, California

Martine Lemaire
Rhône-Poulenc Rorer
Centre de Recherche
Vitry-Alfortville, France

Dezhi Liao
Departments of Physiology and Biophysics
University of Iowa
Iowa City, Iowa

Gary Lynch
Center for the Neurobiology of Learning and Memory
University of California, Irvine
Irvine, California

Daniel V. Madison
Department of Molecular and Cellular Physiology
Beckman Center for Molecular and Genetic Medicine
Stanford University School of Medicine
Stanford, California

Robert C. Malenka
Departments of Psychiatry and Physiology
University of California, San Francisco
San Francisco, California

Roberto Malinow
Departments of Physiology and Biophysics
University of Iowa
Iowa City, Iowa

Toshiya Manabe
Department of Neurophysiology
Institute for Brain Research
University of Tokyo
Tokyo, Japan

Bruce L. McNaughton
Arizona Research Laboratories
Division of Neuronal Systems, Memory and Aging
University of Arizona
Tuscon, Arizona

Eishichi Miyamoto
Department of Pharmacology
Kumamoto Universisty
Kumamoto, Japan

Dominique Muller
Department of Pharmacology
University Medical Center
Geneva, Switzerland

Richard Myers
Center for the Neurobiology of Learning and Memory
University of California, Irvine
Irvine, California

Roger A. Nicoll
Department of Pharmacology and Physiology
University of California, San Francisco
San Francisco, California

Alex V. Nowicky
Department of Physiology
University College London
London, England

Yoichi Oda
Department of Biophysical Engineering
Faculty of Engineering Science
Osaka University
Osaka, Japan

Tim Otto
Department of Psychology
Rutgers University
New Brunswick, New Jersey

Nickolai Otmakhov
Departments of Physiology and Biophysics
University of Iowa
Iowa City, Iowa

David J. Perkel
Department of Pharmacology and Physiology
University of California, San Francisco
San Francisco, California

Odile Piot
Rhône-Poulenc Rorer
Centre de Recherche
Vitry-Alfortville, France

Michel Reibaud
Rhône-Poulenc Rorer
Centre de Recherche
Vitry-Alfortville, France

Pius Renner
Department of Pharmacology and Physiology
University of California, San Francisco
San Francisco, California

François S. Roman
Neurobiologie des Comportements
Université de Provence
Marseille, France

John M. Sarvey
Program in Neuroscience
Uniformed Services University of the Health Sciences
Bethesda, Maryland

Erin M. Schuman
Department of Molecular and Cellular Physiology
Beckman Center for Molecular and Genetic Medicine
Stanford University School of Medicine
Stanford, California

Aneil M. Shirke
Departments of Physiology and Biophysics
University of Iowa
Iowa City, Iowa

Daniel Shulz
Centre National de la Recherche Scientifique
Institut Alfred Fessard
Gif-sur-Yvette, France

Solomon H. Snyder
Department of Neurosciences
Johns Hopkins School of Medicine
Baltimore, Maryland

Bernard Soumireu-Mourat
Neurobiologie des Comportements
Université de Provence
Marseille, France

Patric K. Stanton
Departments of Neuroscience and Neurology
Albert Einstein College of Medicine
Bronx, New York

Jean-Marie Stutzmann
Rhône-Poulenc Rorer
Centre de Recherche
Vitry-Alfortville, France

Libor Velíšek
Departments of Neuroscience and Neurology
Albert Einstein College of Medicine
Bronx, New York

Eric M. Wexler
Departments of Neuroscience and Neurology
Albert Einstein College of Medicine
Bronx, New York

Eric Whelpley
Center for the Neurobiology of Learning and Memory
University of California, Irvine
Irvine, California

David J. A. Wyllie
Department of Pharmacology and Physiology
University of California, San Francisco
San Francisco, California

Index